Successful Contract Administration

The success of every construction project begins with reading and understanding the contract. Contract Administrators and Project Managers for all parties in the construction process must realize the major impact their actions have on cost, schedule, and quality in relation to the contract terms and conditions.

Written in a clear and accessible way from a Constructor's perspective, *Successful Contract Administration* guides the student through the critical issues of understanding contract law and obligations for effective project execution. Through examples, exercises, and case studies, this textbook will:

- improve knowledge and comprehension of key contract elements;
- help the student apply knowledge to real case scenarios;
- improve the student's ability to analyze and create different scenarios for success;
- evaluate critical issues of responsibility and ethics in relation to contract administration.

The text is supported by a companion website featuring additional resources for both students and instructors. Resources for the student include additional case studies, links to useful websites, video commentary and interviews for increased understanding of important chapter material, true/false sample quiz questions, and a flash card glossary to reinforce comprehension of key terms and concepts. Additional instructor material includes a testbank of questions (including true/false, multiple choice, and sample essay questions), website links to contract documents, and PowerPoint slides.

Charles W. Cook has over 40 years' experience in the construction industry, from laborer to President of R. S. Cook and Associates, Inc. He is now an online professor of Construction Management, College of Engineering, at Drexel University in Philadelphia, while presenting programs throughout the United States to construction companies and associations. He has been elected to the Consulting Contractors Council of America, the Carpenters' Company of Philadelphia, the Board of Directors of the General Building Contractors Association, and serves as an arbitrator for the American Arbitration Association.

Successful Contract Administration
For Constructors and Design Professionals

Charles W. Cook

LONDON AND NEW YORK

First published 2014
by Routledge
2 Park Square, Milton Park, Abingdon, Oxon OX14 4RN

and by Routledge
711 Third Avenue, New York, NY 10017

Routledge is an imprint of the Taylor & Francis Group, an informa business

British Library Cataloguing in Publication Data
A catalogue record for this book is available from the British Library

Library of Congress Cataloging in Publication Data
Cook, Charles W. (Contract administrator)
Successful contract administration: for constructors and design professionals / Charles W. Cook.
pages cm
1. Building--Superintendence. 2. Subcontracting. 3. Construction contracts. I. Title.
TH438.C6448 2014
624.068'4--dc23
201420222

ISBN: 978-0-415-84422-2 (pbk)
ISBN: 978-1-315-76579-2 (ebk)

Typeset in Helvetica Neue by
Servis Filmsetting Ltd, Stockport, Cheshire

For my family – the alpha of any good that I have done

Contents

Figures

Foreword

As an author, I approach the subject of contract administration from the perspective of a Constructor with over 40 years of experience in the industry, and I am often reminded that any success I have enjoyed has often been the result of previous lessons learned the hard way. I invite others to learn from my mistakes, because you probably don't have time to make all the mistakes I have made. But I also invite Constructors, and Design Professionals who want to understand Constructors, to study from my perspective for two important reasons:

1. As contract law becomes more complex, too often we find ourselves in adversarial relationships with one or more of the team members on a project. This is not good for any of us, or for our clients, or for our industry as a whole. The more we understand the complexities of delivering the obligations we have agreed to, the more likely we can turn adversaries into allies, which I believe is the potential that technology is making available to all of us if we choose to seize it.
2. Within the past couple centuries, the role of the Design Professional in the total construction experience has diminished from the original "Master Builder" concept that had been the way of construction for millennia. A return to a pure individual Master Builder may not be possible, but we all still need to understand that achieving something incredible is a goal for everyone on each and every project, and to do that requires effective and even extraordinary teamwork. The benefits of Constructors and Design Professionals recognizing the problems and opportunities of contract administration together can only open new and improved methods of delivery, and Constructors, Architects, and Engineers will find new and fulfilling levels of administration and achievement within the construction process as we all move forward together.

I would like to invite you all to a personal perspective and understanding of my world of construction, but I look forward to each of you in the future making it a better world of construction for us all.

Note on the text

In the following pages and during the course of a career in construction, the reader will learn there are many perspectives of what is right and wrong in relation to contract law and administration. What one court might conclude, another might not. Slight subtleties and differences can change the outcome of a situation. This text is not to be used as a legal standard or document. All readers should recognize the need for competent legal counsel familiar with the issues of local, state, and Federal laws pertinent to any specific situation.

The most effective outcome of this text would be to bring about an atmosphere that resolves conflicts effectively and eliminates the need for court or Alternative Dispute Resolution processes.

Student learning outcomes

It's not what you look at that matters. It's what you see.

Henry David Thoreau

Educators realize the true value of a course, a program, and ultimately a degree needs to be measured in terms of what the student is able to take away. I generally consider several steps in a student's education in a particular subject. The first is "walking around knowledge." This is what I believe the student should absorb and remember without referring back to a text, notes, or seeking input from another expert on the matter. Next there should be some recognition by the student of what one does not know. In such a case the student should understand where to seek additional information and help. After a course or program each student should have confidence to know where to find and collect information, organize it, and how to use it and apply it effectively. Finally, the ultimate goal is an ability to analyze situations and create effective solutions.

Working in the mid to late twentieth century, Benjamin Bloom has been widely credited with developing a system of learning objectives related to three primary areas of development for students: (1) Knowing, (2) Feeling, and (3) Doing. Building on the foundation of Bloom's work along with other educators he worked with, accrediting agencies and institutions have created various standards called student learning outcomes, student learning objectives, and student learning priorities. By whatever name, the intention remains the effective growth of the student's abilities from:

1. *remembering* basic facts to
2. *understanding* the information to
3. *applying* the ideas to other circumstances to
4. *analyzing* situations in their basic components to
5. *evaluating* the various issues and challenges to
6. *creating* solutions.

This text has multiple elements to assist the student in the passage from Remembering through to Creating.

Effective use of the "Instant recall" and "Key words" sections at the end of each chapter, along with the flash cards and sample quiz questions on the Companion Website, will help the student with *remembering* the key points of each chapter.

The text itself contains examples and stories to assist in the *understanding* of the issues. At the end of chapters suggestions for additional information are offered. Also, additional video discussions on the Companion Website will help to improve practical understanding of the information.

At the end of each chapter, various exercises including the "You be the judge" exercises and the related discussion on the Companion Website are intended to help in *applying* what has been learned to actual circumstances.

Discussion exercises at the end of each chapter help the student *analyze* the material covered in relation to actual challenges.

In addition, the discussion exercises and the "It's a matter of ethics" discussions at the end of each chapter can help the student *evaluate* the subject matter in relation to real-life issues and *create solutions* before similar situations arise in one's own life.

Of course, the most common means of measuring growth is testing, but this textbook is also meant as a workbook. The growth in confidence and self-esteem of the reader who diligently works through the exercises, shares ideas with others, and comes to conclusions that will affect positive growth in the future is an even better measurement for one's personal success.

There are many priorities for development beyond memorization of facts. Effective communication, creative and critical thinking, self-directed learning, and professional conduct are some of the key benefits a reader will develop by working through the exercises and the additional material at the end of each chapter under "Going the extra mile."

Finally, but certainly not least, Chapter 19 deals with Ethics. Closely related to law, ethics is discussed in this stand-alone chapter, but it should not be considered an "afterthought." For that reason there are ethics discussion exercises at the end of each chapter. The reader is encouraged to open these exercises to group discussion, remembering that the sharing of ideas is an opportunity to find other perspectives. One does not need to adopt a different perspective, but the free exchange of ideas depends on an open atmosphere without fear of ridicule or belligerence.

There are also ethics video segments, beginning with an "Introduction" on the Companion Website, starting with Chapter 13 and continuing through Chapter 19.

If the reader would like to, Chapter 19 can be read first or at any time. Ethics is a difficult subject and a definite challenge for all of us in the construction industry. It will continue to be a challenge well past the completion of this workbook. Under extreme pressures, people face constraints that force decisions about tradeoffs. To gain one issue costs something else, and the choice is not always easy or even obvious.

In the end, there are many right answers for the learning outcomes contained within *remembering* through *creating solutions*. The reader is encouraged to master as many as possible, with the ultimate realization that growth begins by reading on, but it continues each time the book is closed.

CHAPTER 1

Getting started

Starting with this chapter, there will be a list of key learning objectives for each chapter. Many readers might think it will be easier and certainly quicker to skip the learning objectives section. "After all," some might reason, "the points will obviously be repeated in the text that follows." However, an early understanding of the author's intent is actually a positive reinforcement to more effective reading, and it is also a discipline worth acquiring in preparation for understanding the Designer's and Owner's intention within the contract documents.

Here, then, are the key learning objectives of this chapter.

Student learning outcomes

Upon completion the student should be able to. . .

- **recognize there are three phases to contract administration: (1) read the contract, (2) understand the contract, and (3) execute the contract**
- **know the importance of applying lessons learned to future projects**
- **evaluate the importance of schedule, cost, and quality in relation to the contract documents**
- **analyze the other party's needs in relation to the contract documents.**

Figure 1.1 30th Street Station and Cira Center, Philadelphia
Photo credit: Charles W. Cook

The neoclassical columns of 30th Street Station in Philadelphia contrast with the silver glass façade of the Cira Center that at times seemingly blends with the sky it reflects. Built almost eight decades apart, the two structures are a reminder of how different construction has become over the millennia since the ancient Greeks advanced the use of "post and beam" construction. In that time period, the relationship of Designers and Constructors has evolved tremendously. No longer are buildings the products of individual "Master Builders," who took on the task of both design and execution while directing a crew of craft workers. Now most structures are a joint effort of several corporate entities bound together by complex legal documents that require considerable understanding and attention to details.

How well the details of the construction process are handled and the final structure successfully realized requires diligence and commitment by each and every member of a construction team assigned to the task of contract administration.

Whether these are the best of times or the worst of times depends a great deal on how we approach the times. This is not just a textbook. This is also a *workbook*, providing you with a balance of both theory and practice. Case facts and compensation figures may be important to recognize, but how we react to contract conditions and, particularly, how we handle what some term "killer clauses" is of far greater importance to our own success and the successful completion of the projects for which we are responsible.

The Contract Administrator is helping to change the world, one "put in place a unit at a time," but, to do so effectively, he/she must read the contract one paragraph, and even one sentence or clause, at a time. The devil is only in the details that we do not know. Knowing all the details, however, is useless unless we know what to do with them. Knowing what to do with them is helpful, but knowing how we will handle difficult situations is of far greater importance.

Good contract administration depends on three phases: (1) reading the contract, (2) understanding the actual requirements, and (3) executing to a successful conclusion. Successfully doing this on one contract will provide the Contract Administrator with a fourth phase leading to even more success when the knowledge and experience gained from one project is applied to the next contract to be administered in a continuous quality improvement process.

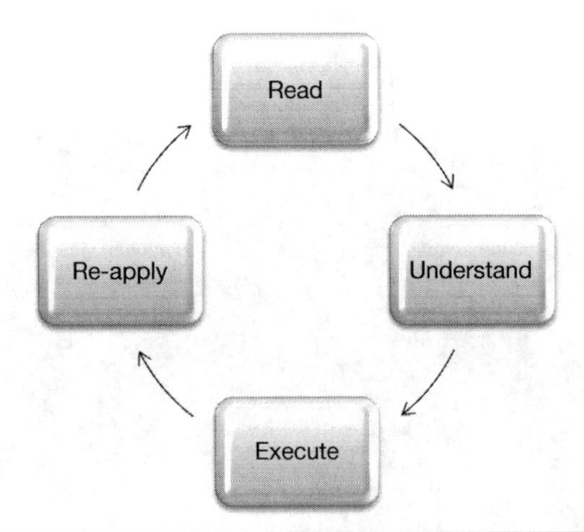

Figure 1.2 Three phases plus one in continuous improvement of contract administration

Figure 1.3 Boilerplate language
Photo credit: Charles W. Cook

Copper, brass, or tin plates have been affixed to fuel storage tanks and boilers since the beginning of the Industrial Revolution. Eventually advertisers developed the practice of sending copper plates of printing to newspapers to include as their standard portion of their advertisements. Seeing the resemblance to the metal "boilerplates," the standard portion of the advertising became known as "boilerplates." From that it became a short jump for printers to call the standard portion of contracts "boilerplates." For many Contract Administrators, the **boilerplate** language may become repetitive and something one is inclined to skip, but within the standard language some important issues might be found, particularly if there have been changes made to the original "boilerplate" contract wording.

Note: Within the text, when an important word is introduced, it will be in bold (e.g. **boilerplate** in the above box). The reader is encouraged to read the text in relation to what is being discussed. At the end of each chapter the key words will be repeated. Definitions can be found in the glossary and the reader can review using the flash card exercises on the Companion Website.

Exercise 1.1

Take a moment
Read a contract

At the back of this book you will find a sample blank contract in Appendix I. The AIA and the ConsensusDOCS contracts are certainly more widely used pre-printed contracts. We are using in this exercise a sample contract the reader has never seen before. It is a contract to be used between a General Contractor and a Subcontractor. The contract contains many basic but important elements of construction contracts we will be discussing in later chapters of this book.

Take a moment now, however, to read the contract before answering the questions that follow on the next page.

After you have read the contract in Appendix I, continue.

You may have thought the questions would be about specific clauses or requirements of the contract, but we first have to understand our own approach to the most fundamental aspect of reading a contract, and that is *reading*, itself.

In answering the following questions, be honest with yourself. This is not a test, but an opportunity to recognize our own challenges. It is through understanding ourselves that we can make choices to improve our value to others.

1. Were you solely concentrating on reading, with no competitors for your full attention (e.g. television, radio, e-mail, text messages, or iTunes)?
2. Were you able to tune out exterior noises in your environment that could be considered distracting or even become interruptions (e.g. e-mail or text alerts)?
3. Did you highlight portions of the contract or make notes?
4. Did your concentration wander to other important things you have to do?
5. If you did not understand something, did you make a note so you could somehow resolve it later?
6. Did you ever feel bored or that reading the whole contract was tedious?
7. Did you read the whole contract (or any of the contract) before you skipped back to these questions?

As you go forward, you might also want to recognize how you "read" this text. As fundamental a skill as reading is for most of us, it can often be neglected. We tend to take shortcuts and even permit distractions to impair our complete concentration. When it comes to the "devil in the details" of a contract or understanding a text, missing or skipping over something important can become critical and even costly for the Contract Administrator.

Based on the answers to the above questions, developing what habit or skill might improve your reading of a contract?

Note: As you do the exercises in this workbook, there will be a temptation to go to the discussion before completing the exercise, but if you actually try to work through the exercise, the lessons will be more fulfilling, provide greater insights into your own challenges and opportunities, and lead to a better understanding of how each of us can be a better Contract Administrator.

Discussion of Exercise 1.1

It is obvious none of us can be a perfect reader 100% of the time.

1. ***Reading***. Let us be honest. Contracts are boring. Additionally, many of us have the attitude "you've seen one, you've seen them all." Unfortunately, it is the part you do not see that will usually cause the most problems. The fun is outside on the project itself, but the devil is in the details we do not know about, so we have to start with the details of the contract by reading them.

Not reading the contract can have disastrous consequences

For example, there are notice provisions within almost all contracts.

1. If a Constructor does not read the contract and know the time frame for giving notice, extensions of time and change orders may be in jeopardy, with the potential of heavy losses.
2. Parties who fail to give proper notice to a bonding company will lose the right to be paid or have work completed as contracted if the entity for whom the bond was written fails to do so.
3. Not responding in a timely manner can put Design Professionals in jeopardy of becoming responsible for extensions to the project schedule.

All these possibilities and many more will be explained in later chapters, but for now it is important to recognize that the consequences of not reading the contract are many and can affect everyone involved in a construction project.

Sometimes the contract can appear identical, but a subtle shift of phrases or clauses can make a big difference.

Example: A major Contractor's staff had failed to recognize a change in the retainage provision of a contract the Owner had made. Instead of paying the Contractor in full at the end of the project, the Owner kept 10% of the total contract for two years after completion as an additional "warranty" provision.

A CASE IN POINT

Reading the contract before signing it is even more important, but sometimes contracts appear similar and are not carefully read.

When companies grow, they often need to take on more and more work to maintain the personnel they have employed. Dollar volume becomes a false means of measuring a successful fiscal year. One such Midwest company that will remain unidentified can serve as an example for many others throughout the nation. Growing in volume and expanding to new territories, it was seeking work from a client that had a very good project under a very tight schedule. Initially, the negotiations were going so well and everyone seemed so eager to obtain an extremely sizeable project that no one was reading the details of the contract. Fortunately, a Contract Administrator in the home office asked to see the contract before it was signed.

Reading through the contract it was discovered that in addition to a very tight schedule there were severe costs (known as Liquidated Damages, to be discussed in a future chapter) to be applied against the Contractor if the project was not completed on time. Had the contract been signed as it was written, losses would have accrued against the Contractor, but by reading and understanding the contractual obligations before signing, negotiations were able to take place so the Owner would pay for all the additional costs that would be necessary to meet the schedule. The CEO of the Contractor told me that incident changed the approach for all future contracts. No matter how familiar the contract might appear, they were always to be thoroughly read and red-lined by the home office, including review by an attorney if necessary.

It is always exciting to think about how much money can be made doing some big contracts. But just as much or even more can be lost. You do not lose money on the job you did not get, but you can on the one you should not have taken.

* * * * * * * * * *

One habit you might want to develop is setting some quiet time aside to really understand the contract. It is often the case that we absorb the most the first time we look at a set of plans. The same can be true of the first time we sit down to read a contract, but it will never happen if we are in a setting that allows others to continuously interrupt us.

* * * * * * * * *

Ultimately, *reading the contract* is a discipline each of us has to develop personally, but careful study of this text will be a nice start. As a workbook, as well as a textbook, make it your own. Highlight, make notes, "dog ear" pages, and do whatever you think will help you in relation to specific passages, concepts, and ideas in the future. Sometimes, merely underlining a portion of the text will be enough to help us remember it, but more often than not, what we read is soon forgotten, so an important aspect of making this text yours is to make it easy for you to refer back to it when you encounter a situation related to contract administration on a future project.

For this to be an effective balance between a workbook and a textbook you will need to work through the exercises while you *read* about the clauses and issues related to contract administration.

2. ***Understanding***. Once we have read the details, it is important to understand the requirements. The text of this book in that sense will be similar to most contract administration texts. It is intended to give a basic understanding of the key clauses and conditions of contract administration for the construction professional. The critical difference of this workbook over a typical textbook will be in phase 3 – Execute.

However, *understanding the contract* is where this text begins. It is not intended to make anyone a lawyer, but recognizing the difference in future chapters between a penalty and Liquidated Damages, as well as many other conditions of contract language, tort law, and the vagaries of statutes and regulatory demands, is important for successful contract administration. As we fully understand the contract, some issues and items in the contract may be in conflict or, at the least, ambiguous, and require resolution: the more immediately the Contract Administrator recognizes and deals with these issues the better it will be for all concerned.

* * * * * * * * * *

Before leaving the Understanding Phase, it is worth emphasizing this text/workbook is for the Contract Administrator. That title is chosen as an umbrella for many other titles, including, but not limited to: CEO, President, Project Manager, Project Administrator, Superintendent, Owner's Representative or Agent, Project Engineer, and Field Engineer. To the extent the success of a project or any portion of a project depends on an individual's performance related to contractual obligations, that person is a Contract Administrator. Constructors, Owners, and Design Professionals all administer contracts. Depending on the size of the company and the size of the project itself, the contract may be administered as a whole or in part by a single individual or several persons, but the obligation to administer effectively will not be less important.

* * * * * * * * *

3. ***Execute***. The third phase of contract administration requires execution. There are some who will basically read the contract simply to understand the fastest way out. Get in and

get out and get paid. That may sound a bit self-serving, but it is the essence of business. The key for all of us is that in the process of doing business, we want to stay in business. To do that, we need to execute effectively, and each of us has different ways of doing that. One is not necessarily better than another. What is best is the best way each of us individually can perform our tasks in relation to the contract. This textbook will help you understand the contract, but the workbook portion will help you understand how you can execute it to the best of your own abilities, which may not be the way the person who purchased this text before or after you will administer contracts.

Ultimately, *executing a project to a successful conclusion* is the goal of everyone who has ever signed a contract or been assigned to manage a project. Achieving that goal often becomes the most challenging aspect of any contract. Safely navigating the "maze" of contract requirements is just as important as erecting the structure or building the bridge, and often just as dangerous from a corporate perspective. No matter how well read and intentioned one might be, between concept and reality are many hazards, and successfully navigating those hazards and ultimately securing final payment can require extraordinary perseverance and attention to detail. But then, that is what makes you valuable.

4. ***Re-apply***. If there is one fatal flaw we all have in the construction industry it is that we run lean and seldom have time to apply what we learned yesterday to what we will do tomorrow. We need to grow from the mistakes of the past and look for ways to even improve what we have done well so we can do even better going forward. Unfortunately, when we get a new contract to administer, we see that it looks similar to the last one, and we put it aside to get to later – but that later never comes. The *curse of knowledge* can lead to disaster if we assume business will be as usual. We need to take what we have learned and apply it to our future, and in doing so apply new strengths to the ever necessary three phases of (1) *read*, (2) *understand*, and (3) *execute*.

Developing a new habit is often difficult, but discipline begins first by knowing what we really want, being true to that desire, and then doing what is necessary to achieve our goal.

Exercise 1.2

Take a moment
To think about yourself and how you think about others

Please note: In all the exercises related to our own behavior there are no right or wrong answers, only paths to understanding ourselves better.

1. Think of any problem you have recently had. Write the essence of the problem here:

____

____

2. Recall who was most responsible for causing the problem and why.

____

____

3. When you started to think or look for a solution, whose needs did you first think about?

____

____

Discussion of Exercise 1.2

In filling out the above three statements it is quite common for people to blame others and to think about their own needs. In arriving at a successful conclusion, however, it is usually more effective when we understand and help the other party to also achieve their needs. Abraham Lincoln pointed out his own method in this regard:

> When I am getting ready to reason with a man, I spend one-third of my time thinking about myself and what I am going to say and two-thirds about him and what he is going to say.

As we are *getting started*, therefore, we must read the contract. Even if it looks familiar we need to read it for what might be different. While we read the contract, we can make lists, highlight sections, and add notes wherever we want. But the contract is not just about what we need to do. It is actually more about what the other party needs.

Look closely for items and issues in the contract that will affect or relate to:

- schedule
- cost
- quality.

These three critical issues and how we initially can have an effect on them will be essential to the ultimate success of the project. Some wise fools have said in relation to schedule, cost, and quality that "you can have any two of the three, but you can't have all three."

But there is no doubt an Owner will expect the Contractor to deliver on all three, and to do it all safely, as well. The Contract Administrator must recognize the contract is not just about what we need to do. It is more about what the other party needs us to do for them.

Of course, the rest of the book will not be as short and sweet as this first chapter, but do not overlook or pass up the opportunities already presented.

In addition to the text, visit the Companion Website for additional commentary on topics related to contract administration.

Ultimately, is it possible that we all could do a bit better when it comes to *reading*, *understanding*, and *executing* the contract? The reading journey we are about to take together is not as difficult as building a project, and hopefully navigating the text will not be as boring as reading a contract. As we learn the cold, hard facts about contracts and the words and clauses that make them a challenge, we want to also understand our own selves in relation to contract administration.

A CASE IN POINT

Denise Scott Brown and her husband, Robert Venturi, are considered two of the most influential Architects of the twentieth century. Denise told me a story of a ritual ceremony they had witnessed in Japan during which the builder was required to demonstrate remorse for his failure. Symbolically, the builder committed Seppuku, a form of ritual suicide, also known as hara-kiri.

It is not hard to imagine decades earlier the ritual might not have been symbolic, but a para-judicial practice in which the offending builder would have actually committed suicide in front of an assembled audience.

Of course, whenever I recollect this story my first thought is that I am glad I did not live and build at such a time in Japan, but upon reflection, the Contract Administrator must always realize that the construction industry is a very dangerous place. Failing to properly administer contractual responsibilities can lead to more than embarrassment. Significant damage to property and even loss of life is always a possibility on a construction project. Casual or careless attention to details can sometimes lead to catastrophic results.

Although today a Contract Administrator's negligence may not require a ritual Seppuku, one's reputation may be the least a Constructor has to be concerned about if tragedy strikes. From losing a client to losing lives, there are many degrees of a haunted conscience.

Fulfilling the contractual obligations of schedule, cost, and quality, while doing it all safely, is an immense undertaking. Paying attention to all the details may seem tedious, but it is what the Constructor is under contract to do, and the Contract Administrator is the key to continuing success.

* * * * * * * * * *

This text/workbook is not intended to substitute for good counsel from attorneys who specialize in specific areas of construction law. Each Contract Administrator will have to make the decision when legal counsel is required. The success of this text for the reader will vary depending upon experience, the complexities of each situation, and the potential cost or consequences of failure.

Ultimately, the value of the Contract Administrator will be measured by how well the project is administered and the contractual obligations fulfilled for all the parties involved. The more that can be done without conflict, the more favorably the Contract Administrator and the Contractor and Design Professionals will be viewed.

So, the key to knowing contract law for the Contract Administrator is, in order:

1. **to both improve your own ability to understand, navigate, and document the results of your company's efforts in regard to contractual obligations;**
2. **at the same time, to recognize when you need help, whether that help be expertise within the company or professional legal input and/or representation;**
3. **to maintain accurate and effective documentation of all efforts to fulfill the contractual obligations, first to avoid disputes, and second, to provide an effective trail of evidence, should another means of resolution become necessary; and**
4. **to effectively assist counsel, in the unfortunate event that path becomes necessary.**

* * * * * * * * *

INSTANT RECALL

- Good contract administration depends on three phases: (1) reading the contract, (2) understanding the actual requirements, and (3) executing to a successful conclusion.
- Re-applying what we have learned from one contract administration experience to the next is important for continuous contract administration improvement.
- No matter how well read and intentioned one might be in relation to contract requirements, between concept and reality are many hazards.
- We need to take what we have learned and apply it to our future.
- Developing a new habit is often difficult, but discipline begins first by knowing what we really want, being true to that desire, and then doing what is necessary to achieve our goal.
- Even if the contract looks familiar, we need to read it for what might be different.

- Look closely for items and issues in the contract that will affect or relate to:
  - schedule
  - cost
  - quality.
- The contract is not just about what we need to do. It is more about what the other party needs us to do for them.
- The key to knowing contract law for the Contract Administrator is, in order:
  1. to both improve your own ability to understand, navigate, and document the results of your company's efforts in regard to contractual obligations;
  2. at the same time, to recognize when you need help, whether that help be expertise within the company or professional legal input and/or representation;
  3. to maintain accurate and effective documentation of all efforts to fulfill the contractual obligations, first to avoid disputes, and second, to provide an effective trail of evidence should, another means of resolution become necessary; and
  4. to effectively assist counsel in the unfortunate event that path becomes necessary.

YOU BE THE JUDGE

A contract to install 109.5 miles of pipeline from Bethany, Missouri to Mason City, Iowa, was typed with the exception of paragraph 2.03, which was handwritten:

"2.03 All settlements by company [Wood River] of claims in the name of contractor [Willbros] shall be based on substantial evidence of contractor's liability that the claims are valid and are reasonable in amounts.

"Contractor shall not be liable under any circumstances or responsible to company for consequential loss or damages of any kind whatsoever including but not limited to loss of use, loss of product, loss of revenue or profit. RW 7/27/80 GB 7/27/80"

(The initials RW and GB show that representatives from both Wood River and Willbros acknowledged the change.)

However, another typed paragraph clearly stated the Contractor would be responsible to the Owner for damages to the Owner's property due to the Contractor's work.

A rupture of the pipeline, spilling significant amounts of oil, cost the Owner (Wood River) over a million dollars in damages and repairs. Wood River entered a claim against the Contractor, stating the rupture was due to negligence on the part of the Contractor (Willbros).

How do you rule (for Willbros or for Wood River)?

In what amount?

IT'S A MATTER OF ETHICS

(**Note:** All "It's a matter of ethics" topics can be considered by the individual student or be opened to classroom discussion.)
When contracts were pre-printed, changes to the "boilerplate" were easily recognized. Today, many contracts are no longer pre-printed. They are generated from computer software. This has made it possible for some contracts to be printed with changes in the typical language that are not immediately obvious. Many such contracts do call the changes out through margin notations, but others do not. How ethical is it to change typical computer generated "boilerplate" language on a contract without advising about or noting the change to the other party signing the contract? Does it make a difference if the change in the language is for a first-time contract signer rather than a company that has signed many contracts prior to the changes in the standard language?

Going the extra mile

1. Go to the Companion Website at www.routledge.com/cw/cook for Part I of a video discussion of that subcontract you did or did not read thoroughly in Exercise 1.1
2. **Discussion exercise** (with friends or as part of a classroom discussion): Often when people are confronted with a failure to achieve the required standards or do not meet an obligation they resort to distorting the facts or outright lying. Some believe a first lie leads to a "bottomless pit." One lie will lead to another deception, and so forth, at least until the truth is discovered, and then that can be a pit from which a relationship cannot escape. Do you agree that a single lie, caught or uncaught, is the beginning of a limitless downward trend in a relationship? How does it change a relationship if the lie is discovered? If it never is discovered?

3. Complete the "You be the judge" exercise above and then go to the Companion Website for a discussion of the actual case. There will also be more "You be the judge" opportunities on the Companion Website for you to work on after some chapters.

4. Students are encouraged to consult the Companion Website for a complete glossary of words used in the text, and practice the flash cards to study the key words and concepts discussed. For this chapter there is one key word the reader should be familiar with when reading or discussing contracts.

Key word

Boilerplate

CHAPTER 2

An introduction to contracts, torts, statutes and regulations

One key to learning is repetition. Repeating letters and sounds eventually enabled us to read, and the constant rehearsing of numbers formed the basis of all our mathematical abilities. In this chapter we are going to identify some fundamental elements of contract administration related to law that will be dealt with in more detail in later chapters.

Student learning outcomes

Upon completion the student should be able to. . .

- **recognize some historical foundations for our existing legal system**
- **identify AIA and other pre-printed and commonly used contracts**
- **understand the importance of tort law and statutory/regulatory laws in relation to construction**
- **know the importance of ethics in the construction workplace**
- **analyze fundamental challenges of contract administration.**

As reinforcement to future contract administration, the reader will note that the numbering of headings and subjects within the chapters will follow the numerical sequencing format of many contracts.

2.1 How we got to where we are

Those who cannot remember the past are condemned to repeat it

George Santayana

The above quote is sometimes stated as "Those who do not learn from history are condemned to repeat its mistakes," and used to expose the efforts of misguided individuals or groups, even nations, going in the same wrong direction others have gone before and then winding up with the same disastrous consequences. The same could be said of many construction companies that have sought relief in litigation when they were not able to find it in their project and contract administration.

However, should not the opposing corollary also be true?

Those who remember the past can avoid repeating its mistakes.

2.1.1 Criminal and civil law

We are about to discuss several aspects of our legal system. We will start by separating law into two main areas: (1) *criminal law* and (2) *civil/common law* (including *contract law, tort*, and *statutory and regulatory law*), and already that may seem confusing. So, taking a brief look at how we got to where we are might help us benefit from the lessons of the past.

Since this is a construction text, imagine a great, two-tower suspension bridge that will carry several lanes of law and legal traditions from Europe to the United States.

The strength of the bridge depends on the foundations and construction of the two towers.

The east tower's foundation comes from the eastern Mediterranean, and construction was first begun over 3,000 years ago. A "milestone event" in construction of the east tower's foundation was when Moses presented the Ten Commandments at Mount Sinai. Primarily the contribution of the Ten Commandments as we know them gave rise to criminal law in our legal system. In criminal law the state exacts forms of retribution such as imprisonment, and even death in extreme cases. Such remedies and retribution go beyond making the injured party whole from the result of whatever wrong was committed.

Presenting the list of crimes to the people, Moses then established a group of individuals empowered to enforce the laws. Eventually individuals as wise as Solomon would find ways to expand and effectively enforce nuances of the Commandments, but the basic laws of "Thou shalt not" had formed the basis of criminal law.

Mosaic law, however, also codified several other wrongs in addition to what has been termed the Ten Commandments. These were not criminal but civil in nature, dealing with disputes between individuals rather than with crimes against the safety and security of the entire group.

* * * * * * * * * * *

This, then, becomes the important distinction for our purposes: Criminal law seeks to punish the offender, civil law attempts to restore or make whole the injured party.

* * * * * * * * *

Such laws were often quite detailed and contained precise penalties:

> When someone borrows an animal from another and it is injured or dies, the owner not being present, full restitution shall be made. If the owner was present there shall be no restitution; if it was hired, only the hiring fee is due.
>
> *Exodus 22.14*

This method of describing a wrong and exacting penalty would eventually evolve through Roman law and spread throughout the Roman Empire, becoming the foundation of civil law in Europe. Dating back to the simple "eye for an eye" concept of everyone is equal under the law, European civil law is based on codified punishments for specific offenses.

It may seem difficult for us in the twenty-first century, so used to police forces and a legal system, to imagine people just on the cusp of civilization. Thousands of years ago, such small steps were required to establish peace among a people wandering through landscapes made up of vast and barren wildernesses occasionally interrupted by villages and tribes that often did not speak the same language. A change in power was slowly taking place. Justice would be measured through doing what is right for everyone rather than the concept that what is right belongs to whoever is the stronger.

The Code of Hammurabi

Establishing laws and peace among the civilized was a slow but steady process. Over three centuries before Moses brought the Commandments to the wandering Israelites, the sixth King of Babylon established his code of conduct that would become the rules for all within his realm. Named after him, these 282 rules have become known as the *Code of Hammurabi*. In it we find the first written laws related to the construction industry. These "eye for an eye, tooth for a tooth" type laws undoubtedly represented the broad concept later established in Judaic law.

228. If a builder builds a house for someone and completes it, he shall give him a fee of two shekels in money for each sar of surface.

229 If a builder builds a house for some one, and does not construct it properly, and the house which he built falls in and kills its owner, then that builder shall be put to death.

230. If it kills the son of the owner the son of that builder shall be put to death.

231. If it kills a slave of the owner, then he shall pay slave for slave to the owner of the house.

232. If it ruins goods, he shall make compensation for all that has been ruined, and inasmuch as he did not construct properly this house which he built and it fell, he shall re-erect the house from his own means.

233. If a builder builds a house for someone, even though he has not yet completed it; if then the walls seem toppling, the builder must make the walls solid from his own means.

The other tower of the suspension bridge carrying laws and traditions from Europe to America comes from Greece and the great Athenian scholars. Their contribution to law came from the simple concept that an **adversarial system** of resolving disputes would be better than the "stronger take all" approach that had prevailed for centuries. In the adversarial approach, one party would oppose the other not with violence but before a neutral judge. To do this effectively, the rule of law had to involve more than criminal actions. Disputes over who promised to do what for whom had to be settled in a "civilized" way. *Civil* laws had to be written and enforced. Socrates and others began with the premise each individual can know right from wrong. Aristotle moved that concept even farther by acknowledging that what is important for the whole supersedes what is good for the individual, and he concluded that disputes should be settled through adversarial means within a legal system. Thus, the basic "day in court" was confirmed. He pointed out:

> A man is the most noble of animals if he be perfect in virtue, so is he the lowest of all if he be severed from law and justice.

However, the European traditions of civil law that still exist today are only the start of the foundation of the law as recognized in the United States. Most of the foundation and support for our laws came from England to the original colonies, and this heritage has been termed common law. Common law is civil law with a very different method of determining guilt or innocence, as well as restitution.

Common law is not as codified as civil law (such as Hammurabi's codes). Civil law has legislated rules to be applied to each case, and in civil law the codes are continuously updated for each possible offense or situation and the punishment exactly described. Conversely, common law relies upon precedence in previous cases and goes back to the concept that we can know right from wrong without it having to be specifically legislated. Of course, this common law system requires an immense amount of record keeping to document what has gone before, and it also gives extraordinary powers to judges to interpret the law. This has

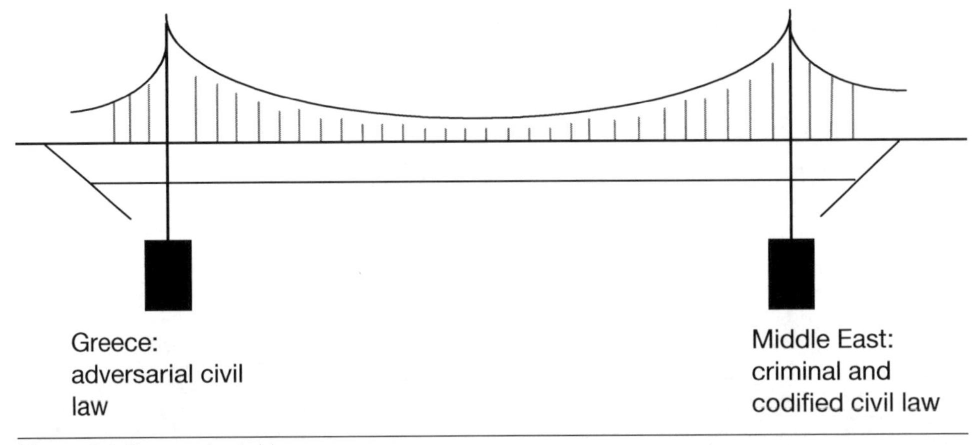

Figure 2.1 Suspension bridge

given rise to some criticizing the system, because judges with the ability to enforce what they believe the law should be have the power to "legislate" from the bench. Common law also carries with it the burden of a long and often tedious appeal process if the parties feel improper interpretation of precedent cases has been imposed. This can mean that recovery or restitution for a claim can be further delayed. Appeals are also available in civil law courts of Europe, but the conditions are often more restricted.

There are, of course, still many legal issues in common law that are similar to and often identical to civil law. Whether we want to view them as civil or common, in construction the vast majority of our legal issues fall under the non-criminal law system. However, in accordance with the typical use of language today, we will often in this text refer to the non-criminal issues of law as civil law, even though, for the most part, the lanes that have carried legal traffic to the United States came from ramps that originated in British common law.

* * * * * * * * * *

This discussion has so far involved centuries- and millennia-old historical background related to our legal system. A new habit you might want to develop in relation to studying history is studying recent and current local history. Specifically, make it a habit to go back and study past projects. Know what went wrong, and even more importantly, study what went right. "Those who remember the past can avoid repeating it." Additionally, those who study the past can repeat its successes.

* * * * * * * * *

It is important to point out that just as it is possible to change lanes on a bridge, it is also possible that civil actions can be prosecuted on a criminal basis, and the opposite has also been true.

In civil law, we recognize the right of each person to protect their own self and their property. Civil law, therefore, usually involves the individual seeking his or her own justice. Criminal law involves a superior entity such as a state or the Federal government enforcing the peace for the whole community.

Example: If a Foreman and a Superintendent are arguing on the project site and the Foreman in frustration throws a scrap 2x4, and it happens to strike the Superintendent's truck, that would result in a civil case so the Superintendent could recover damages done to the truck. However, if the blow was interpreted as actually intended for the Superintendent and not his truck, then assault would be the issue, especially if the 2x4 actually hit the Superintendent. In such a situation the case could become a criminal one if someone called the police and the Superintendent decided to press charges.

A CASE IN POINT

Most of us understand that a person cannot be tried twice for the same crime. However, the distinction between criminal and civil acts can change that considerably. O. J. Simpson was a former NFL football star, the first ever to rush for more than 2,000 yards in a single season. On June 12, 1994, Simpson's former wife, Nicole Brown, was brutally murdered along with her friend Ronald Goldman. The murderer in an apparent violent attack used a knife. A bloody glove led to O. J. Simpson's arrest, but during the controversial trial the glove was produced as evidence, and when he was asked to put it on his hand, it did not seem to fit, leading to a famous phrase from the defense's closing argument that, "If it does not fit, you must acquit." Simpson was found not guilty in the criminal trial of the horrendous murders.

The parents of both victims were certainly not convinced of Simpson's innocence and they pursued and won a civil action against him for wrongful death. The parents were awarded $33.5 million in damages, of which Simpson has paid very little; and, of course, the money could not bring back the two victims.

Some believe the difference in the two judgments was due to differences between criminal and civil procedures of evidence as well as the experience gained by the new team for the prosecution in relation to what had not worked when the State of California's attorneys had tried the criminal case.

The importance of recognizing the potential to prosecute in either "lane," (criminal or civil) for those in construction is to understand the danger of our own industry. Sometimes New York City prosecutors are taking what used to be unfortunate and tragic civil actions in which crane collapses killed construction workers as well as innocent pedestrians and turning them into criminal cases where the penalties involve more than money. In most cases a corporation and even individuals can insure against civil actions due to accidents, but criminal actions carry much more severe penalties, including imprisonment, that cannot be covered by an insurance policy, and the defense of such claims is also not covered by the typical insurance policy.

Figure 2.2 Construction cranes
Photo credit: Charles W. Cook

Regulations on construction cranes have become increasingly strict, but too many accidents still occur. Failure to comply, or inadequate inspection, as well as lack of updated training for operators, can lead to fines, at a minimum. Should an accident occur, severe civil restitution for damage to property, injury, or loss of life can be imposed through tort claims. In addition, some prosecutors are finding reason to press criminal charges against Constructors failing to perform crane operations safely. The Contract Administrator involved in a project with one or more cranes must recognize the extreme need for continuous caution and diligent compliance with all regulations.

* * * * * * * * * * *

The key point to recognize is that there are possible multiple consequences of doing something wrong – particularly if your actions can be construed as "criminally negligent." Even while our workers are exposed to extreme dangers, we must also recognize the potential for disaster to the general public. We work next to speeding traffic on our highways, or near passengers and freight trains carrying commerce over rail lines, or above pedestrians walking on sidewalks as beams are carried to the top of skyscrapers. We cannot dismiss our responsibilities. We are "our brother's and sister's keeper." We do not need laws to tell us that. Socrates knew we all could know right from wrong, but with so many distractions and requirements competing for our attention it is too easy to put off vigilance when we most need it.

* * * * * * * * * *

Exercise 2.1

Take a moment

Think about how you view criminal and civil fault

Below is a scale with two opposites – Criminal wrong and Civil wrong. At what point or under what circumstances would you consider yourself closer to being criminally wrong rather than merely civilly wrong.

Criminal ____ Civil

Circumstance: A beam being hoisted by a tower crane falls twenty-two stories, bounces off the trailer that had transported the beam to the project site, and lands in the street.

Does it matter how you were involved? Does it matter what or who was injured? Or, why the accident took place?

Discussion of Exercise 2.1

There are probably several issues you might consider before you would think you might be criminally liable:

1. What was your position on the project? Were you on the site as a Superintendent or were you off site as a Project Manager?
2. Were you part owner of the crane?
3. Did you have anything to do with choosing this crane and/or operator over another company or operator?
4. Did you give any direction to those operating the sling or the crane?
5. Had you known of previous difficulties on the project related to the crane, the operator, or the iron workers wrapping the sling?
6. Had you any choice in selection of the staging area and placement of the crane related to traffic in the street?
7. Was it only property that was damaged (e.g. the trailer, some fencing and scaffolding, and maybe a parked truck or car)? Or were there fatalities?
8. If there were fatalities, were they workers on the jobsite? Were any pedestrians or drivers killed?

9 Could there have been alcohol or drugs involved in any of the crew working on the lift? Did you have suspicions before the accident that drugs or alcohol were being used? Did you actually know drugs and alcohol were being used?

The list of possibilities could go on. One important conclusion you might draw, even if you find yourself completely innocent of any criminal negligence, is what you think and what a prosecutor might think could be two very different things. Defending a criminal action can be costly and emotionally draining. The protection that insurance often provides for a civil wrong is not available for a criminal wrong.

As with many of these exercises, there may be no right or wrong answer, but it is better to think about them before you are in the middle of a disaster, and hopefully, by doing so, you will avoid such circumstances.

Examples: At this point let us take a moment to differentiate between criminal, tort, and statutory/regulatory law:

1. Someone steals tools or equipment from a project site; this is a crime.
2. A concrete truck backs into another vehicle while delivering concrete; this will bring a tort case, usually to be settled by insurance companies outside of any court proceedings.
3. Driving being considered by each state a privilege rather than a right, someone driving without a license or insurance will be violating the statutes and regulations of the state in which they are driving.

Just as we began this section of Chapter 2 with the warning from George Santayana about history, perhaps it is wise to conclude this section with another observation, hopefully one that is less dire. Outside the National Archives building in Washington, DC, there is a statue by James Earl Fraser of a young female figure holding a child and a sheaf of wheat in her right hand as symbols of the future and hope, while in her left hand she cradles an urn with the ashes of past generations. The quotation on the base is from a speech by the abolitionist Wendell Phillips, who spoke about taking the lessons of the past into a successful future.

The heritage of the past is the seed that brings forth the harvest of the future.

2.2 Contract law

I shall fulfill my contract. No more nor less.

Lillie Langtry

Another chapter will deal more specifically with two fundamental issues related to contract law: (1) what is a contract? and (2) what elements and/or documents are included as part of the contract?

For now it is important to recognize that a contract creates obligations on the parties making the contract. If one or another of the parties fails to meet the obligations agreed to, the other party can have the contract provisions enforced or remedies (usually of a financial nature) determined by a court. A **contract** in its most fundamental definition would be an agreement between two or more parties for which if there is a **breach** of the agreement a court can determine a **remedy**.

Figure 2.3 Heritage Statue at National Archives Building
Photo credit: United States Government

2.3 Contract forms

All sensible people are selfish, and nature is tugging at every contract to make the terms of it fair.
Ralph Waldo Emerson

The most widely used pre-printed contracts are the ones published by the American Institute of Architects (AIA). They consist of over a hundred separate contracts that deal with different relationships within the industry from standard agreements between Owner and Architect to a standard agreement for *Pro Bono* (for free or for the good) work by the Architect for the Owner. The most used and perhaps best known for Constructors is "A101 Standard Form of Agreement Between Owner and Contractor," where the basis of payment is a stipulated sum.

These contracts are copyrighted and sold for use through the AIA. For a fee access can be obtained and the documents downloaded on a computer.

Representing Architects, the AIA is obligated to protect its members within the legal provisions of the contract, and for that reason Contractors have found some provisions of the AIA contracts onerous. In 2007, the Associated General Contractors of America (AGC), representing its Contractor members, refused to endorse the AIA contracts and also published a series of over a hundred contract forms separated into categories of potential need or use:

- 200 Series – General Contracting
- 300 Series – Collaborative
- 400 Series – Design-Build
- 500 Series – Construction Management

- 700 Series – Subcontracting
- 800 Series – Program Management

To gain popular support a coalition of leading design (but not the AIA) and construction industry associations have been included in the drafting or endorsing of the documents.

A third major series of standard contracts is by the Engineer's Joint Contract Documents Committee (EJCDC). These come from a joint input of the American Council of Engineering Companies along with the American Society of Civil Engineers and the National Society of Professional Engineers. Civil projects, such as bridges and highways, are areas where these contracts are particularly useful and often used.

In addition to these three major contract forms, Owners often have their own contract forms. In general these forms heavily favor the Owner, and the Contractor has to be careful of some of the potentially burdensome clauses that shift risk from the Owner to the Contractor, often when the Contractor has no control over the potential risk.

AIA and ConsensusDOCS

The AIA has pointed out that the reason the AGC did not endorse the AIA forms was its intention all along to come out with its own documents. The AGC counters that its contracts are fairer to all the parties involved in the construction process, pointing to DOCS in the title itself referring to **D**esigners, **O**wners, **C**ontractors, and **S**pecialty contractors, **S**ureties, and **S**uppliers. In 2013 the list of associations involved in the DOCS included:

Air Conditioning Contractors of America
American Society of Professional Estimators
American Subcontractors Association, Inc.
Architectural Woodwork Institute
ASFE/The Geoprofessional Business Association
Associated Builders and Contractors, Inc.
Associated General Contractors of America
Associated Specialty Contractors, Inc.
Association for Facilities Engineering
Association of the Wall and Ceiling Industry
Construction Financial Management Association
Construction Industry Round Table Construction Owners Association of America
Construction Specifications Institute
The Construction Uses Roundtable
Door and Hardware Institute
Finishing Contractors Association
Independent Contractors Association
Independent Electrical Contractors
Lean Construction Institute
Mechanical Contractors Association
National Association of Construction Auditors
National Association of Electrical Distributors
National Association of State Facilities Administrators
National Association of Surety Bond Producers
National Association of Women in Construction
National Electrical Contractors Association

National Electrical Contractors Association
National Ground Water Association
National Hispanic Construction Association
National Insulation Association
National Roofing Association
National Subcontractors Alliance
National Utility Contractors Association
Painting and Decoration Contractors of America
Plumbing Heating Cooling Contractors Association
Sheet Metal and Air Conditioning National Association
The Surety & Fidelity Association of America
Water and Wastewater Equipment Manufacturers Association
Women Construction Owners and Executives

Once again, the devil will be in the details, and the Contract Administrator cannot stop simply at the definitions clauses, or the scope of work and compensation paragraphs. The journey through any contract is often convoluted and tedious. If Emerson is correct that all sensible people are selfish, we will have to look after ourselves as others are looking after themselves. The key will be, as with nature, finding a fair balance.

Exercise 2.2

Take a moment
What's in a word

Contracts are obviously made up of words, and words can be quite powerful. Changing a word or even some punctuation can change the entire meaning of what is being conveyed. The challenge for many people today is txtng. Certainly, the elimination of vowels can be a challenge to understanding unless you know the shortcuts from practice or experience.

In ancient times, those who wrote the text for the Bible dropped vowels quite frequently to make the writing quicker. That has led to some debate by scholars about certain words.

More importantly, within our own language we have confusion. Imagine someone learning American English and being faced with the conundrum that we *park* on our *driveways*, but we *drive* on *parkways*.

As communication becomes faster and more ubiquitous, we have to be quite careful. The wrong word here or there can make a big difference in contract administration.

Here is a quick example of how future scholars, or maybe a project partner, might be stumped in some instances because they are not sure you used the correct vowel on two otherwise identical statements. What is the difference if a Supplier e-mails you:

"That will affect our warranty"

or

"That will effect our warranty"?

Either one could be the case, but the issue and potential challenge for the Contract Administrator will be quite different.

Discussion of Exercise 2.2

In the first example, affect is used as we normally understand it, meaning "to have an influence upon the warranty." In this case the Contract Administrator might need to inquire what the issues regarding the warranty might be. If it will alter or decrease the warranty in some way the Owner needs to know.

In the second example, "effect" is being used in a way we do not usually use it, but it might be the way the Supplier does. Basically, the Supplier is telling you the warranty has started as a result of whatever "That" was. This is very important, especially if the Owner does not think or know the warranty period has begun.

It could have been a typo, or maybe not. Words do matter, and the Contract Administrator has to know how much and to whom they matter.

2.4 Tort

No man is above the law; and no man is below it.

Theodore Roosevelt

Other than contract law, we want to recognize, at least briefly for now, that there are a couple other "lanes" of construction law important to the Contract Administrator in construction – tort and statutory/regulatory law.

These will be discussed in more detail in later chapters, but generally we might think of tort law as related to the responsibility to act reasonably toward each other. Courts consider tort law to involve three broad issues; (1) a civil wrong, (2) between two parties that do not have to be contractually obligated, (3) in which the wronged party is entitled to compensation.

Examples:

1. A pedestrian is injured by falling debris from a construction site.
2. A car is damaged by a construction vehicle that was on route to make a delivery to a project.
3. In order to gain favorable consideration, another Contractor falsely tells an Owner that a competitor failed to properly fulfill a contract for another Owner.

Note: Torts can be negligent or intentional. In the first two examples there was negligence. The third example, however, is an intentional tort.

As we recall, civil law is different than criminal law. In tort law, which is part of the civil law system (founded on common law principles), it is usually the wronged party that initiates the prosecution of the civil wrong. Federal, state, or local governments such as municipalities or townships have court systems to hear cases regarding civil wrongs, but the initiation of a case is usually through a lawyer or representatives of the wronged party. In essence, tort law is in place because each person within a community is entitled to equal protection from harm to his or her person or property. We will deal with many of the specifics of tort law in Chapter 15, but it should be restated that some tort actions can also be transferred to criminal prosecution, depending on the actions and the injuries sustained by the wronged party.

2.5 Statutory/regulatory law

If you have ten thousand regulations you destroy all respect for the law.

Winston Churchill

Even Hammurabi's 282 codes are minor in comparison to the volumes of rules we now have to comply with or face the consequences. Not all of the rules we must follow, however, are laws in the strictest sense of the word. Some of them are statutes or regulations passed by governing bodies empowered to set standards for the conduct of everyone within their jurisdiction.

Sometimes statutes and regulations do not affect an individual. For example, most of us benefit from the rules and regulations that set standards for restaurants, but we certainly do not expect an inspector to visit us in our own kitchen or dining room at home.

Within construction, however, there are many statutes and regulations each corporation must follow. Statutory and regulatory laws need to be recognized up front, since they may cause the greatest "hidden" factors in contract administration. There will be clauses within contracts and obligations that are not stated in the contracts that each Constructor must comply with in order to do business and successfully complete a project.

Examples of statutes and regulations:

1. Distances between sprinkler heads in a ceiling becomes a requirement as part of building codes.
2. The required amount of compensation by craft is regulated on government-funded projects.
3. Providing runoff protection to insure construction waste does not pollute streams and other waterways is part of environmental regulations.

There is an important adage related to law enforcement, and that is:

> Ignorance is not an excuse.

Never try to tell an inspector, an officer of the law, or the judge and jury you did not know you were doing wrong. In order to do business you are expected to understand the rules that protect everyone. From speed limits to clean air, we all have to do our part, and sometimes the toughest part of doing one's part is knowing what part is yours. Following the obligations of the contract is a good and very important beginning. Not doing harm to others through some tort action outside of the contractual relationships is also important. Doing everything within the statutory/regulatory boundaries set by the various governments and agencies associated with construction is also very important and usually monumental. No one person can know all the statutes and regulations, or that person would lose more than respect for the law; he or she would lose all time for doing anything else. It is here that one must not only work as a team with the Owner and Designers, but we all must create an atmosphere in which everyone can contribute their expertise within their area of project control. Excavators will need to help with proper runoff protection. Electricians will need to make sure lockout tag-out practices are correct. Accountants will need to monitor proper payroll reporting. In short, on most projects no one person can do everything, but each Contract Administrator must know there are those who can cover all that is needed to be done. If you know, for instance, the electrician is unfamiliar with lockout tag-out procedures, then you will have to make sure someone correctly advises the electrician and monitors compliance.

Figure 2.4 Ulysses S. Grant
Photo credit: Brady-Handy Photograph (Library of Congress)

Ulysses S. Grant, as President of the United States, would often take a break from the stress of the office and walk a couple blocks to the lobby of the Willard Hotel. There he would intend to have a drink and perhaps smoke a cigar, and there he would often find several people wanting to grab his attention about one issue or another. He never looked forward to these encounters, but he enjoyed the break from the White House more, so he derogatorily termed these individuals "lobbyists."

Today many of the statutes and regulations have been "rewritten" or influenced more by lobbyists than by the President or members of Congress.

2.6 Above and beyond the law – ethics in the workplace

Grub first, then ethics.

Bertolt Brecht

Whereas the Federal, state, or local governments will enforce laws, personal ethics are enforced solely by one's conscience, and therefore do not necessarily have a penalty associated with them. Group ethics can be enforced through the "governing" association, such as the American Bar Association, made up of over 400,000 lawyers and law students. Penalties for violation of any group's ethics code might include sanctions within the group or even expulsion from the group.

We will have a lengthier discussion of **ethics** in the last chapter of this text. However, the subject is something that always needs to be present within all successful management of projects. Just as with contract law, tort, and statutory/regulatory law, ethics is introduced early in the discussion of Contract Administration. As the playwright Bertolt Brecht implies, however, knowing what is right is sometimes quite different than our actions turn out to be when the stakes are personally high. The worldwide hit musical *Les Miserables*, based on Victor Hugo's novel by the same title, follows the life of ex-prisoner Jean Valjean, who was released after almost twenty years' imprisonment for stealing a loaf of bread to feed the starving son of his sister. Hugo was certainly writing against the civil law practices of excessive set penalties for certain crimes (France was under the civil rather than common law system of the English courts) as being too harsh on the low and "miserable" of our society, but he was also acknowledging that there must be a higher law above human law.

For some, this knowledge of right and wrong is not just "set in stone" Commandments of morality. Knowing what is right is personal ethics, but we also must recognize now that ethics can also be determined by a group as a code of conduct they determine is appropriate for themselves. Professionals, such as doctors and lawyers, often do this as standards for their actions.

In such a case ethics as a standard that a group of people set for themselves is important, but, of course, such standards may not agree with the way the individuals conduct business or agree with the standards other groups set for themselves.

For example, **bid shopping** is widely practiced in the construction industry. Sometimes, an Owner "shops" bids, because the Owner wants a particular Contractor to do the work even though that Contractor was not low. Some Owners and Architects will combine forces in post-bid-submission meetings to discuss the various parts of the job and costs of particular elements. Good ideas or low **Subcontractor** costs of one party are sometimes shared with other Contractors who then reduce their higher prices.

Some Contractors are also notorious bid shoppers, acting as though every bid day is Black Friday with lots of shopping bargains and the thrill of a deal is in each Subcontractor's bid. On bid day, some estimators "leak" low prices (sometimes fictitious low prices) with eager Subcontractors who want to have the lowest price "on the street" to get the job, particularly if key Subcontractors have to be named on the bid form.

Other Subcontractors or Suppliers will generally wait until after the contract award, when the successful Contractor will contact them and try to "shop" a lower price to make up for money they may have "left on the table" (the difference between their successful low bid price and the second bidder's price).

And, of course, bid shopping can also proceed from Subcontractors to Sub-subcontractors and Suppliers.

As many within the industry acknowledge, "It is easy to be honest for $100, but the stakes are different when there are thousands, even hundreds of thousands, or sometimes millions of dollars involved." Then we are talking some real grub.

Ultimately, then, ethics must be considered individually, but it is far better to understand one's ethical position before the situation arises or is thrust upon us. Although the grub may change, Brecht's observation is appropriate. From Jean Valjean's loaf of bread to the needs each of us feel we cannot live without, we all will face challenges to do the right thing.

When you know your values the right choice is obvious, but not always easy.

Exercise 2.3

Take a moment
Think about your own emotional intelligence

Sometimes our own human nature can be a challenge when it comes to acting ethically, or even effectively, when confronted by one of the many challenges in construction or our lives. There is an important aspect to each person's successful contract administration, and that is what scientists call **emotional intelligence**. How we react emotionally to the daily challenges of project management, as well as how we handle the seemingly constant pressure of new demands that can impact schedules or affect the quality and conformance to design requirements, is essential to successfully administering contracts.

You may want to go back and think about how you react to conflict. If possible it would be nice to go back to our first "primal cry," when a doctor smacked each of us on our bottom right after we were born. Was your reaction "Hey, it's not fun out here. Put me back where I was nice and warm," or was it "Hey stop that, or I'll smack you one as soon as you turn me around!" Of course we had no words, only emotions, but what were they? They were probably the foundation of how we respond to conflict and challenges today.

We all have a "fight or flight" reaction to things that happen to us. Think about how you handle challenges and conflict. Do you go right at them for a solution or do you put them aside for a better time? If an issue comes up do you solve it then and there or do you say "I'll get back to you." In a meeting do you confront opposition, or do you wait until after the meeting to assemble allies for your position?

As with any of the personal development exercises, there is no right or wrong answer, but just thinking about your own style can be important. When it comes to the issues and challenges you face, where do you place yourself on the compete (fight) or avoid (flight) scale? Be honest with yourself, because the most growth will come from understanding who you truly are emotionally.

Compete ____ Avoid

Discussion of Exercise 2.3

There may be excellent reasons to either compete or avoid any given situation. If the opposition has a cannon and you only have a sling shot, unless you are David, it might be best to avoid Goliath. On the other hand, if someone is refusing to do something safely and is endangering others, can we afford to look away? Sometimes we can compete best by getting help (a bigger cannon or someone with the authority to stop an unsafe action). Police are seldom the karate experts seen on television. They more often overwhelm a suspect with numbers. So getting help makes good sense.

The key to the exercise is to get us thinking about our emotional response to issues and challenges, because it actually comes before our "intelligence" response.

How do we respond to the barking dog? Do we tell it to be quiet or do we walk on the other side of the street? Again, it may depend on how big the dog is. And that is also true of how big the problem is for each of us as a Contract Administrator. If we do not perceive the danger, then we react differently. The question then is what is our tolerance for problems – what is our threshold of crisis?

Whatever our reactions are, they are not wrong. They are who we are. This helps us to know how we are going to best handle the future. Whether we first tend to compete or first tend to avoid, we now can look up to the horizon and see which way we need to go to get to that bright future.

2.7 How to get to where we want to be

The good thing about the future is it comes one day at a time.

Abraham Lincoln

If, as we have discussed, each member of the project team is expected to contribute their expertise, and if we intend our contribution to be the realization of the Design Professional's vision and the fulfillment of the Owner's needs, then our task is monumental, the risks are many and often unknown, and the prospect of failing is usually hidden but ever present. To succeed it seems only logical that working together will minimize hazards.

To find the Designer's vision and the Owner's needs we can start with the contract. In each contract is our promise or set of promises, but true success comes from "under-promising and over-delivering." You become the Contract Administrator champion by doing just that – going beyond the promises.

The contract is the base from which we start. And the nice thing about contract law is that the obligations are contained within the documents that have been signed and made a part of the project obligations. When we study tort and statutory/regulatory law in more depth, that will not be the case.

We must always keep in mind that the subject of construction contract law is a work in progress for every Contract Administrator. The lessons will continue long after this text is finished. As we take each step (often repetitive ones), it is sometimes good to look up and see the brightness of the future ahead of us. We need a vision to refresh us along this journey that will not end so long as we are involved in the industry.

The construction industry has a great diversity of talents and needs among the many contributing parties. Satisfying the specified needs of each contract is the first and crucial step to meeting one's obligations. It does not, however, assure ultimate success.

Client satisfaction, repeat business, rapport leading to ease of flow for information and scheduling, and going above and beyond are all factors that turn a job into a career, and a project into a success.

Seen on a bumper sticker:

Go the extra mile – it's never crowded

INSTANT RECALL

- For our study, law can be separated into two main areas: (1) criminal law and (2) civil/common law (including contract law, tort, statutory and regulatory law).
- Where criminal law seeks to punish the offender, civil law attempts to restore or make whole the injured party. The European traditions of civil law that still exist today are only the start of the foundation of the law as recognized in the United States. Most of the foundation and support for United States laws came from England to the original colonies, and this heritage has been termed common law.
- Common law relies upon precedent in previous cases and goes back to the concept that we can know right from wrong even without it having to be legislated.
- The dangers of our own industry can bring profound difficulties upon Constructors, even changing civil wrongs to criminal prosecution.
- A contract in its most fundamental definition would be an agreement between two or more parties for which, if there is a breach of the agreement, a court can determine a remedy.
- Generally we might think of tort law as involving three broad issues: (1) a civil wrong, (2) between two parties that do not have to be contractually obligated, (3) in which the wronged party is entitled to compensation.
- Within construction there are many statutes and regulations each corporation must follow.
- The subject of construction contract law is a work in progress for every Contract Administrator.
- How we react emotionally to the daily challenges of project management, as well as how we handle the seemingly constant pressure of new demands that can impact schedules or affect the quality and conformance to design requirements is essential to successfully administering contracts.

YOU BE THE JUDGE

In most of our "You be the judge" exercises, you will read about actual cases, but for this chapter let us practice a bit on determining what are contract cases, what should be a tort claim, and what are statutory/regulatory issues. Make your determination and then go to the Companion Website for a brief discussion of each case.

Circle your choice:

1. The Owner refuses to make the final payment because the building inspector has found a code violation.

   Contract claim *Tort claim* *Statutory/regulatory*

2. The Owner refuses to make the final payment until a full-year warranty is delivered on the mechanical equipment.

   Contract claim *Tort claim* *Statutory/regulatory*

3. The Owner refuses to make the final payment because the Contractor is seeking additional compensation for work the Owner did not show on the drawings.

Contract claim *Tort claim* *Statutory/regulatory*

4. A driver was struck by a construction vehicle that was returning to the shop for refueling at the end of the work day.

Contract claim *Tort claim* *Statutory/regulatory*

5. Another driver was struck by a construction vehicle while the driver of the construction vehicle was heading home after the work day.

Contract claim *Tort claim* *Statutory/regulatory*

6. A Contractor uncovers asbestos above the ceiling. The Owner had no knowledge of its existence, but the Contractor stops work immediately and informs the Owner.

Contract claim *Tort claim* *Statutory/regulatory*

7. A Contractor discovers asbestos above the ceiling and removes it as quickly as possible in an attempt that no one will notice or report it

Contract claim *Tort claim* *Statutory/regulatory*

IT'S A MATTER OF ETHICS

Different cultures have different value systems. For instance, there are many stories about how "you can't get anything done in XYZ country" if you do not bribe everyone along the way. What would you do in such a situation? Why?

Going the extra mile

1. Go to the Companion Website www.routledge.com/cw/cook and view the video discussion Part II of the Subcontract discussed in Chapter 1.
2. Go to www.constructingleaders.com and view the video "Moving the Gruber Wagon Works." The video will be a prelude to the text in a future chapter related to alternative project delivery methods, such as Design-Build.
3. After you have made your own rulings on the above "You be the judge" cases, go to the Companion Website www.routledge.com/cw/cook for a discussion of the example cases.

4. **Discussion exercise I** (with friends or as part of a classroom discussion): In a capitalist society, we generally figure the lowest price is the best way to go. Construction, however, is a very complex industry. From the moment an Owner considers what Design Professionals to use through the final completion by the Constructor, there is a tension to balance cost with quality. Included in the quality is the service, responding to needs, meeting schedules, and communicating effectively. Additionally, a safe project and jobsite, along with longevity of the structure past warranty, is not something that is determined by lowest cost. Without lowering one's price, what would you use to promote your superior services? Do you think your approach would be effective? As a Contract Administrator, how would you ensure what you say you can deliver?

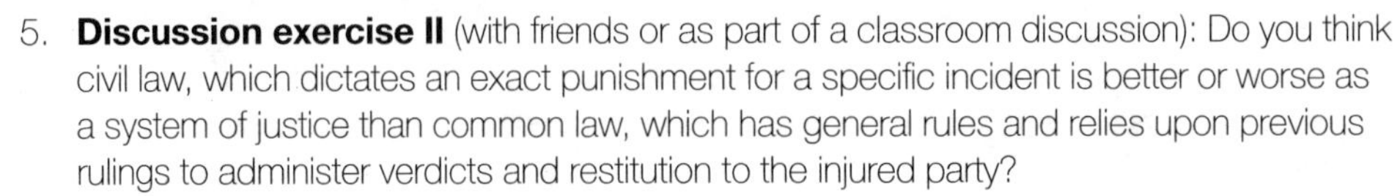

5. **Discussion exercise II** (with friends or as part of a classroom discussion): Do you think civil law, which dictates an exact punishment for a specific incident is better or worse as a system of justice than common law, which has general rules and relies upon previous rulings to administer verdicts and restitution to the injured party?
6. Effective use of the correct vocabulary improves communication and promotes respect among colleagues. The reader should become familiar with the key words listed below that have been used in the chapter. Definitions can be found in the Glossary. Additionally, the student is encouraged to use the flash card exercise in the chapter's Companion Website as practice for possible quiz or exam questions.

Key words

Adversarial system

Bid shopping

Breach

Contract

Emotional intelligence

Ethics

Remedy

Subcontractor

CHAPTER 3

Owners, Designers, and Constructors

For millennia there was the builder and the Owner. Recently (in the past two and a half centuries) the design process has for the most part separated from the construction process, creating three primary entities involved in a construction project: The Owner (with ultimate authority), the Designer(s) to create vision and insure quality, and the Constructor(s) to make the vision a reality.

Student learning outcomes

Upon completion the student should be able to. . .

- **know the three basic entities involved in the traditional Design-Bid-Build construction project**
- **recognize the basic relationships and responsibilities of the Owner, Designer, and Constructor**
- **distinguish between single and Multi-prime Contracts**
- **analyze how he or she might relate to the challenges of contract administration.**

The foundation of our legal system is based on enforcement by a superior force (local, state, or Federal) and representation through an adversarial system. The Greeks found the adversarial system also a model for government, and we can often see today the problem of two parties in adversarial relationship working together for the good of the country.

The same can be true if we approach a project on an adversarial basis. Unfortunately, based on a Design-Bid-Build project contract, an adversarial relationship can quickly and easily be the rule rather than the exception. The Design-Bid-Build sequence sets up a contractual attitude in many Constructors to do the least while being paid in full. Mistakes in bidding are common, so finding relief through change orders becomes a means of making a losing project or estimate profitable, or at least break even. In such an atmosphere, the adversarial relationship destroys project harmony and often disrupts progress.

Figure 3.1 U.S. Capitol, Washington, DC
Photo credit: United States Government

In some respects the founders went back to Aristotle's adversarial system of conflict resolution when designing the legislative branch of the United States government. They divided it into the Senate and the House of Representatives. To further "fuel" the adversarial relationship, however, there are basically two parties, Democrat and Republican. This dueling duality was not part of the founders' vision. As much as Aristotle's system seemed to work in ancient Greece, where violent methods of resolving conflict had prevailed for centuries, the adversarial relationship of the two parties and the two chambers (Senate and House of Representatives) has supported the public perception that Congress is the least effective branch of the United States government.

Keep that lesson about adversarial conflict resolution in mind when choosing how to approach ambiguities, controversies, and conflicts in project management and contract administration.

Learning in the previous chapter how we got to where we are was a beginning. Now that we have crossed that bridge, however, it is the task of the effective Contract Administrator/Project Supervisor to eliminate the need for adversarial relationships. To do that requires a thorough and objective understanding of the contract.

Henry David Thoreau, in commenting on the Greek meaning for the word *kosmos* (our cosmos or universe) noted to the Greeks the *kosmos* meant beauty and order. The universe was the beauty and order they could see in the world around them. Each project we administer is our *kosmos* – to which we are to ensure beauty and order.

3.1 The project parties

The will of the people is the best law.

Ulysses S. Grant

For all of their existence, birds, bees, and beavers have not changed their designs, and they continue to function perfectly. Humans are never fully satisfied with what has been. We strive for what comes next, and sometimes it comes with initial failure. It can be an unforeseen field condition or a simple piece of something that won't fit where it was drawn that delays a whole project crew from moving forward effectively. It can be as small as an epoxied bolt that ultimately fails and brings a precast concrete panel in a tunnel crashing down, killing a motorist, or it can be as large as the whole design of a bridge based on previous bridge designs, but incapable of resisting the sustained forty-five mile per hour winds that are common in the Tacoma Narrows. Whatever the cause, failure is possible. The only way to minimize the potential for failure will be open and effective communication among all the parties in the construction process. To the extent the contractual relationships prevent that, the task of the Contract Administrator becomes an even greater challenge. And to the extent contracts are written to shift risks from one party to another, the challenges become the prelude to disaster.

The vast majority of the Contract Administrator's experience and effort will be satisfying the requirements of each project's contract, so in the heading above "parties" does not refer to "a good time." It refers to the individual companies and entities that have to work together effectively. If they do, then there will be good times. If they do not, then discord and difficulties will follow, with the potential endgame of individuals totally unassociated with the construction industry making a determination or judgment about work they may have little experience or knowledge of or of how things are done in the industry. Judges and juries will try to be fair, but in the end, the message is clear:

> If at all possible, avoid remedies imposed by courts.

Fortunately, there are a few steps before a problem becomes a claim involving courts or arbitration:

1. A problem arises that requires a solution.
2. Solutions are sought.
3. Disagreements are "aired."
4. Appeals are made for a reasonable resolution.
5. Parties fail to reach a resolution, and one or both parties determine to either drop the matter; or
6. to escalate it to litigation or another means of conflict resolution outside the project team.

How you handle these stages along the way usually determines whether the next step is necessary.

Avoiding litigation is what could make you indispensable, but the task is not always easy. If one approaches Contract Law with the attitude of "What can be gotten away with," the likelihood that eventually something will not be "gotten away with" becomes greater. There is, however, a fundamental challenge every Contract Administrator faces in a typical Owner–Contractor relationship that has developed out of the Design-Bid-Build system of construction project delivery. (Other methods of project delivery will be discussed in a later chapter.)

In general there are three parties to the project, but there are usually two separate contractual relationships, immediately setting up challenges for communication and responsibility:

3.1.1 Owner in the middle

From Figure 3.2, it immediately becomes obvious the Owner is in the middle. This three-party relationship first emerged about two centuries ago and will be discussed in a little more detail in another chapter. At first this separation of the Designer and the Constructor may have seemed a good idea. Owners might have thought the idea of "playing" one against the other would be an effective means of getting lower costs while maintaining higher quality. In reality, it has led to more and more finger-pointing disputes and often lengthy litigation. The Owner's main concern after contract award is usually schedule, and that too has often been affected by the separate parties hardening their position on disputed items.

As construction has become more complex and the techniques more technical, Owners have had difficulty finding within their own staffs enough individuals with sufficient expertise to handle the complexities of the construction process, particularly related to the schedules. The Owner does not want a space that is being built simply because it is easy. The Owner wants a space for what can be done in or on the space after construction. Those on the Owner's staff that are put in charge often have other tasks within the company to manage. Construction becomes a distraction. Contract management was not what they thought they were going to do for a living, and often for effective management of any construction process the Owner either has to employ more staff specialized in handling construction or hires an outside **Construction Manager** to oversee each construction project. Either way, while the Owner might have believed a couple centuries ago that the three-party relationship might have saved money, it usually does not work out that way.

What the Constructor in this relationship must keep in mind is that the Owner still needs the space completed on time, within budget, meeting standards of quality, and doing it all safely for all concerned. To do that, each Owner, often with the assistance of the Architect, who generally is the first person contacted regarding a new construction project, will provide the Constructor with a contract detailing as best as possible what needs to be done in order to achieve the results the Owner expects.

The contract between the Owner and the Constructor therefore becomes the essential document to achieving harmony between these two critically important parties. As easy as that may sound, the task can be complicated simply because the contract documents that the Owner has presented were not created by the Owner, but rather by the Designers. If there are ambiguities, discrepancies, or even worse, defects within the design documents, the Owner will expect the Designer and the Constructor to find a solution. However, there is

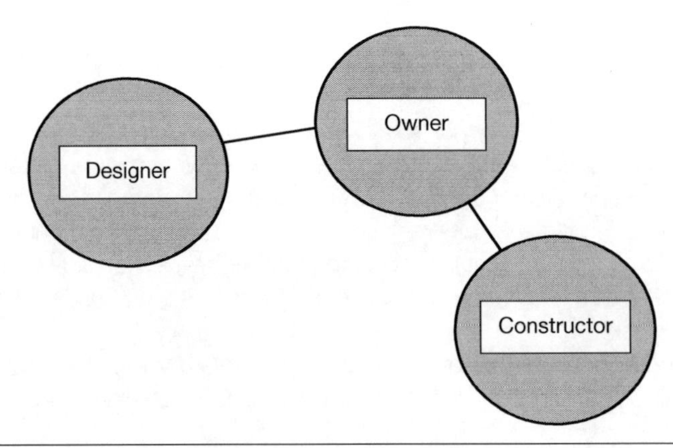

Figure 3.2 Typical Design-Bid-Build relationship of Designers–Owners–Constructors

usually no direct contractual relationship between the Designer and the Constructor. Therefore working together to find solutions that keep the project on time and within budget can lead to deterioration of project rapport very quickly as one party blames the other for delays or discrepancies.

Of course, the Owner has contracted with the Design Professional(s) with the understanding that the design and the construction as ultimately built and approved will function perfectly. In many construction projects today, Designers and Owners are relying more and more on specialty Contractors to provide details on the best erection practices and details for particular parts of a project. These are often submitted in the **shop drawing** stage of the construction process, and the Designers approve the construction, usually with some stamp that declares approval is for purposes of conformity with the design intent only (thus leaving out acknowledging structural integrity and responsibility). Meanwhile, the Constructor, in approaching the project, will rely upon the contract documents to perform the work as shown and be paid for delivering what was designed. If something goes wrong after the project is completed, the Owner is going to expect one or both parties are responsible. This can lead to further problems for the Owner as well as the Designers and Constructor.

Designs are becoming more and more complex. Often the Architect and Engineers rely upon the expertise of **Specialty Contractors** for input regarding specific features of certain materials. Whether in New York or another city, how many of us walk city streets never thinking a piece of glass might suddenly fall on us? We do hear about the frightening exceptions and sometimes the rare tragic accidents, but in order to insure buildings stay whole and bridges do not collapse, a team of individuals, including the Owner, the Designers, and the Constructors, have made vertical and horizontal construction safe. That takes a great deal of effort, and, of course, there can be some unfortunate failures, but it is the Contract Administrator's responsibility to try to eliminate any such issues before they occur.

Figure 3.3 Skyscrapers on a city street
Photo credit: Charles W. Cook

A CASE IN POINT

As was mentioned earlier, much of our legal system is based on English common law, which relies on previous rulings. How courts accrue responsibility and allot damages varies. Sometimes, however, the actual settlements of cases are sealed, and we do not know the results of what was ultimately agreed to by the parties in dispute.

Such is the case with the sixty-story John Hancock Tower in Boston. Shortly after installation, five 100-pound 4 x 11 foot panes of blue mirror glass went crashing to the pavement below. The police closed off the surrounding streets. The problem appeared to be due to the oscillation of the building, combined with thermal stress from the repeated heating and cooling of the glass. The design had called for inner and outer panes of glass, but the thermal expansion of the inner pane seemed to be forcing the outer pane out, causing it to fall. The solution was to install one pane in lieu of two. That was done at a suspected cost of five to seven million dollars. That was just the cost of repairing the problem, and not the actual settlement for damages. Who finally paid for all the costs will never be known, since the issue was settled by the parties and then sealed.

After the considerable delays, including moving final completion from 1971 to 1976, the John Hancock Tower won several awards for architectural excellence, including the AIA's National Honor Award.

There is also a story, unconfirmed, that when the Owner of the building, the John Hancock Insurance Company, threatened to recover the costs of repair from the Contractor, the Contractor quickly cautioned against that strategy – the Contractor's insurance company being John Hancock. At least that is how the story goes "on the street."

3.1.1.1 Multi-prime Contracts

One other issue an Owner often creates is multiple contracts for both the Designers and the Constructors. These **Multi-prime Contracts** will be discussed in another chapter, but once there is more than one Constructor (e.g. a General Contractor for the overall work, and then separate Electrical, Mechanical, and Plumbing **Prime Contractors** for those areas of work), then the coordination and the scheduling of work becomes even more difficult. If the Owner, for example, has also elected to contract separately on the Design side of the project, and has a separate mechanical engineering firm doing work outside of the Architect's responsibilities, then more issues can arise as to who is responsible for what. The finger pointing can erupt into an endless circle of blame – what some people call "Who struck John?"

Once again, for you as a Contract Administrator, it may not be your fault, but it will be your problem.

3.1.2 The Designer

If we divide construction into two broad types: (1) horizontal (highway and bridge construction) and (2) vertical (building construction), then we have two main categories of Design Professionals. In highway and bridge construction the Engineer is often foremost, and sometimes the only Design Professional required. In building construction, however, the Architect is almost always foremost and usually subcontracts the engineering of the building if there are not Engineers on the Architect's staff.

Figure 3.4 The Benjamin Franklin Bridge, a suspension bridge across the Delaware River connecting Philadelphia, Pennsylvania, and Camden, New Jersey
Photo credit: Charles W. Cook

Figure 3.5 The Camden side of the Benjamin Franklin Bridge
Photo credit: Charles W. Cook

When crossing a bridge we trust in the engineering more than the architecture, but the Ben Franklin Bridge prime Designer was Paul Cret, an Architect. With over a 9,000-foot span, and 135 feet above the Delaware River, the bridge was to be the largest suspension bridge ever constructed at the time (completed in 1926), and there was general and genuine concern that it was being designed by an Architect. To answer the concerns, Leon Moisseiff was brought on as the design Engineer, and Ralph Modjeski was the chief Engineer. Moisseiff was considered the leading bridge Engineer in the country, and although some wished him or Modjeski to head the project, generally everyone was satisfied with Moisseiff assisting Paul Cret, who remained the chief Designer. The structural tower, suspension, and cabling designed by Moisseiff and the supporting stone towers by Cret suggest an imbalance in responsibilities, but today the "Ben Franklin" is a beautiful bridge joining two cities and states, and few remember the controversy over the design, or that Leon Moisseiff was also the Engineer who designed the ill-fated Tacoma Narrows Bridge that failed so dramatically.

3.1.2.1 The Engineer as Designer

Whether they are directly employed by the Owner or part of the architectural package, the chief responsibility of civil and structural Engineers is ultimately to defy gravity. From the very beginning of civilization, engineering probably involved a trial and error method. Considering Hammurabi's Code, once something did work, the Engineer was not too eager to try something else that might not. So for centuries building engineering progressed slowly.

But still today, the civil and structural Engineer's chief task and ultimate contribution is the continued free standing of whatever has been designed. We all defy gravity every time we get out of bed or take a step, but the Engineer has to ensure the work will continue to defy gravity after everyone who has built the structure has left.

Although certainly buildings and bridges are easily seen defying gravity, it may be a bit more difficult to see the effects of gravity on highways, but place a car or tractor trailer truck on top of a thin layer of concrete or asphalt, and it will soon become obvious that gravity is once again playing an active role in how successful the Engineer's work has been.

Taking the vision of the Architect for a building masterpiece or the infrastructure needs of a Federal or state owner for highways and bridges, and ensuring it can be built is a considerable responsibility. Guaranteeing it will last for the duration of the Owner's needs adds to the burden of the engineering. The civil or structural Engineer has to be sure it will initially work, and then that, through decay such as oxidation, or even structural fatigue, it will last.

There is, however, a barrier to engineering perfection, and that is human error. We all recognize there have been great disasters in structural engineering. If a reader has not already seen video of the Tacoma Narrows Bridge collapse, it would be worthwhile to view one of the many videos available on the internet. Contractually, the Engineer is expected to foresee what could happen under ordinary use of the structure, and then add some more protection for the possibility of extraordinary circumstances, but there are still limits. The Empire State Building was the first skyscraper hit by an airplane in New York City. In 1945 an Army B-25 bomber collided into the building while trying to land at Newark airport during a heavy fog. Although the crew members were killed, the Empire State Building survived that accident. The towers of the World Trade Center used new design methods that would withstand what the Empire State Building had, and even more. What the engineering did not account for was the tremendous heat from planes with nearly full tanks of gas burning after the collision.

The responsibility of Engineers, therefore, is enormous. The chance for errors in unforeseen conditions can bring great pressure on the design process. Over several decades construction has become more litigious as the separation of the design and the construction has led to more and more finger pointing. This has led to even greater pressure on the Engineer. In many instances now, particularly in vertical construction, the Engineer, being subcontracted to the Architect, is not allowed to communicate directly with the Constructor. Architects are merely protecting their own exposure to responsibility if something is communicated for which the architectural firm will be liable, but this often adds time to the project management process.

Added to the Engineer's burden is the challenge of achieving the vision of the Architect while finding the most cost-effective but still safe means of doing so. The John Hancock Tower in Boston was the first of its kind, and the engineering was not tested in the field. It only seemed possible on paper. More and more Architects are designing their masterpieces, and the Engineer is expected to find ways and means to make the vision a structurally sound reality. This also means the Constructor will sometimes be building in some previously unexplored "territory." This is not new. Architectural design has always pushed engineering to new limits. But with new limits come new understandings of what does not work, and this can once again seem to be through the original engineering method of trial and error, something neither the Architect nor the Engineer wish to risk.

Additionally, where engineering design for centuries had been primarily concerned with structural issues, the use of electricity and mechanical equipment has created new design fields. The complexities of electrical, mechanical, and plumbing designs, for example, require specific engineering to meet the demands of both the Owner and the public as the cost of using resources becomes an important factor in construction design.

Figure 3.6 The Fort Lee Bridge under construction, New Jersey
Photo credit: Robert S. Cook

Figure 3.7 The Fort Lee Bridge under construction, New Jersey
Photo credit: Robert S. Cook

These rare 1929 photographs of the Fort Lee Bridge under construction were taken by Robert Cook, then a young, would-be Engineer student, before he entered the University of Pennsylvania. In the class history, the comment read that they had entered the Towne School of Engineering, "itching to learn how to design the George Washington Bridge."

Few people realize that the design for the Tacoma Narrows Bridge was not unique. Other bridges before and after had used similar designs. The Fort Lee Bridge was using similar design principles as the Tacoma Narrows Bridge. Today, now known as the George Washington Bridge, the structure carries heavy traffic on a daily basis between New York and New Jersey. The fault in the design was in part corrected with the addition of the second level on the bridge.

Note the "Roebling Cables" sign atop the tower. John Roebling (father) and his son Washington Roebling are best remembered as the Designer and Constructor of the Brooklyn Bridge in 1883, but the family name and reputation continued long after in bridge construction.

3.1.3 The Architect as the Designer

For millennia the Architect was also the builder. Today we usually see the Architect as the chief Designer. Even Architects are acknowledging that much of what they used to do has been taken over by the Constructor building the project, or, more often, a separate Construction Manager hired by the Owner. If this trend continues, the Architect runs the risk of becoming less significant and perhaps being absorbed by the Constructor in a Design-Build construction environment.

Whether a trend towards traditional functions of the Designer slowly being absorbed within the builder's responsibilities will continue is uncertain, but Architects recognize the heavy responsibility and even burdens that can be placed upon them in the beginning of a project.

Creating the largest artworks known to humans, Architects have far greater responsibilities than just having a vision of something grand or nice. Lines on paper have to become a reality. Bricks and sticks have to work together. The Owner has to be able to use the space for its intended purpose, not just admire it for how nice it looks. A factory floor finish is more important than how the color of the walls complements the forklifts. Compliance with ADA (American with Disabilities Act) standards is more important than how the cubicles fit so nicely together. Installation of the entire sprinkler system is more important than how nice the ceiling grid looks on paper.

Between achieving an effective and safe building designed according to codes and creating a work of art there are many challenges to the ultimate success and satisfaction of both the Owner and the Architect. The Constructor will find it his or her responsibility to meet and even exceed those expectations.

A CASE IN POINT

Figure 3.8 The PSFS Building, Philadelphia
Photo credit: Charles W. Cook

There was a time, not very long ago, when architectural firms retained the construction supervision of the project, often handled today by the Construction Manager or the General Contractor.

The PSFS Building is now a historic and iconic landmark on the Philadelphia skyline. During design and construction it was both controversial and challenging. Architects George Howe and William Lescaze had never done such a large project. In fact the project, being the first of what is termed "International Design" buildings in the United States, with several unique features, including central air conditioning and electric push-button elevators, was beyond their usual experience. The Owner was the Philadelphia Savings Fund Society – the first savings bank established in the United States. Howe had done several branch banks, but this was a total departure from anything previously conceived. In the middle of Philadelphia's retail district, retail shops would occupy the first floor and banking would be done on the second floor, accessed via escalators. Above the banking would be a skyscraper series of floors for offices. In fact, the building would become the nation's first modern International Style skyscraper. George A. Fuller would be the Builder, bringing onto the site up to a hundred Subcontractors and Suppliers to perform many firsts in architectural design. Recognizing his own limitations in regard to project management of this magnitude, for his "clerk of the works" George Howe did not go to someone for his design knowledge. Instead he chose a carpenter Superintendent, Frank Armstrong Cook, who had a great deal of construction field experience, including supervising the construction of the Philadelphia Fidelity Bank "skyscraper." Having previously and successfully coordinated craft workers, Subcontractors, and Suppliers, Frank Cook was relied upon to do for the Architect what Owners since the last half of the twentieth century hire an entire construction management firm to do separately, thus diminishing the architectural firm's role and responsibilities in managing the actual construction.

The PSFS Bank stands no longer, but the image remains atop the historic structure that was built below it to such elegant European Modernism standards. The sign remains an iconic beacon on the skyline above the renovated structure, which was converted into a hotel in 1997.

For a few, it has become a symbol of an earlier time when an Architect took a Constructor onto his team and thus assumed contract administration to an extent no longer assumed by Design Professionals. At least, that is the way my father told with pride the story of what his Uncle Frank had accomplished for the famous Architect, George Howe.

Within the contractual relationships, however, is the potential for struggle and disappointment. The Constructor's main motive is profit. To achieve that, a quick in and out is in order. Satisfying the contract requirements is an obligation, but interpreting just what those obligations are can lead to disagreements and, ultimately, litigation. It is reasonable to assume that the Architect believes the contract documents clearly indicate the final intent of what is expected. Changes outside the original documents would be and should be the Owner's responsibility, and therefore the cost and effect on the schedule are not part of the Architect's concerns. Changes brought about through design ambiguities or errors, however, can quickly become the Architect's responsibility, and we just suggested that the Architect believes he or she has clearly stated without ambiguity or error what is expected from the contract documents. In other words, no one wants to be told they are wrong. The Architect wants an A+ for the work, and the Constructor might be suggesting C- is more appropriate.

Although not a party to the Owner–Architect contract, the Constructor, through the Owner–Constructor contract, may carry many of the burdens and responsibilities one would otherwise think the Architect should bear. These are done through what are known as **risk shifting clauses**. Once again, the task of the Constructor is to fulfill the contract requirements, but the ultimate challenge may be to keep all parties communicating effectively so teamwork and mutual responsibility for the overall success can be maintained.

The rest of this text will be discussing the issues and obligations associated with the Constructor contractually. As the discussion above has shown, the Constructor is at one end of the contract chain in a typical Design-Bid-Build project. This is not an ideal place to be if the "everyone else" associated with the work does not feel a sense of genuine teamwork. Whether you, as the Contract Administrator, can foster such an atmosphere is not guaranteed, but the effort should be made.

3.1.4 The Constructor

Someone has to build it – make the vision a reality. Bringing the project in on time, within budget, meeting standards of quality, and doing it safely is a formidable task. Within the obligations of the Constructor, as well as everyone else on the project, is the duty to cooperate. It is an implied duty of all contracts. Courts will always recognize this obligation even when it is not written into the contract itself. How the Contract Administrator improves and fosters that cooperation is essential to the ultimate success.

Think of those needs and vision as the "will of the people," and recognize, as Ulysses S. Grant did, that their will is the best law. Fulfill the will of the parties, and that will be the best path to success.

3.1.4.1 Subcontractors and Suppliers

Very few projects are ever done by a single Constructor. With so many specialty trades required for a single project, it is almost certain one or more Subcontractors will be needed. Additionally, Contractors and Subcontractors will purchase material from Suppliers. In some cases the supplies will be from a "full service" facility that carries the entire range of materials, from nuts and bolts to heating, ventilating, and mechanical equipment. At other times, the Supplier will be very specialized for a particular product. Coordinating space and schedule for all the different parties involved in the project becomes a major concern for the Contract Administrator.

Exercise 3.1

Take a moment
Check your language

Below is a list of words and phrases. Think about how often, if at all, you use them in conflict resolution situations. Check the ones you do use, and circle the ones you think you should use more often. Copy the page and keep it near you until the use of the words or phrases becomes a new habit for moving potential conflicts to resolutions:

Could we try this?	What do you need from this?
Good thought	Are we staying on track?
Is there a better way?	Your turn, then let me respond.
Thanks!	So what would be best for everyone?

I keep thinking there's something we both would like (or could do) better.

Are we effectively using all of our combined strengths?

Discussion of Exercise 3.1

Not all changes need to be dramatic. Sometimes a little change, even in a virtue or positive character trait we are already very good at, can have dramatic results and improve our success. How we bring others "onto the team" is crucial for the success of the team.

INSTANT RECALL

- In general there are three parties to the project, but there are usually two separate contractual relationships, immediately setting up challenges for communication and responsibility.
- Judges and juries will try to be fair, but in the end, the message is clear: *If at all possible, avoid remedies imposed by courts.*
- In Design-Bid-Build, the Owner is in the middle between the Designers and the Constructors.
- The Owner still needs the space completed on time, within budget, meeting standards of quality, and doing it all safely for all concerned.
- If we divide construction into two broad types: (1) horizontal (highway and bridge construction) and (2) vertical (building construction), then we have two main categories of Design Professionals. In highway and bridge construction the Engineer is often foremost, and sometimes the only Design Professional required. In building construction, however, the Architect is almost always foremost and usually subcontracts the engineering of the building if there are not Engineers on the Architect's staff.
- Although engineering design for centuries had been primarily concerned with structural issues, the use of electricity and mechanical equipment has created new design fields.
- For millennia, the Architect was the builder.
- Within the obligations of the Constructor, as well as everyone else on the project, is the duty to cooperate.

YOU BE THE JUDGE

M. J. Oldenstedt Plumbing appears before you as a Plaintiff on appeal of a previous judgment favoring K-Mart. The issue revolves around whether or not the contract form has to be returned in order for the contract to exist. In March of 1991 M. J. Oldenstedt (Plaintiff) received drawings and subsequently submitted a bid for work in the amount of $709,500 via telephone.

The bid was accepted and a written contract was received by the Plaintiff about May 21.

The contract apparently called for more items to be done than the Plaintiff had originally considered in the plumbing scope. Although the Plaintiff signed the contract and told the K-Mart representative that it had been signed and would be returned, he never returned the contract. M. J. Oldenstedt personnel did, however, proceed with the work.

In the absence of an actually signed contract form returned to K-Mart, was K-Mart entitled to withhold monies from the Plaintiff based on the terms of the contract and any failures by the Plaintiff to fulfill the terms of the unreturned contract?

Your ruling:

Find for M. J. Oldenstedt or K-Mart:

Why?

IT'S A MATTER OF ETHICS

The Owner has told you they want you to do the project if you can cut just $25,000 from the $1,000,000 bid proposal you have already submitted. Would you do it, knowing the Owner has told you another Contractor had a lower price when the bids were originally submitted?

Yes No

Going the extra mile

1. Go to the Companion Website www.routledge.com/cw/cook and view the video discussion Part III of the subcontract discussed in Chapter 1.
2. Go to www.constructingleaders.com and view the video "Moving the Gruber Wagon Works." The video will be a prelude to the text in a future chapter related to alternative project delivery methods and Design-Build.
3. After you have made your own ruling on the "You be the judge" exercise above, go to the Companion Website for a discussion of the actual case rulings.

4. **Discussion exercise I** (with friends or as part of a classroom discussion): Compared to other artists, it might be said Architects are more like composers than painters. An artist completes the painting and puts his signature on it. For a composer, the final work of art will require several different additional artists, each with a specific task. Perhaps the most important individual will be the conductor, and that is where you, as the Contract Administrator, must come into the process. It will be up to you to make it all happen as planned, not as composed on a set of music sheets, but in accordance with the plans and specifications.

   a. How can you successfully bring so many different "artists" together to complete the construction project? Is it wrong if you do it differently than someone else? How, ultimately, will you be judged by the audience (the Owner)?

5. Effective use of the correct vocabulary improves communication and promotes respect among colleagues. The reader should become familiar with the key words listed below that have been used in the chapter. Definitions can be found in the Glossary. Additionally, the student is encouraged to use the flash card exercise in the chapter's Companion Website as practice for possible quiz or exam questions.

Key words

Construction Manager

Multi-prime Contract

Prime Contractor

Risk shifting clauses

Shop drawing

Specialty Contractor

CHAPTER 4

The basics of construction contracts

In this chapter, basic concepts of construction contracts will be discussed in more detail. To be successful, a Contract Administrator must begin with the fundamentals, understanding who is going to do what in return for what.

Student learning outcomes

Upon completion the student should be able to. . .

- **understand what is and what is not a contract**
- **recognize what constitutes the entire contractual agreement**
- **identify what might not be considered part of the contract**
- **differentiate between Unit Cost, lump sum, and Time and Material or Cost Plus Contracts**
- **understand contract issues regarding changes**
- **analyze which contract type is better under different circumstances**
- **recognize the importance of cash flow and how contract language can affect payments.**

4.1 What is a contract?

A verbal contract isn't worth the paper it's written on.

Samuel Goldwyn

Co-owner of MGM, Samuel Goldwyn was famous for such "Goldwynisms." A bit like Yogi Berra's, his expressions often had some point to them, even though initially they made no sense whatsoever. No doubt the days that our great-grandfathers knew in Construction, when a "man's word was his bond," have slipped away with more and more documentation and paperwork. Everything has to be in writing. The paperless jobsite is still a fantasy, and if the paperless jobsite ever does become a reality, it will no doubt still be lost in the fog of "the cloud."

When our great-grandfathers (a time when our great-grandmothers were not on a project site and seldom in the office) did so much on a handshake, there was a reason most of what they said was not reduced to writing. Without e-mail, faxes or copy machines, in our great-grandfathers' day formal writing required a manual typewriter, and someone who could use it (not likely our great-grandfather). Job minutes would be kept by hand at best, and copying anyone on a letter meant tediously including carbon paper for each person to be included (and don't make a typo because correcting it will take longer than writing the letter in the first place).

Today, all that has changed, and with it comes the responsibility to "Document. Document! DOCUMENT!"

The Project Administrator is almost certainly one's own typist. Electronic media sends things instantaneously, and sometimes regrettably. We are well aware of how we had better have it in writing, but the time pressures we face on a project often require us to move without written final approval, so we try to document our understandings along the way. And even then, if we have not done it well, it may not be worth the paper it is or is not printed on.

Still, our indispensable contribution to the project is not about just doing what is legal. We are not intending to become attorneys through this text. In fact, the more successful we are, the less we will need attorneys, but to be successful in a world where we need to know what we need to have in writing requires us to go back again to the contract, and ultimately understand just what is a contract.

We have previously defined a contract as "an agreement between two or more parties for which if there is a breach of the agreement a court can determine a remedy."

If we go back to understanding contract law, we recognize that there are two critically fundamental issues: (1) what is and what is not a contract and (2) what elements and/or documents are included as part of the contract.

This second issue becomes vitally important for you as a Contract Administrator, so let us "build" a contract with these two elements in mind.

4.1.1 The Contract Administrator's contractual obligations

In essence, a contract begins with a promise or set of promises by one party (in this case the Constructor). As we all know, not all promises are in writing.

> **Example:** Marriage vows are spoken verbally, and although a marriage certificate and license may formalize the union, the vows themselves (usually stated as "I promise to. . ."), are the true marriage "agreement." A breach of those vows is basically what the court will enforce in a divorce settlement.

Although many contracts do not have to be in writing, in most cases of construction law, contracts have been reduced to writing. At the same time, Contract Administrators, along with authorized personnel representing the company, must recognize that some verbal commitments made in the field or during meetings may still have the force of a verbal contract in the form of a **change order** or what is sometimes referred to as a **contract modification**. It is important to put these agreements and the extent of the "promise" into writing as soon as possible. In fact, contractually there may even be a time limit to do so.

That said, then, where does one look for the written promises? In construction, the written promise is often quite extensive and does not simply involve a simple promise of "I will do this for you." As long and complicated as the contract to which both parties have put their signatures might be, there is almost always more to the contract than just the "signature form." These additional documents include:

- plans and drawings (usually supplied by the Designers)
- specifications detailing the standards for each area of work
- addenda to the bid documents
- eventually, change orders or contract modifications to the original contract.

All of these become the "promise" of what the Constructor is obligated to do. In essence each of us promises to do "this" (in Latin that would be "*quid*.")

4.1.2 The Owner's contractual obligations

However, no promise can be without what is termed under the concepts of common law to be **consideration**. The courts will not recognize that a contract exists if the party making the original promise is not given something in return. If someone promises to do "*quid*," then the party benefiting from the "*quid*" needs to do or provide that which was expected in return, and "that" in Latin translates as *"quo."* Thus, we have the expression ***quid pro quo***. The courts will expect the two parties will have agreed to do "this for that." Without consideration, no contract exists.

As part of this *quid pro quo* is the obvious element of **acceptance**. No one is required to accept an offer. Acceptance of the *quid pro quo* promises begins the point at which the contract exists.

Example: If I promise to give you a present for your birthday, but I fail to do so, there is no contract. However, if you agree that if I promise to give you a simonizing kit you will polish my car in addition to yours, and you fail to simonize my car after I give you the kit, then a contract has been "breached." In such a case, however, we will probably only find a Judge Jack or Judge Jill type court drama to accept such a case if we cannot work it out ourselves. Courts are heavily backlogged, and the expense of taking such a case before a judge would almost surely be prohibitive or senseless compared to going down the street and paying for a complete detailing of both cars.

The *quid pro quo* in construction is almost always "I promise to complete this structure (highway, bridge, building, etc.) and in return you will pay me (almost always as installment or progress payments eventually amounting to total compensation) the originally agreed-upon price plus or minus any add or deduct change order or contract modification costs incurred during the construction process."

That represents the fundamental "promise for a promise" basis of the contract. *Quid pro quo* is basically the "tit for tat" relationship between two parties, but there are additional issues that must be observed.

4.1.3 Other issues for a legal and binding contract

In addition to the "*quo*" consideration, all contracts have to also comply with some other important issues.

4.1.3.1 The promised acts have to be legal

There cannot be a contract to commit an illegal act. We hear the term "contract killer," but such a "contract" could never be taken to court to be enforced or remedied. That is obviously an extreme example, but the same would be true if an illegal act was contracted for in construction. An Owner cannot contract for removal of asbestos without the proper environmental safeguards. Although someone might agree to remove asbestos against the rules and regulations established for such an abatement of hazardous materials, a court would not enforce a breach of this non-contract. Another example would be that a Contractor cannot require a Subcontractor to perform a change order that they know would result in a structurally unsafe condition or be done in a manner contrary to the building codes.

4.1.3.2 Both parties have to be mentally competent

Age and/or mental capacity at the time the contract is signed can invalidate a contract.

In most states individuals under the age of 21 are considered minors and a parent or guardian will be required to execute a contract for them. Also, individuals with diminished mental capacity, whether from a permanent mental disorder or a temporarily self-induced diminution through drugs or alcohol, cannot be held responsible for contracts signed when they were mentally incompetent.

Related to the mental competence is also the subject of duress. Persons cannot be forced to sign a contract under the "make you an offer you can't refuse" of *The Godfather* movies type. In construction, duress would be a bit different, but still does occur even though it can and should void a contract.

Examples:

1. A Contractor could not require a Subcontractor to take on a new contract or even change order for a reduced rate on the basis that a currently due payment on an existing contract will otherwise be significantly delayed.
2. It would be an invalid change order (a change order is made a part of a contract through the change order or contract modification clause found in most contracts, and to be discussed later) if the Designer threatened an abusive punch list unless the Contractor reduced the price submitted for the change.

4.1.3.3 Oral understandings

Once a contract has been reduced to writing, the two parties should not rely upon any previous non-written agreements they may have made prior to the signing of the contract. Almost all contracts will have language similar to this found in the Appendix I subcontract:

> This Subcontract embodies the entire agreement between Contractor and Subcontractor. Subcontractor represents that in entering into this subcontract it does not rely on any previous oral or implied representation, inducement or understanding of any kind or nature.

Any understanding either party might have had prior to the written agreement will be overruled by what has been agreed to in writing. Attempting to incorporate something outside the written documents (the signed contract and also all the drawings, specifications, and addenda) to include something that was said or even written before the contract is very difficult. The courts will want to interpret the signed contract as the whole agreement.

Oral evidence, known as **parol evidence**, might not even be allowed to be admitted. A court attempts to understand the contract strictly based on what is written in the signed contract. Written statements not incorporated into the contract or oral agreements before or at the time of signing that have not been confirmed by the contract documents are usually considered not to have been made and are not part of the duties or responsibilities of the parties to the contract.

4.1.3.4 Implied understandings

Contrary to previous understandings, oral or written, there are implied standards one should expect within each contract. For example, the parties to a contract have the implied duty to cooperate. An Owner cannot ask for a renovation to a property to be completed and then keep the builder from entering the premises. Similarly, the builder is expected to take all standard precautions not to interrupt the Owner's typical flow of work in areas that are not under renovation.

4.1.3.5 Uncertainly or impossibility

There can also be situations of **uncertainty** or even **impossibility** that will nullify a contract.

Uncertainty can sometimes arise in fast-track construction when the Constructor is working with a Guaranteed Maximum Price but no final drawings. It would be best not to rely upon such an issue saving the company from potential disaster, but if an agreed-upon price is based on vague parameters of what the final structure is expected to be, and through construction the scope changes significantly, the Constructor can possibly rely upon a change to the Guaranteed Maximum Price. This all should have been negotiated at the time the increases were occurring, but it is also possible an Owner with a vague concept of what was needed would have pushed forward changes (perhaps not recognizing the potential for a significant overrun), and not until the last minute refuse an increase in the Guaranteed Maximum Price.

As always, communicating clearly on a timely basis will make the Contract Administrator invaluable.

Related to uncertainty is the doctrine of impossibility. There have been some major examples of impossibility in Department of Defense contracts. Such contracts are usually based on **performance specifications** where the Contractor is required to meet certain performance criteria for a new military weapon (e.g. an aircraft). At first the requirements seem reasonable, but they later prove impossible.

In construction this could happen in some performance specification situations where the contract calls for specific conditions to be met for certain units that are being installed. For example, a Mechanical Contractor might be required to meet certain cooling and heating levels using specified heat pump units. The Designer may have specified these units in a particular climate before and expected them to perform the same again, but for the new contract the location was changed to the Arizona desert and achieving the contractual levels would be impossible.

Escaping from contractual obligations under the doctrine of impossibility, however, is not as simple as it sounds. It has to be shown that not only was it impossible for the existing Contractor, but *no* Contractor could have accomplished the contractual obligations.

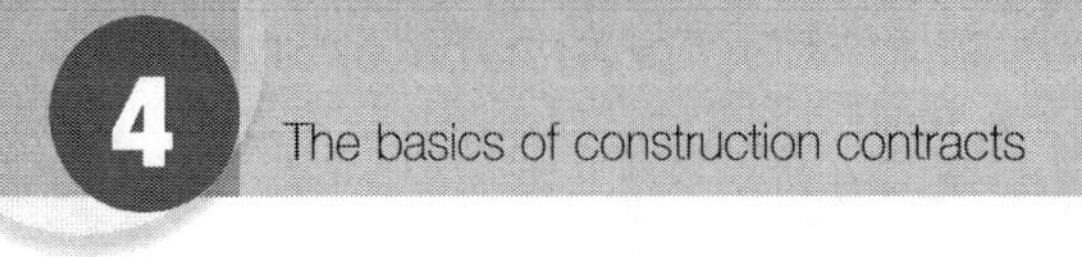

A CASE IN POINT

What would you do?

Figure 4.1 The Empire State Building, New York City
Photo credit: Charles W. Cook

Figure 4.2 Communications tower at the top of the Empire State Building
Photo credit: Charles W. Cook

Today we look at the top of the Empire State Building and cannot imagine it without a communication tower. Architecturally, it seems to complete the majestic rise of the building, and structurally, the engineering to resist lateral forces of wind has been effective, but it still had to be put in place after the building itself had already been completed. The 200-foot extension, 1,250 feet above the ground, had to be erected on top of the mooring mast that was first used in downtown New York City to dock blimps. The task of erecting the antenna fell to Ed Thompson, a gentle, quiet man who thought a lot about not only doing things correctly, but also doing them safely.

"Uncertainty" at 1,250 feet or contractual "impossibility" was not an option. The contract was to put the antenna in place, but how to get it to the top of the building when construction had been completed and the tenants were occupying the building was the issue. No one could walk the long antenna pieces up the 102 floors even if they would have fit in the stairwells. There were no cranes in place to do it. The setbacks of the building and the height made pulling the tower pieces up by ropes or even with a block and tackle impractical and dangerous. In fact, during construction of the Empire State Building one pedestrian had been killed when a masonry block fell to the street, so endangering the public was out of the question.

Between a signed contract and the actual accomplishment of the "*quid*" so you get paid the "*quo*" can be the challenge, but a Contract Administrator must rise to that challenge – even when it has to be 1,250 feet above the street and top off at 1,454 feet.

How would you do it?

Discussion of Empire State Building Story

Sometimes the simplest solutions are the best, and often right in front of us. Ed Thompson had the long antenna pieces loaded on top of the elevators and rode them up the elevator shafts the 102 stories.

But for now, the lesson is hopefully that experience counts. The Contract Administrator does not always have all the answers. The younger or newer one is to construction, the more we must depend on others who have done it before. We can know all the details of the contract, but we also have to recognize the experience of those who have been doing it for longer than we have. Between managing the specific demands of the fine print on a contract and leading others who surely know better than ourselves how to get it done and have been doing it their way seemingly forever can be one of our biggest challenges. Flaunting book learning or contract knowledge will not be the answer.

* * * * * * * * * *

Open communication and soliciting ideas from all the team members is crucial to joining knowledge with experience, while mitigating our own inexperience and eliminating the tension between a "know it all" and a "been there done that" attitude.

* * * * * * * * *

4.1.3.6 Mistakes versus errors of judgment

Although you will not make friends and your reputation might suffer, if you do make a mistake in your offer (submission of your bid proposal) you can be released from your obligations and the contract would be voided. Note, that to do this, however, one must be alert and up front. Too late into the project, and the complications of releasing from a half-performed contract can be enormous.

On the other hand, errors of judgment are not reason to be released from a contract. If your company has decided to "gamble" on some situation (e.g. that there are no underground obstructions), and the bid documents clearly stated, and the contract confirms, that all underground obstructions are your responsibility, then you have to bear the costs of removing the obstructions. The gamble that there were no obstructions would be considered an error in judgment.

4.1.3.7 Promissory estoppel

There is one form of oral contract that should be recognized.

The bid process can be a time of great stress and pressure. Large dollar amounts with possible exclusions and vague scopes of work often lead to bid mistakes. Generally, the low bidder has the challenge to validate the estimate with the team that is going to build the project, and more and more Contractors are relying upon Subcontractors to perform most if not all of the actual work.

What obligations do the Subcontractors have who bid the work to the successful Contractor? Under the premise that to be valid a contract requires an offer and an acceptance, if there was no agreement at bid time between the Contractor and Subcontractor the assumption would be there is no contract.

In such a case courts recognize what is known as **promissory estoppel**.

Estoppel is a legal term that prevents one party from acting contrary to the way they have been consistently acting. Thus promissory estoppel is a bar preventing someone from making a promise and then backing out from it. These are not those "I promise to give you a birthday present" type promises. These refer to such standard practices of submitting bids (often verbal) to a Contractor, who relies upon that price when submitting a proposal to an Owner, and then the Subcontractor backs out of the bid once the Contractor contacts the Subcontractor to award the contract.

Based on promissory estoppel, one party cannot withdraw its obligations if it knows the other party relied upon its actions or offer.

Examples:

1. A **General Contractor** receives an $8,000 bid from a Plumbing Subcontractor. He relies on that bid and has no reason to believe the Subcontractor made an error. The General is awarded the project and sends a contract to the Subcontractor, who now has more work than people to do the work, so he refuses to hold to his original price, and insists he will need an extra $3,000 to hire more workers at a premium to do the work. The General can go out and get another price, and sue the Plumbing Subcontractor for the difference.
   Note: The cost of the process of litigation should be considered in determining whether it is worth the General Contractor's effort or needs.
2. However, if the General sends a subcontract containing burdensome clauses to the Subcontractor, and the Subcontractor had never seen these clauses before, or the General changes the scope of work, then the Subcontractor can refuse, on the basis that there was no formal agreement, and the General Contractor has changed the conditions the Subcontractor had originally relied upon.

Another important point to recognize is that the previously discussed "bid shopping" will negate promissory estoppel. Once a counter-offer is made to "shop" down the price of the Subcontractor, the Subcontractor is no longer obligated to maintain the original offer, even if the Contractor comes back and states it will now take the original offer. The act of "shopping" negated the promissory estoppel, since the original offer was initially not accepted.

4.1.3.8 Breach of contract

Through the entire process of administering the contract, and in some instances attempting to defend actions taken or not taken, the Contract Administrator must be aware that not fulfilling any part of the contract can be considered a **breach of contract**. The contract will usually have specific remedies described within the contract for any breach, such as withholding payments or even removing part or all of the remaining contract and having it performed by another party. Otherwise a breach of contract might be resolved during a court process, but the Contract Administrator should always recognize that fulfilling the contract terms and conditions is the best way to avoid litigation and keep relations among all parties positive.

4.2 Types of compensation

Rule no. 1: Never lose money. Rule no. 2: Never forget Rule no. 1.

Warren Buffett

As we mentioned, the typical "*quo*" in a construction contract is money. There are basically three ways the money of a contract can be determined or distributed.

4.2.1 Lump sum contracts

By far the most common form of payment in construction is **lump sum**. In such an agreement the Owner is contractually obligated to pay a total cost agreed to, usually through a bid submission or negotiated after discussion of the scope of work. This does not mean the payment is withheld until the completion of the project and then paid in one final large amount. Contracts call for **progress payments** throughout the entire length of the contract.

The flow of these progress payments becomes extremely critical, and the Contract Administrator must pay particular attention that everything does flow smoothly. There are those who consider cash flow the life blood of business, and in construction there are many challenges.

4.2.1.1 Cash flow under lump sum

The contract will generally state the General Contractor is to submit an invoice for work done during each month by the last day of each month. Note that this means each Contractor working during that month has already expended payroll costs for that month without being able to invoice for them until the end of the month. In addition, material and other costs such as rental equipment may have already been paid for by the Contractor or others on the project.

The contract will then usually call for a review period by the Owner or Owner's representative (usually the Design Professional), and payment will be made within thirty days of submission of the approved invoice. This, of course, is if all goes according to contract and there are no problems with the billing or the work as reviewed by the Designer or Owner.

In order to expedite processing, a "pencil application" is often submitted several days before the actual invoice due date. This "pencil application" is then reviewed and any adjustments are recommended before the formal submission.

Exercise 4.1

Take a moment
Some more words to resolve by

Similar to a previous exercise, how often do you use the words below, and which ones would be worthwhile including in your oral and written communication related to conflict resolution?

accommodate	appreciate	assist	benefit
collaborate	cooperate	enable	expedite
guarantee	hope	satisfied	successful
win/win			

Take a moment to compose a short note to accompany or to be sent as a e-mail advising of submission being forwarded of your first payment request using three or more of the words above.

Discussion of Exercise 4.1

There are many ways to write a letter. The key is, whether the letter helps to move the reader to my position. In order to improve the success of communication, it is best to keep three things in mind:

1. Is the reader the person who can make happen what I want?
2. Have I given the reader a "feel good" reason to want to help me?
3. Does it make reasonable sense for the reader to do what I am asking?

For a sample letter, go to the Companion Website after you have written your own.

4.2.1.2 Retainage

The typical cash flow, then, is that construction on the project proceeds for a month, and then another month will transpire before that work is paid for. However, under almost all contracts today the full amount of the work is not paid for on each progress billing. The Owner will keep back up to 10% of the requested amount as **retainage**. Holding money on the Contractor is certainly one way of guaranteeing the Contractor will remain on the project and complete it as soon as possible. The amount the Owner retains is usually 10%, but it can be changed part way through the project and is sometimes reduced to 5% after 50% of the project is complete. The assumption by the Owner and Designer is that work done early in the project such as excavation, footings, and slab work should have shown any problems by the half-way point, and those Subcontractors should not have to wait to receive their retainage until the painter or the ceiling installer have completed their work. In practice, the General Contractor does not always distribute the reduced retainage back to the early Subcontractors, and this can lead to some difficulties on future projects between such Subcontractors and General Contractors.

4.2.1.3 Owner's need for 10%

Most Contractors believe the typical reason for holding retainage is so there will be money held by the Owner in case there are problems in the work discovered after it had been initially paid for. This does make sense, but it also brings into question how effective the Owner and Designers are in monitoring the work, considering that the work continues and payments usually lag actual construction by thirty to sixty days or more.

Another and more immediate reason for most Owners to hold 10% retainage is that banks are holding that amount on their construction loan. Few Owners do have the cash needed to finance the entire project. At best they have it invested, but more likely on large projects they have taken a construction loan. The banks will hold 10% of each draw the Owner makes to pay for the work done to date. Obviously, the Constructor has no control whatsoever over the agreement between and Owner and the bank making the construction loan, and sometimes issues between the Owner and the bank do delay the payment request made by the Constructor past the thirty days from invoice submission agreed to in the contract. At the least this can lead to some embarrassment, but it can also cause friction and affect the harmony of the project team, through no fault of the Constructor.

One remedy to keep in mind to alleviate some of this problem is that some contracts (and sometimes it is not a contractual obligation but a job-meeting request) call for the Constructor to submit to the Owner an estimate of cash needs for contract billing ten to thirty days in advance. If this request or contract obligation is made, it is very important to provide that information as accurately and early as possible. Such an Owner is not adding to the Constructor's burden but is actually trying to ensure effective cash flow on their part.

As technology improves, the accuracy of these "pre-estimates" of billings will become even easier and more accurate. **Building Information Modeling Technology (BIM)** will effectively be used to move drawings from 2-D to 3-D. Owners, Designers, and Constructors can view not just the elements of the structure, they also can view the completed building in three dimensions, from all angles or perspectives. Through this process the technology will be aiding all parties in the Design-Build team to find **crashes** between different trades (e.g. sprinkler pipes being run through rather than under a structural beam) that do not easily appear in two dimensions. However, the fourth dimension of BIM is cost, and once this is fully incorporated effectively into the software and projected along with the fifth dimension (schedule), the ability to predict what the cost for work will be on any given date will be easily foreseeable and much more accurate.

4.2.1.4 Front-end loading

With as much as 10% being held on the work that has already been done, some Constructors practice what is known as **front-end loading** the costs. In fact, this practice is so widespread that many Owners and Designers expect it, and for that reason will challenge the initial cost breakdown submitted by the Constructor. The way front-end loading works is that the Constructor will submit a breakdown of all the costs for the project, but the costs for the early portion of the project will be artificially inflated, while the cost for the last part of the project will be similarly decreased.

Example: Using some round numbers, let us take a project that costs $1,000,000. Let us say included for the first half of the project are costs for initial excavation and concrete footings, slabs, steel, and masonry of $400,000. For the rest of the project there will be costs from all the other trades of $600,000. The contract calls for 10% retainage for the length of the project. A total of $100,000 will be held from the Constructor and the Subcontractors on the $1,000,000 project. That means the Constructor as well as Subcontractors will not have a critical amount of the money they need to cover their overhead and profit, as well as some direct expenses, on the project until the project is completely finished.

Constructor's front-end load solution: Change the value of the initial costs to $500,000 and reduce the costs of the other back-end trades to $500,000.

	As bid	**As invoiced with front-end loading**
First half	$400,000	$500,000
Final half	$600,000	$500,000
Total	$1,000,000	$1,000,000

The practice of front-end loading is quite extensive, but it is critical that the Contract Administrator understands the ramifications of the practice. Accounting of the project that is front-end loaded produces a myriad of difficulties toward the end of the project, particularly if that project is in trouble. Without sufficient cash flow at the end, the project can become a cash drain for the company. If the false "profit" from front-end loading the job has already been spent for some reason, the prospect for recovery is difficult and will create tensions and affect relationships at the end of the project.

Frighteningly, the practice of front-end loading at the beginning of a recession can be the major cause of Contractor bankruptcies at the end of a recession. Companies that thought they were doing well are suddenly faced with no backlog and more debt than cash flow, along with the contractual obligations of finishing back-end work that was undervalued because of earlier front-end loading.

In the example above, the amount of the project was $1,000,000. However, some mega projects can reach much larger amounts. Front-end loading such a project may bring the Constructor enormous amounts of money early in the project, but the Owner will be severely burdened financially by paying for what has not yet been delivered, and if the Constructor does default for any reason, the cost of completing the project for the Owner will be much more than remains unpaid in the Owner's funds or construction loan.

For any Constructor there is an additional burden for front-end loading a project, and that is the necessity to keep two sets of books to understand what is actually profit and what is just advanced billings. This can be critical for tax purposes, so that money that is not truly profit is not taxed. And it can also be critical related to bonding, so that the company has a true and accurate record to present when seeking more bonding capacity. Bonding will be discussed extensively in a later chapter.

4.2.1.5 Pay when Paid/Pay if Paid Clauses

One way Constructors protect their cash flow is with clauses in their subcontracts that are known as **Pay when Paid** or **Pay if Paid**. These are critical to almost all Constructors, since they do not have the amount of cash available to "carry" all the Subcontractors on the project until they themselves are paid. Both of these clauses are known as a **condition precedent**, clauses in which something has to take place before another action of the contract is required.

> **Example:** Inheritance by a minor of money from an estate might have a condition precedent that the minor reach the age of maturity before the money is made available.

4.2.1.5.1 Pay when Paid

The purpose of the Pay when Paid Clause is to allow the Contractor to wait for payment by the Owner before being obligated to pay the Subcontractor. The clause usually also contains a period of time past the actual receipt of payment from the Owner to allow for depositing the money and clearing the funds into the Contractor's checking account.

4.2.1.5.2 Pay if Paid

The Pay if Paid Clause is not as widely used, and it is not as acceptable even by courts. Owners can sometimes withhold payments for reasons that are not the fault of a particular Subcontractor. For example, there might be a difficulty that has been discovered on work paid for on a previous invoice, and now the amount being held is confused or entangled with the current application for payment. The money withheld could be withheld for a considerable amount of time and it was never the result of what the Subcontractor did. Subcontractors rightly object to such a clause, because it can only lead to more confusion than necessary.

It is important to note that even with such a clause (or the Pay when Paid Clause), a Contractor is not protected against never paying the Subcontractor. Courts generally recognize that the purpose of these clauses is to affect the timing but not the obligation to pay the Subcontractor.

Again, in a later chapter, we will discuss bonding, and the inclusion of either a Pay when Paid or Pay if Paid Clause does not change the notice requirements of a Subcontractor to alert a bonding company that they have not been paid. Also, filing of Liens against an Owner's property for lack of payment will be discussed in another chapter, and the timing of this needs to be governed by the legal requirements and not the duration of any "pay if or when. . ." clause. The Contract Administrator, therefore, needs to be alert to the timing of notices related to such filings.

4.2.1.6 Payments to disadvantaged businesses

When taking on public work, many Constructors assume responsibility for the **Disadvantaged Business Enterprises** that are required to be included in the construction process. Municipal, state, and Federal contracts will require certain percentages of Minority and Women Owned Businesses be a part of the construction workforce, Subcontractors, and Suppliers. Fulfilling this obligation in the bidding phase is just the beginning of the effort to fulfill what will become important contractual requirements with significant accounting and paperwork submissions.

The construction industry has been justly criticized for its lack of diversity. Correcting that lack does present additional challenges for the Constructor. One of these is directly related to the lump sum contracting method and the payment provisions of most contracts. Without receiving payment for sixty or more days from the time that a Disadvantaged or any Subcontractor with severe cash flow needs begins work is devastating. For this reason many **joint ventures** between a Minority or Disadvantaged Business Enterprise and a more established Contractor are created to help with both cash and technical issues related to the industry or, more specifically, the project. In the 1960s and 1970s, when the requirements were first appearing in public contracts, many of these joint ventures were merely fronts to allow the "majority" Contractor to gain more work. Now more scrutiny has put a stop to most of that practice.

In lieu of a joint venture, cash flow usually becomes a considerable challenge for start-up or disadvantaged Subcontractors, and the Contract Administrator faced with long payment terms from the Owner needs to recognize the possibility that some critical Subcontractors will not be able to survive under a Pay when Paid subcontract, and cash flow will have to be assisted from the Contractor's own funds.

We have to recognize that lump sum contracting, although the most common form of payment today, is not perfect or ideal. In addition to putting burdens on Disadvantaged Business Enterprises, it also challenges all Contractors and Subcontractors from a cash flow basis. One of the reasons the industry has remained closed to disadvantaged businesses is the huge cash flow requirements under lump sum contracting, and to overcome such challenges companies either face failure or a very slow, steady, and lengthy growth process.

4.2.2 Joint ventures

It is important to recognize that not all joint ventures in construction are with disadvantaged enterprises. Two or more companies will form joint ventures to gain a strategic advantage each would not have separately. Such advantages will often include bidding larger projects, better expertise in a specific type of construction, and/or moving into a new region or locale. If a joint venture is awarded a contract for a project, the Contract Administrator(s) will have an additional challenge to monitor and maintain the balance in contract requirements and compensation for each entity of the joint venture.

Example:
The contract is much larger than anything Contractor A has ever done before

Contractor A provides all field supervision, project craft personnel and materials, tools, and equipment to complete all the work not subcontracted by Contractor B. Contractor A is well known in the area and will have access to key personnel and better pricing, based on an excellent local reputation.

Contractor B provides all project management of the steel, masonry, mechanical, electrical, and plumbing Subcontractors. Contractor B, being a larger Contractor, provides all initial funding and insurances for the joint venture, as well as assurances for the Owner that the project can be completed, based on the size of past projects performed by Contractor B.

A joint venture could also be formed between a Constructor and an Architect where the Owner wishes both the Design and Constructor to be provided under one contract. Such a combined delivery method will be discussed in the next chapter.

Example of cash flow on lump sum:
Where's the money?

Before we leave lump sum contracts and cash flow, let us take a simple example of the lack of payment on a contract that is staying within the typical payment guidelines.

Let this be a simple ten-month job worth the round number of $1,000,000, and to keep it very simple, let us say each month $100,000 of billable work is put in place. Let's jump ahead to the last day before the end of month eight and see where the money has not gone. Remember, month eight's work has not yet been billed and month seven's work has not yet been paid. Only payments through month six have been received by the Constructor:

Month eight:	
Total put in place	$800,000
Total previously paid (retainage deducted at 10%)	$540,000
Withheld from the Contractors for work already put in place	$260,000

Certainly, with the Pay when Paid Clause all of this money is not being put out by the General Contractor. Much of it has been put out by Subcontractors that will actually have to wait a little bit longer than the Contractor for the money to trickle down to them.

The lesson for the Contract Administrator is that large amounts of money are spent on projects, and the cash flow is slow. Contractors and Subcontractors work on narrow profit margins on Design-Bid-Build lump sum projects that were competitively bid and awarded to the low bidder. Contract Administrators have to be vigilant that payment clauses are adhered to and money flows effectively. A few days (and sometimes those few days turn into a month delay) of not being paid can create serious financial challenges for Contractors and Subcontractors.

Even good Contractors and Subcontractors can get into financial difficulty, perhaps on a bad project, and that can create even more difficulties for the Contract Administrator as crews run thinner and less productive or deliveries are delayed because Suppliers are waiting for overdue payments. Meeting the schedule can become a major challenge for the Contract Administrator simply because, somewhere, somebody needs to be paid.

4.2.3 Time and Material or Cost Plus contracts

Time and Material (T&M) payments, often referred to as **Cost Plus** contracts are based on the actual costs of the workers (Time) and the cost of materials (Material) that are put in place. In addition to the actual costs, the Contractor is then allowed to include a markup for overhead and profit, thus the term "Cost Plus." Other costs such as Subcontractor costs are also included, provided the Owner has not contracted separately with the specific Subcontractor. Rental costs for equipment are usually invoiced under Time and Material. Some equipment, however, is expected to be part of the tools and equipment the Contractor brings to the project. For instance, rolling scaffold for a Ceiling Contractor might be considered part of the tools and material expected to be included in the Contractor's overhead and markup, whereas a mechanical scissor lift to reach the top of a warehouse ceiling would be considered rental equipment, and the additional cost would be invoiced along with the labor and other material.

One issue that may become a point of contention is material that is used but not incorporated into the project, such as temporary shoring. One rule often cited is, "Was the material consumed in the construction, or has it been removed for use at another project?"

Time and Material contracts are not as common as lump sum contracts, partly because of the "trust factor." Owners believe Contractors will work more diligently if they are focused on bringing the project in under a budget. Time and Material may take away the incentive to work hard to "get in and get out." For that reason most Time and Material contracts are between parties that have a long-standing relationship. They also usually involve facilities or work requirements that a Contractor has become familiar with and performs effectively for the client's needs. Bringing in another Contractor might entail a burdensome learning curve and additional monitoring the Owner would prefer to avoid.

Markup for overhead and profit is usually negotiated and will vary depending on the work and the amount of effort the Contractor is expected to put forth monitoring or managing the field conditions. For instance, the Contractor's own labor might have a 15% to 20% markup, because this is the largest immediate cost to the Contractor, and it is probably the expertise of the labor that the Owner most appreciates. Material is also significant, but not quite as important, and it might have a 5% to 10% markup. Finally, any Subcontractors involved might have a 2% to 5% markup.

One of the critical issues for most Time and Material contracts is the Contractor's safety record. If accidents have been serious for a Contractor, then insurance costs can increase significantly. High costs of insurance will tend to make a Time and Material contract unattractive for most Owners.

4.2.3.1 Time and Material for change orders

Although Time and Material contracts are far less common than lump sum, they do enter into some lump sum contracts to resolve change orders. In ConsensusDOCS, Article 8 is the Changes Clause, first establishing the right of the Contractor to request and/or the Owner to order changes to the existing contract. The Article goes on to list ways the change can be priced, which first includes a negotiated lump sum. However, if the parties cannot resolve what the cost should be, based on lump sum pricing or, perhaps, Unit Costs (to be discussed next), then in accordance with Article 8.3.1.3 or.4 a form of Time and Material can be used (Note: the blanks for the percentage of overhead and profit would have been filled in prior to contract signing.)

8.3.1.3 costs calculated on a basis agreed upon by the Owner and Contractor plus _______% Overhead and _______% profit; or

8.3.1.4 If an increase or decrease cannot be agreed to as set forth in Clauses 1 through 3 above, and the Owner issues an Interim Directed Change, the cost of the change in the Work shall be determined by the reasonable actual expense and savings of the performance of the Work resulting from the change. If there is a net increase in the Contract Price, the Contractor's Overhead and profit shall be adjusted accordingly. In case of a net decrease in the Contract Price, the Contractor's Overhead and profit shall not be adjusted unless ten percent (10%) or more of the Project is deleted. The Contractor shall maintain a documented, itemized accounting evidencing the expenses and savings.

4.2.3.2 Compare contracts

The reader is encouraged to compare ConsensusDOCS 8.3 Determination of Costs with the AIA A201 7.3.7 for some subtle but significant differences. Most significant is the absence of the Architect involvement in changes in the ConsensusDOCS, whereas price increases or decreases in A201 involve the Architect, Owner and Contractor together. Although the AIA process seems more team worthy, it is important for the Contract Administrator to recognize Architects may not have the authority to direct Constructors to perform contract changes. By contract, that usually requires written authorization from the Owner.

Exercise 4.2

Take a moment
Overhead and profit add or deduct

In ConsensusDOCS 8.3.1.4 paragraph above, there is a very different method of handling "Overhead and profit" for an increase in the contract price than there is for a decrease in the contract price. This method also differs from the AIA A201 contract. Owners have often wanted overhead and profit to be included in the overall accounting of a decrease in the contract. In fact, some Owners not using the ConsensusDOCS contract still want to take away overhead and profit on all deduct change orders.

If you are faced with such a contract provision, and knowing your argument has to be succinct, in fifty words or less, what would you state is your justification for not including overhead and profit in the deduction price?

____

____

____

____

____

____

Discussion of Exercise 4.2

With the introduction of ConsensusDOCS, more and more Owners are recognizing the overhead on a deduct seldom changes. The actual workers will not be required to perform the work, and that is what the costs of the deduct will include, but the personnel involved in the project as "overhead" managers, accountants, and officers do not work any less. In fact, in order to fully document the deduction, some will probably have worked more than they would have if the deduction had not taken place. Additionally, unless the length of the project is significantly reduced (e.g. 10% less to be accomplished), then some of the overhead items such as trailers, trucks, or phones will still be used the same amount of time.

As for profit, the deduction does not really change the need to be in business for a profit, but the project, itself, represents a fixed target for the company. Construction is a service industry. The Contractor does not make more profit by sending more items across an assembly line. The personnel working on one project cannot jump to another because a small item has been eliminated. Construction uses teamwork to obtain profit from one project at a time. Profit can come only from successful completion of work on hand. Taking on a project usually means the Contractor will not be taking on other projects with that staff, based on the workload of these available personnel. They are employed to service the project, and unless that project were to disappear, they will not influence profit on any other jobsite during their time on the project. In addition, even bonding capacity if the Contractor does bonded work can be capped and will not be reduced by a small deduct change order. So, again, unless a significant deduction takes place, the prudent Contractor is basing revenues on the amount of work on hand.

Now, of course that is more than the fifty-word limit set for you, but in a job meeting, confronted by an Architect or Owner, you want to be able to keep your position or presentation short. Success is not guaranteed, but being prepared is a good start.

4.2.4 Unit Cost contracts

Unit Cost contracts for the most part are restricted to heavy and highway work as well as itemized units within lump sum contracting for Building Contractors.

Although today we generally see Unit Costs as only a method of resolving costs within the more generally used lump sum contract, Unit Cost contracting was for millennia considered the most fair and effective method of contract payment.

If you go back to the Code of Hammurabi and refer to Code 228, you will see the Builder is to be compensated on a Unit Cost basis. In the case of the Babylonian Builder it will be two shekels for each sar of surface. A sar was a basic unit of measurement in ancient Babylon. It was used more for land area, covering about 36 square meters.

Although individual trade and craft workers may have been paid on a daily basis (their day consisting of sunrise to sunset), the Builder was usually paid by the units that were put in place. The fairness of this was that every Builder would be paid the same for the same work. How quickly and well they did the work was to their advantage or detriment, particularly if the work had to be redone. Passing from Europe to the colonies in America, the Unit Cost form of payment still was the most commonly used method of payment until the early decades of the nineteenth century. The reasons for this will be discussed shortly under Design-Build in this chapter.

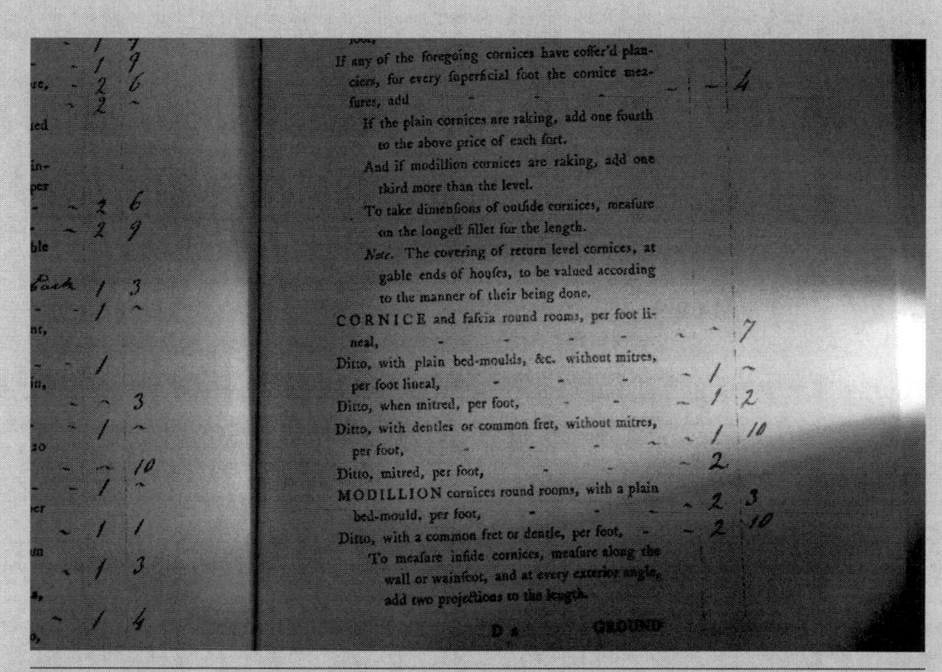
If any of the foregoing cornices have coffer'd plan-ciers, for every ſuperficial foot the cornice mea-ſures, add — 4

If the plain cornices are raking, add one fourth to the above price of each ſort.

And if modillion cornices are raking, add one third more than the level.

To take dimenſions of outſide cornices, meaſure on the longeſt fillet for the length.

Note. The covering of return level cornices, at gable ends of houſes, to be valued according to the manner of their being done.

	s.	d.
CORNICE and faſcia round rooms, per foot li-neal,	–	7
Ditto, with plain bed-moulds, &c. without mitres, per foot lineal,	1	–
Ditto, when mitred, per foot,	1	2
Ditto, with dentles or common fret, without mitres, per foot,	1	10
Ditto, mitred, per foot,	2	
MODILLION cornices round rooms, with a plain bed-mould, per foot,	2	3
Ditto, with a common fret or dentle, per foot,	2	10

To meaſure inſide cornices, meaſure along the wall or wainſcot, and at every exterior angle, add two projections to the length.

D 2 GROUND

Figure 4.3 Page from a reproduction of the secret Book of Prices kept by the Carpenters' Company of the City and County of Philadelphia, 1786
Photo credit: Charles W. Cook

Through colonial times and up until the early nineteenth century, the Carpenters' Company of the City and County of Philadelphia kept a secret Book of Prices. The page shown above is from a reproduction of the 1786 copy of the book.

The Carpenters' Company members were able to totally control all construction in the City of Philadelphia, because no construction would be paid for until it was "measured" by a member of the Company, and the price of the work then figured on the basis of the various units in the Book of Prices. In fact, the prices were so secret that when Thomas Jefferson asked for a copy to check that what he was paying in Virginia for work was comparable, the Committee on The Book of Prices turned down his request.

As an example, cornice and fascia around rooms would cost the Owner 7 pence (pennies) per lineal foot. However, with plain "bed-moulds, [molding immediately under a projection as a sort of rest for the molding above] &c. without mitres," the price increases to a shilling per lineal foot, or 5 more pence per lineal foot.

Note: English currency denominations were still being used – the book's date being one year prior to the adoption of the United States Constitution.

One advantage today to including Unit Costs in a contract is when an Owner or Designer who is not exactly certain how much they intend to do can get a competitively bid Unit Cost from Contractors during the bidding process, and these will generally be less expensive than if negotiated after the proposals have been submitted and a final Contractor chosen.

Example: The project requires a considerable amount of excavation over several miles for a new highway, including a new clover-leaf exchange. The Engineers know there are differences in the soils, which include rock, clay, and dirt. Not knowing the exact extent of any of the subsurface conditions, they ask each Contractor in their bid to submit Unit Costs for removal of each possible material type. Usually such contracts will be based on a minimum quantity for each type, since to remove just ten cubic yards of rock would be enormously more expensive than to figure the continuous removal of hundreds of yards of rock.

Also, although the Designers do not know the exact amount of each subsurface material type, they have certainly made some preliminary estimates, and they will then use those quantities to determine from the submitted proposals who has in fact submitted what would appear to be the low bid based on the quantities expected to be excavated.

Unfortunately, there is also a potential for tremendously wasteful spending in Unit Cost contracting. Federal, state, and municipal governments often subscribe to a "pool" of Unit Cost Contractors who have been asked on an annual or periodic basis to submit Unit Cost pricing for various work, often of a maintenance nature. Contractors have to protect themselves against pricing too low for a single item, and so a higher overhead and mobilization is usually added. The result is one unit will cost as much as perhaps ten would if bid as a lump sum project. This is how the public discovers it is paying $500 for a toilet seat in the military.

Obviously, each of the three methods, lump sum, Time and Material, and Cost Plus, has certain advantages and disadvantages. In some cases the situation will dictate which method will be best for just compensation to the Contractor while protecting the Owner from paying more than the work should be worth. Within a lump sum contract both Time and Material and Unit Cost may be used, based on previously agreed costs and markups for certain items or possibilities.

Exercise 4.3

Take a moment

Which compensation method would be best?

A Lump sum

B Time and Material

C Unit Cost

1. ——— . Unknown quantities of certain types of work that will be required to be done on the project.
2. ——— Time to fully detail vision of structure.
3. ——— Need to move quickly without knowing final extent of costs.

Discussion of Exercise 4.3

1. In most cases when Designers are not sure of the quantities they will encounter or need to complete the work, but they do know specific elements of work will need to be accomplished (e.g. excavation in cubic yards), they will ask for Unit Costs to be submitted for those items of anticipated work along with the initial bid proposal.
2. When Designers and Owners have the time to fully detail on plans and in the specifications what is envisioned for a particular project, they will seek lump sum bid proposals, usually from more than one Contractor.
3. If the situation requires a quick response, particularly if the completion schedule is important, then authorizing the work on a Time and Material basis will be faster than waiting for a lump sum proposal, or, even worse, waiting for a lump sum proposal to be resolved.

One final issue regarding Time and Material or Cost Plus change orders within an existing lump sum contract is to realize the costs of the Time and Material must be clearly delineated and lump sum contract work cannot be included in the hours or material of the invoice for the Time and Material change order. This may sound elementary, but in practice it can become a major challenge for the Contract Administrator.

Example: The difficulty can come from two directions. Those supervising the work of your own forces may have difficulty separating the times when individuals are working on the actual contract and when they are working on the change. The same will be true of Subcontractors (and unfortunately there are some Subcontractors, just as there are Contractors, who might take advantage of this situation to inflate their T&M work while actually doing contract lump sum work).

Let us imagine the crews are working on a hospital wing with twelve private rooms and twelve semi-private (two-patient) rooms, and the Owner now recognizes that two of the private rooms will have to be converted to semi-private two-patient rooms. The work has to continue immediately, so all the Subcontractors and you are told to do the work on a T&M basis. For the semi-private room there are minor changes in the drywall, ceiling, and some additional fixtures such as curtain track and television brackets, which your forces will do. For the electrical and pipe-fitter trades there are several more wires, outlets, and significant additional piping for medical gases to be installed.

No one has the time to stand over each worker to determine whether they are working on extra or contract work. And at the same time, just cutting the hours in half for each room is probably not feasible or even accurate.

This is when good rapport and open communication with all parties on the contract is essential. Some may view the Contractor's intentions negatively and object to the invoicing submitted as inflated. Depending on due diligence, accurate records, and detailed breakdowns, the Contract Administrator can succeed where many others initially fail to justify inflated T&M change order requests. Fortunately, once a successful rapport is achieved, trust is established, and the likelihood of more T&M work is greater.

4.2.5 Risk and compensation

When weighing compensation methods, all parties usually look to how the risk can be appropriately shared or distributed. Ultimately, the best method for whoever is paying for the work will usually determine how compensation will be made. Unfortunately there are times when some Constructors will take a lump sum project when the risk is too great for that method. In such a case the stress on the Contract Administrator can be tremendous. Although it will almost always benefit the Owner to have a set lump sum price before any work begins, there are times when no one really understands the entire scope or, perhaps, the time something will take. It is then that some other form of compensation becomes effective. An Owner might "ball-park" what a project should cost based on experience or input from others, and then direct a Constructor to proceed on a Cost Plus or Unit Cost basis. In such circumstances, the Contract Administrator should recognize that in addition to being a team member looking out for schedule, quality, and safety, now the Contract Administrator is going to be even more closely scrutinized on how well he or she manages the costs.

INSTANT RECALL

- In essence, a contract begins with a promise or set of promises by one party and requires an acceptance by another party with some form of consideration for the original promise.
- There cannot be a contract to commit an illegal act.
- Any understanding either party might have had prior to the written agreement will be overruled by what has been agreed to in writing.
- The parties to a contract have the implied duty to cooperate.
- Although you will not make friends and your reputation might suffer, if you do make a mistake in your offer (submission of your bid proposal) you can be released from your obligations and the contract would be voided.
- By far the most common form of payment in construction is the lump sum. In such an agreement the Owner is contractually obligated to pay a total cost agreed to, usually through a bid submission, or negotiated after discussion of the scope of work.
- The practice of front-end loading is quite extensive, but it is critical the Contract Administrator understands the ramifications of the practice.
- One way Constructors protect their cash flow is with clauses in their subcontracts that are known as Pay when Paid or Pay if Paid.
- Time and Material (T&M) payments are based on the actual costs of the workers (Time) and the cost of materials (Material) that are put in place.
- Unit Cost contracts for the most part are restricted to heavy and highway work as well as itemized units within lump sum contracting for building Contractors.

YOU BE THE JUDGE

A Contractor (Christian Bionghi, doing business as Abacus Technical) appeals a case to you in Southern California on the basis she had been promised prior to signing the contract that language within the contract stating the Owner, Metropolitan Water District of Southern California (MWD), could terminate the contract without cause would not be applied. Abacus was able to demonstrate MWD had made assurances prior to signing the contract that it would not terminate the contract without good cause. MWD pointed to language in the mutually revised contract stating "Both parties have participated in the drafting of this Agreement."

The lower court ruled that MWD could terminate the contract.

Circle your decision:

Affirm the Lower Court's Decision Reverse the Lower Court's Decision

After finishing this chapter you can go to the Companion Website and check your decision with the actual court decision.

IT'S A MATTER OF ETHICS

Think to yourself or discuss with others: If you promised someone before signing a contract to do something above and beyond what the contract actually required, and the contract contained language that negated or superseded your promise, what would you do?

Going the extra mile

1. Go to the Companion Website www.routledge.com/cw/cook and view the video discussion Part III of the Subcontract in Appendix I.
2. Optional assignment: go to www.constructingleaders.com and view the video "Carpenters' Hall" for a better understanding of construction in early America, and its importance in the founding of our nation.
3. **Discussion exercise** (with friends or as part of a classroom discussion: Do you believe that, the longer and more specific contracts become, the better the relationship and results will be for all concerned? Why?
4. Do "You be the judge" above and then go to the Companion Website to read how the actual case was resolved, and try a couple more "You be the judge" cases there.
5. Effective use of the correct vocabulary improves communication and promotes respect among colleagues. The reader should become familiar with the key words listed below that have been used in the chapter. Definitions can be found in the Glossary. Additionally, the student is encouraged to use the flash card exercise in the chapter's Companion Website as practice for possible quiz or exam questions.

Key words

Acceptance
Breach of contract
Building Information Modeling Technology (BIM)
Change order
Condition precedent
Consideration
Contract modification
Cost Plus
Crash
Disadvantaged Business Enterprises
Estoppel
Front-end loading
General Contractor
Impossibility
Joint venture
Lump sum
Parol evidence
Pay when Paid/Pay if Paid Clauses
Performance specifications
Progress payments
Promissory estoppel
Quid pro quo
Retainage
Time and Material
Uncertainty
Unit Cost

CHAPTER 5

Different construction contract delivery methods

In the previous chapter we discussed some essential and basic elements of a contract. Today, more and more Owners are looking to different delivery methods to secure the best final product in the most cost-effective approach. Whether a Contract Administrator has the opportunity to choose the delivery method or is handed the terms committed to by someone else, knowing how to deliver is as vital as knowing what to deliver.

Student learning outcomes

Upon completion the student should be able to. . .

- **recognize different types of construction contract delivery methods**
- **understand the duties and expectations of Designers and Constructors in the different contract delivery methods**
- **analyze which method of delivery is more effective in a particular construction situation**
- **determine how best to apply Unit Cost, lump sum, and Time and Material or Cost Plus pricing within a contract**
- **understand the expectations for contract administration within the various delivery methods of construction contracts.**

5.1 Design-Bid-Build

Coming together is a beginning.
Keeping together is progress.
Working together is success.

Henry Ford

For almost two centuries most construction contracts have relied upon a lump sum Design-Bid-Build scenario where the Owner hires a Designer to complete plans and specifications for what is desired, then puts the project out for builders to submit lump sum bids to construct the entire project, and finally awards one Builder the project to complete, usually based on the lowest price.

One of the greatest challenges, but often the reason for great success, is the teamwork required for the Design-Bid-Build method of contract delivery. On the surface, Design-Bid-Build sets up the potential for adversarial relationships between the three main parties to the process – the Designer, the Owner, and the Builder. It is not easy to achieve Henry Ford's "Working together success."

Decades of Design-Bid-Build have led to extensive and numerous refinements in the written contracts that bind the parties together, and there will undoubtedly be future revisions to such contracts. Often changes are made not because of the vast majority of teams that work well, but as a reaction to the few that do not.

The chief advantages of the Design-Bid-Build sequence are:

- The entire project is usually complete in design before the bidding phase so the Owner has a clear understanding of final costs once the bid proposals are received.
- Through the bidding process the Owner should get a very competitive price.
- The parties to the contracts have a clear understanding of their roles and the requirements of achieving project closeout.

Unfortunately, there is also one major disadvantage, and that is that the contractual relationships can lead to adversarial relationships. Shortly, in another chapter, we will discuss major "killer clauses" that place risk and burdens on the Constructor. These clauses are written because some Constructors in the past have used ambiguities in the plans, delays in the project, or other challenges of the project administration to their financial advantage.

Human nature being what it is, each party under contractual obligations within the project will first look to see no harm will be done to oneself. With the clear separation of responsibilities and tasks in the Design-Bid-Build environment, it is easier for one party to blame another than assume some major financial responsibility themselves.

5.2 Construction management

Leadership is the art of accomplishing more than the science of management says is possible.

Colin Powell

The typical Design-Bid-Build relationship of Designer–Owner–Constructor is shown in the three-circle model (Figure 5.1).

For decades the Owner relied upon the Designers to oversee the execution of the Constructor's work. In addition to being the visionary of what a structure could be, the Designers were also the managers of the project – and they would supply what became

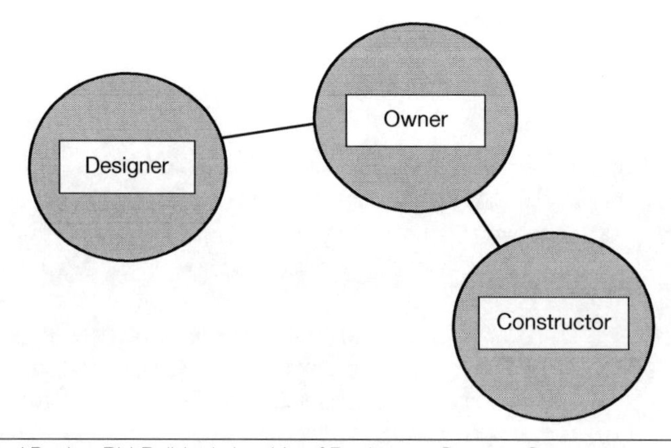

Figure 5.1 Typical Design-Bid-Build relationship of Designers–Owners–Constructors

known as a "clerk of the works" to see that the vision became a reality (cf. PSFS Building story in Chapter 3).

Over several decades the role of the Architect has faded from the Owner's representative in all construction matters to the expert on all design matters. Part of this may certainly be due to the potential litigation that seems to follow many projects. Part is also due to training, education, and experience. And, just as the PSFS Building was moving the nation into the modern era of design, it was also moving construction into greater complexities, slowly separating the vision from the details of execution. What did become obvious to Owners was that more and more expertise in the actual construction management of the project was necessary. Just as George Howe had found his expertise in the construction side of the Design-Bid-Build spectrum when he and William Lescaze designed the PSFS Building, Owners were looking to Constructors for management expertise.

Figure 5.2 The PSFS Building under construction
Photo credit: Hagley Museum and Library

Ushering in the new era of International Style design in America, George Howe worked in the waning decades of Designers who assumed full control of managing the construction phase of Design-Bid-Build contracts. More and more responsibility for Contract Administration would slowly evolve to Constructors and Construction Managers.

The diminishing input of the Architect in totally managing the construction phase and the multitude of Contractors, Subcontractors, and Sub-subcontractors, along with Suppliers, gave rise to a new partner in the construction process, the **Construction Manager (CM)**. CM companies were formed based on their previous construction experience. Many of these companies had previously been **General Contractors**, performing the basic carpentry trades of the project and subcontracting specialty trades such as the Mechanical, Electrical, and Plumbing (MEP) trades. They found a way to increase their volume of work, often without the risk they had previously assumed when they did so much of the work themselves.

Initially this placed an additional layer of cost on the project, but quickly Owners looked to the Construction Manager to eliminate the "middle" General Contractor, and all trades, including the lead carpentry trades, would be subcontracted. The model was changing, and many thought for the better.

Note: Early in the development of the CM concept, it was recognized that supervision and coordination of the Subcontractors could be from the CM, but the contractual relationships with the Subcontractors could either be directly with the Owner or with the CM, potentially shifting some risk and responsibility.

Owners, however, saw the chance to improve on the business relationship, and so contracts were drafted that put the Construction Manager at risk, known today as **CM at Risk**. Basically this was putting back on the Construction Manager the previous burdens of guaranteeing the project costs that the General Contractor had formerly assumed when submitting bid proposals. The main difference was the Construction Manager set a Guaranteed Maximum Price and charged costs plus a fee, similar to a Time and Material plus fee contract, rather than a lump sum that the General usually worked under.

The evolution, however, was not complete. With more and more risk, the Construction Manager sought more and more input into the design phase of project development. If they were to assume the risk of a **Guaranteed Maximum Price (GMP)**, they wanted to be part of the design process. This, of course, was in part due to the "uncertainty" principle that can negate a contract. Moreover, it gave the Construction Manager a better negotiating position to influence the design for practical considerations.

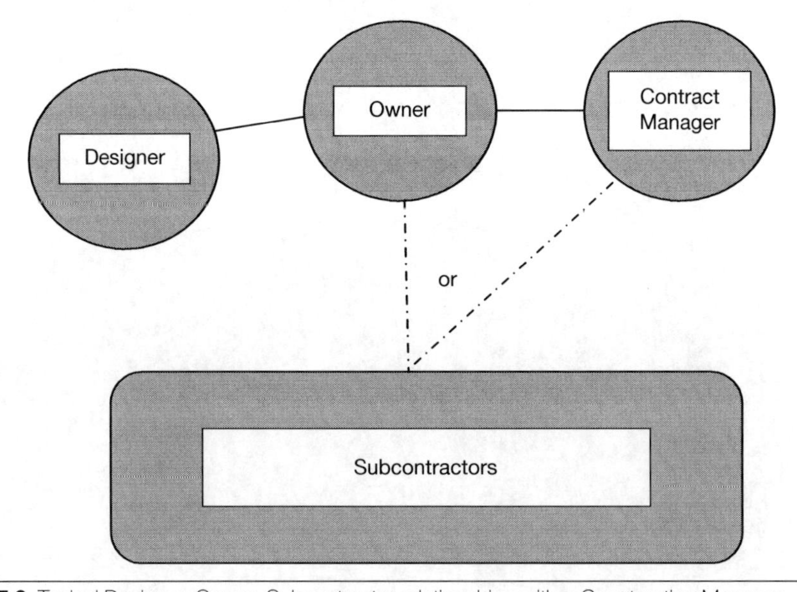

Figure 5.3 Typical Designer–Owner–Subcontractor relationships with a Construction Manager

It is important to note that depending on when and how effective the Construction Manager was made a part of the project team would not only determine the accuracy of the Guaranteed Maximum Price, but it also would determine how changes to the contract would be priced within or above the GMP.

> **Example:** A Construction Manager who has been brought in after the foundations have already been designed would naturally assume the structure that would go above the footings would be within the structural limits of the design. Submitting a Guaranteed Maximum Price based on the existing design would be appropriate. Changes of office configurations would remain within the GMP. If, however, the Owner decided to add one or more stories to the structure, thus requiring a reworking of the foundation, then not just the additional stories would increase the GMP, but also the cost of any significant changes in the foundation would be added to the GMP.

As the Construction Management process continued to evolve through the late twentieth century, more and more General Contractors did less and less of their own work, becoming *de facto* Construction Managers, even if the contract did not read so. As it now stands, Construction Managers can be purely CMs (also sometimes known as agency Construction Managers) that do nothing but act as the Owner's agent and do not carry a risk or burden of a Guaranteed Maximum Price. They might do this on one or more projects for the Owner simultaneously. Or the Construction Manager can be a CM at Risk, solely supervising the work, or some Construction Managers have evolved back to the General Contractor model, acting as a CM at Risk, but also assuming the role of General with the possibility of performing some work with their own forces.

As more and more members of construction teams began to see the successes and failings of the CM process, many began to see new opportunities to improve the entire process. The typical Design-Bid-Build method was being re-examined.

5.2.1 Prime Contractor

A **Prime Contractor** is an entity that has a direct contractual relationship with the Owner to build the project or a portion of the project. There can be multiple Prime Contractors on some projects where the Owner contracts separately with several different key trades such as a General Contractor to do most of the work related to foundations and the structure, while separate Mechanical, Electrical, and Plumbing Contractors also have a direct contract with the Owner. This type of construction relationship with the Owner is referred to as **Multi-prime Contracting**.

When separate Contractors are involved in Multi-prime Contracts, the potential for delay and lost productivity waiting for one or another of the primes to complete sections of work is increased. Unilateral communication through the Owner or agent of the Owner can be a real impediment to progress. The Contract Administrator should attempt to open lines of communication in a 360 degree manner. There will be no contractual relationship between primes, but cooperation starts with knowing what the other party needs. Fulfilling needs is the start of working together. It also is the end-product of a good contract – that *quid pro quo*.

5.2.2 Subcontractor/Supplier

Subcontractors are a contract removed from the Owner's contract with the prime Contractor. Subcontractors can also subcontract work to another entity, and these are known as Sub-subcontractors.

A Supplier provides material or equipment to a project, but a Supplier does not provide labor on the project (other than perhaps during the manufacturing, shipping, and delivery to the project). The Supplier is not responsible for the "put in place" labor on the project.

The good Contract Administrator recognizes not only can we not do it by ourselves, but we are extremely dependent upon others. They may need the work, but we need them to perform it often under difficult circumstances, on our schedule, not their convenience, and achieve the standards of quality the Owner and Designers expect. How we treat Subcontractors and Suppliers will almost always have a greater impact on the success of their performance than the details and clauses within a contract.

A CASE IN POINT

As an eager Project Manager intent on pleasing the client by staying on schedule, my first question at every job meeting or conference for a Subcontractor coming onto the project for the first time was, "How soon can you get done?"

For a while, it seemed to work, but eventually I realized there was often a great deal of difference between the answer and action.

Then one day, while away at an industry convention, I was talking with a good friend from Indiana about the frustration of keeping all the Subcontractors on schedule. He said he had had the same problem, but told me he learned to change the question. In short, I was asking the wrong question.

I appreciated his advice, and when I got back I tried his strategy. I asked the next Subcontractor starting on one of my projects, "How can I help you make money?"

My friend was right!

If I really want to succeed, it's not about me.

5.2.3 Multi-prime Contracting

One outgrowth of the Design-Bid-Build contracting method that can also be reason for an effective Construction Manager is the use of **Multi-prime Contracting**. This method is most often used by public Owners, such as school boards, townships, or for state or Federal projects. With or without a CM the usual primes will be a General Contractor for the overall structure, along with separate Electrical, Plumbing and/or Mechanical Contractors. The chief purpose is to save money from the markup a General Contractor would put on the MEP trades.

In theory that is probably a savings, but in practice it probably does not offset many of the scheduling and coordination difficulties that arise. Without a contractual relationship between the General and the other primes, communication is usually indirect, except during joint project meetings, and the result can be delays and often ineffective production, due to failure to complete certain areas of the project in a manner timely to the work of the other prime trades.

A good Construction Manager can help in this regard, but some government agencies do not hire a CM and attempt to coordinate the various primes themselves.

5.3 Design-Build contracting

Build a better mousetrap and the world will beat a path to your door.

Ralph Waldo Emerson

The inclusion of the Construction Manager into the design process has brought about new thinking in what has been the traditional practice of construction for the past two centuries. Separating the design from the construction phase of the work has been very recent in terms of the millennia of civilization.

Before the separation of design from construction, the whole project was undertaken for the Owner by the Master Builder. The Master Builder was the Architect, Engineer, and Constructor all in one. In the beginning this was a single person. By the Hammurabi Code we can see the person in charge was called a builder.

The Master Builder was generally the Mason, and from the construction of the Egyptian pyramids through the castles and cathedrals of Europe one can understand the prominence of the masonry trade in the construction process. Unlike today, engineering was not a mathematical certainty but a trial-and-error undertaking, working off the successes of the past handed down through the craft guilds.

Hidden within the construction process were the ambitions of the carpenters. Masons could not accomplish their tasks without the support structures first put in place as temporary shoring by the carpenters. When William Penn invited all craft workers in the construction industry to his colony in the New World, everyone was certain Penn would want masons to again take the lead. William Penn had witnessed the devastating Great Fire of London that destroyed so much of the city, and he made fireproof construction a top priority for Philadelphia. Obviously, no seventeenth-century construction could be entirely fireproof, so when the carpenters arrived in Penn's colony, Pennsylvania ("Penn's Woods"), they saw things much differently. With such huge quantities of lumber and virgin forests, they were determined to become the Master Builders, and within a century, through marriages and political intrigue, they had positioned themselves to be the Master Builders and then subcontract work to the masons.

They created their own monopoly of construction in Philadelphia through their Book of Prices (see Figure 4.3), but they also failed to improve their design capabilities as quickly as those who had been trained in Europe, who were learning new engineering and design techniques. A few decades after the American Revolution a new influx of European Engineers convinced Owners in the young United States of America that the carpenters were good at building but not designing, and by the early nineteenth century, the division of Designer and Constructor had taken over in Philadelphia and spread throughout the new nation.

Figure 5.4 Carpenters' Hall, Philadelphia
Photo credit: Charles W. Cook

Robert Smith, Master Builder of Carpenters' Hall, was the first person actually referred to in America as an "architect." He was the Designer and/or Builder of structures from the colony of Virginia through the colonies of Rhode and Massachusetts. Designer and Builder of Carpenters' Hall, he also served on the Committee of 24 (he was #25), writing letters to individuals throughout the colonies to come to a Continental Congress. From that Congress would eventually come the Revolutionary War and the separation from England.

The Carpenters' Company, founded in 1724, continues to operate and opens Carpenters' Hall to visitors on the first floor. The Company remains the oldest extant trade guild in America.

Today, Design-Build is making a major resurgence in favorable contract methods. Three reasons for using Design-Build over Design-Bid-Build as a project delivery method are:

- time
- cost
- quality.

Of course, today the Master Builder is not a single individual. The Master Builder's task and responsibilities are shared by specifically trained individuals. One individual would not be able to handle all the complexities of the various trades, but under one company the Design-Build team no longer has an adversarial relationship that exists when one company designs and another builds. They are the same entity.

This is a great advantage for the Owner, also. What might appear to be effective monitoring of the construction process with separate entities has not proven to be so in many cases. Having the same party responsible for execution that created the design forces harmony on what is often disharmony under Design-Bid-Build procedures. Since the Design-Builder is liable for all design ambiguities and defects, there is less arguing and fault finding and more solutions are forthcoming more quickly.

Of course, the concept of scrutiny and effective monitoring of standards of quality has to be addressed. Owners need to interview and select the Design-Build team based on high integrity and service. Hiding ambiguities and flaws is unacceptable. The Owner wants a Design-Builder with an effective track record of accomplishment.

For a Constructor, there are several issues that are important to resolve even before entering the Design-Build relationship:

- Are the parties (Owner, Architect, and the Contractor, along with critical Subcontractors) ready to participate in a Design-Build relationship?
- Will some still want all the advantages and not contribute to the whole?
- Will there be a satisfactorily defined relationship (usually through a contract) of the responsibilities and expectations of each of the parties?
- If there is a joint partnership evolving, who will be responsible for such items as permits, licenses, insurance, bonding?
- Who will take the lead on the critical decision-making process for specific items (e.g. schedule, budget, changes or extras, and standards of quality) important to the Owner and project completion?
- How certain are all the parties that there are no hidden or subsurface conditions that could significantly change the work?
- Does the Owner have a significant contingency set aside for unforeseen conditions or changes to the scope?
- How will disputes be resolved?

The popularity of the Design-Build delivery method for construction seems to be increasing.

One major caution for most Constructors, however, entering their first Design-Build contract, is Errors and Omissions Insurance. Insurance will be discussed further in a later chapter, but the Constructor must recognize, as part of the Design team, that the responsibility and liability for design flaws can be placed also on the Constructor. This type of insurance is not typical for most Constructors, although it is becoming a bit more common as Contractors and Subcontractors take on more "design" functions through the submission of shop drawings that detail what the Architect's drawings did not. In a Design-Build project the Constructor should definitely alert the insurance company to make sure there is adequate coverage.

Whether the industry will fully return to Design-Build is not certain. Smaller projects may continue with Design-Bid-Build on the basis that smaller Constructors will not have the capacity to bring Architects and Engineers "in-house." In such circumstances some Constructors have taken the design firm on as a Subcontractor or even created joint ventures. And the same would be true for smaller architectural and engineering firms. Regardless, the trend to find a better "mousetrap" continues and has led to other project delivery methods.

5.3.1 Performance specifications

One form of Design-Build that has been around longer than the trend to return to Design-Build is the performance specification contract. It can also be a portion of or within a Design-Bid-Build contract. Contractors and often specialty Subcontractors are asked to install equipment to achieve a certain level of performance. This can be another important reason a Constructor should consult an insurance agent to make sure they are covered, since performance specifications carry with them the burden of design to some extent. Typically the choice of material or equipment will not be included in a **performance specification**. What will be stated are the standards or parameters of the performance the final product is to achieve. If a product is specified, it might negate the performance specification under the doctrine of "impossibility," provided it could be shown that the specified item could not achieve the performance required.

Example: The specifications call for a specific air filter system to be installed in a hospital operating suite. The Minimum Efficiency Reporting Value (MERV) that is to be achieved is 20. The specified system is more than capable of doing that in a single room, but in multiple operating rooms in this hospital, the system will not work. The Mechanical Subcontractor generally would not be held responsible for following the specifications. The possible exception to that would be if the Mechanical Contractor who installed the equipment should have known the installation would not work.

Exercise 5.1

Take a moment
Decide who is responsible?

In the above example, if the Mechanical Contractor had known the equipment would not be adequate to perform what was specified under the performance specification portion of the contract, and knowing this, the Foreman of the Mechanical Subcontractor went to the General Contractor and told the Superintendent about what would happen, then who would be responsible?

Designer Architect and/or Mechanical Engineer
Owner
General Contractor
Mechanical Subcontractor

Discussion of Exercise 5.1

The easy answer might be the General Contractor, provided the Superintendent had not passed the information on to the Owner or Architect. There might have been some extenuating circumstances why the information was not passed on.

- Perhaps the Superintendent never understood exactly what the Mechanical Foreman was saying.
- Perhaps it did not seem reasonable to the Superintendent that the Foreman would know better than the Architect.
- Perhaps the Superintendent did not want to embarrass the Architect on this, since there were other issues that were also in need of clarification due to design flaws.
- Perhaps the Superintendent got busy and forgot about the issue until it was too late.

There might be many reasons, but none of them is a good one. In a "shoot the messenger" atmosphere, the bearer of bad news may not want to come forward. That is why it is important the Contract Administrator creates a collaborative atmosphere for all parties to contribute. Project meetings are an excellent time to bring up issues, but if the time is not right for a meeting when one learns of a problem, then a note, an e-mail, or a call should be made. Seek an answer through a Request for Information, or, at the least, document the concern through some form of communication.

Obviously, this is essential not just for performance specifications, but for all conditions and situations that arise on the project that could affect the schedule, the cost, the quality, or the safety on the project.

Successful performance in any contract, but particularly related to a performance specification, can be quite exciting. Challenges aside, the Contract Administrator must recognize the potential liability related to failure, and, if necessary, insure against it. Recognizing the potential for failure is one of the best ways to not fail.

Exercise 5.2

Take another moment
What's in a word?

Look up the Greek origins of the word "architect," and write down their meanings.

*Arch*__________

*Teckton*_______________

Were you surprised?

Discussion of Exercise 5.2

Most people would think the word "architect" is about a Designer, artist, drawer, but the actual origin of the word and the profession is in building. In fact, it is quite probable the first Master Builders were more concerned about the challenges of engineering and defying gravity than they were with how attractive something looked. Making sure something did not fall down was far more important in Hammurabi's Code than that it looked nice. Moving back to Design-Build, therefore, is bringing the Master Builder concept back. Rather than a single individual, however, the entity of the Design-Builder brings a team together to perform the tasks the single Master Builder had in the past. The advantage, of course, is that there are far more disciplines, trades, and, in our immediate case, codes and laws the Master Builder must be concerned with than in the past. (For an insightful and amusing perspective, read *The Bricklayer* portion of Rudyard Kipling's poem "A Truthful Song," about a reincarnated Pharaoh commenting on nineteenth-century construction practices.)

5.4 Joint venture

As mentioned in a previous chapter, a joint venture contract binds two or more individual entities to the performance of a specific project together. The partners to the agreement share the management and costs of the project, and they also will split the revenue in accordance with the percentages agreed to in the written contract. Joint ventures are often used to aid

Minority or Disadvantaged Business Enterprises, but they have sometimes been abused by the majority Contractors in such joint ventures, by merely using the joint venture as a front for their own benefits.

Other legitimate uses of joint ventures will combine two strong Contractors that need each other for improved bonding capacity or other financial concerns. Also some joint ventures are formed because one party has expertise in an area critical to the project, and the other has assets (e.g. personnel available, equipment, finances and/or bonding) that will make the joint venture more successful.

Joint ventures are tremendous opportunities to bring new teams together for great accomplishments. Usually the terms set the hierarchy for team leadership, but in successful joint ventures, everyone contributing to the best of their ability usually means present and often future success.

5.5 Integrated Project Delivery

No man is an island, entire of itself; every man is a piece of the continent.

John Donne

Different than a joint venture, and often more widespread, is a growing concept in cooperative project delivery. In lieu of one Design-Build company, **Integrated Project Delivery** or IPD has emerged as an effective method of pulling many diverse elements of the project team together while maintaining established individual entities. There are in fact a separate group of IPD contracts published by the AIA in which the Architect does maintain some significant autonomy.

However, the trend for parties to work together is something the Owner is driving, and technology is also helping. **Building Information Modeling** (BIM) is making the 3-D potential of finding crashes extremely attractive in the design phase of project delivery. Many architectural firms, however, need the input of key construction trades to fully realize the 3-D models. As construction details have become more complex, they have been assumed by the actual trades rather than detailed on the architectural drawings. This has led to a dependence on key Suppliers and Specialty Contractors to include their input during the design phase.

Somewhere between the intention to perform Design-Build contracting and the need to fast-track a project has come the Owner's acceptance that including more than the Designers in the design phase of a project can actually save time (and usually time is money) as well as create a harmonious project atmosphere where the final product – the Owner's usable structure – is the primary focus of everyone involved.

Self-interests of each individual or party to the project are secondary, and some Owners have even provided financial incentives to the parties that sign on to the IPD agreement by offering a share in savings if the parties can work together to meet the requirements while saving money above the project budget, including any contingencies the Owner may have allocated to the budget. If this arrangement is made, it is important that the percentages of sharing and commitment be in writing to avoid misunderstandings or disputes afterward as to who did more and therefore deserves more reward.

The IPD team can involve just the Owner, Constructor and the Design Professional(s), but the Constructor will usually want to include Specialty Contractors and even some key Suppliers for their specific expertise in key areas. Once this method has been chosen it is also important to recognize to some extent that the parties will be waiving liability for the input and

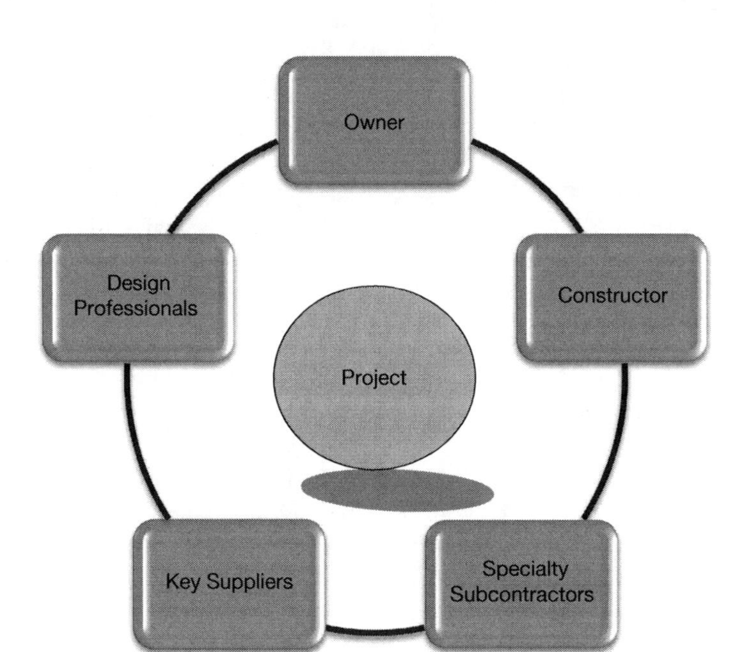

Figure 5.5 Parties that could be involved in an IPD project

acceptance by the other parties of contributions made to the project. Just as in a Design-Build contract, the team is now assuming errors and omissions, and the Owner will be expecting no problems from the expanded expertise of the team.

Figure 5.5 is a simplified model of the actual parties that could be involved in an IPD project. For instance, the "Design Professionals" rectangle could include the Architect, a Civil Engineer, Electrical Engineer, and a Mechanical Engineer, or even more within that one category of "Design Professionals."

Within the Integrated Project Delivery model, the Owner can bring one or more of the parties into the process at different times, but usually the sooner each entity is made a part of the process, the more time and savings the project team can bring to the final product.

The Owner, therefore, can benefit from IPD from conceptualization through design, permitting, construction, into project closeout and facilities management phases of the structure's lifespan.

Example: Hospitals are very expensive structures to build. The equipment costs for hospitals are often only part of the challenge. Delivery of the equipment can also be important. Also, providing the required space for the equipment does not mean having the right dimensions. Wiring is often just the beginning. Providing shielding and other requirements for safe operation is important. Bringing manufacturers onto the team early can improve schedule and final quality of the installation. In such a team the Owner can rely upon the Architect, the Constructor, the Installer, and the Manufacturer to coordinate the issues of design and final operation of the equipment from beginning to end.

Just as there are "implied" commitments within a written contract, there are expectations for all the parties within the IPD project partnerships. These include:

- full and open communication among all the partners
- confidence that each individual brings expertise to share
- trust that all the parties intend to work for the benefit of everyone else.

Still, there are some challenges yet to be worked out among the IPD method. Just as technology has assisted in making IPD more effective, it also has limited some potential. For years there has been anticipation of a "paperless" jobsite. This would not mean the elimination of documentation, but merely the conversion of paper documentation to electronic filing of all material. Although the possibility exists, and technology is moving toward it, some obstacles remain.

- Firewalls between partners and firewalls within the files of a single partner are issues that create challenges for IPD partners.
- Who can "read only" and who (and under what circumstances) can make changes to documents has to be resolved, and the solution to this creates a myriad of other challenges related to responsibility and errors and omissions.
- Who ultimately is copyright protected for the final documents?
- Some information that might be shared on a project between partners to this one IPD project could be proprietary to a specific firm and place that firm in a less than competitive position in future bid submissions.

For this reason IPD may be considered a work in progress or a method in transition to a better "mousetrap." No one method of delivery has been cited as perfect and none of the developments has had a pure cause and effect on each other method. The discussion is merely intended to suggest that technology, combined with the universal intent of all parties that better cooperation can lead to a more successful future for everyone involved in construction, is a goal worth pursuing.

5.6 Private Public Partnerships (PPP)

Combining parts of Design-Build and IPD, the evolution of project delivery that has brought another form of contracting to public works is **Private Public Partnerships** (also known as PPP or P3). In these contracts a private entity contracts with a public authority to produce a product that benefits the "community." Ultimately, the risks associated with the design and construction are borne by the private party or parties involved in the contractual relationship, and ultimately the maintenance of the project may be the responsibility of the private party. Whatever income produced after the project's completion will usually be used to pay for the cost of construction. A real estate developer might put such a deal together, expecting a return on investment over a twenty- or thirty-year period.

> **Example:** A state-owned university might lack the funds to build a new dormitory, but a private developer, along with a builder, and an eventual entity to manage the space, could build the dormitory and then either lease it back to the university or collect room and board fees from the students for the life of the dormitory. The structure being on university property, the agreement might contain a stipulation that after so many years (the period of time the private developer had planned for recovering all costs and profit) the structure would become the property of the university.

An additional legal level to such agreements is what is known as a **Special Purpose Vehicle (SPV)**. A Special Purpose Vehicle is often created through a contract to form the entity that will design, build, and maintain the structure through the length of the contract. Such an SPV might include a Design-Builder, along with a bank to fund the project, and some maintenance company to continue to look after the property after final completion.

If a Constructor does become involved or intends to set up a PPP, it is very important that the Contract Administrator seek competent and experienced help in creating the contractual agreement binding the parties to the project. Monitoring the contract and the obligations after the PPP has been formed will be a challenge, but attempting to put one together without advice of counsel would almost certainly lead to difficulties and potential disaster.

5.7 Turnkey

One final method of project delivery requires the Constructor to do everything. Although this may seem to be an evolution of Design-Build, it has many differences, often involving a smaller Contractor that has particular expertise in certain installations. The Owner expects a complete product installation from beginning to end, with no other party giving input or supervision. It becomes a contract where the Owner expects the Contractor to do everything so all that the Owner has to do when the project is completed is "turn the key" and open the door for "business."

Certainly this can seem like a great way to do business, but recognize it carries with it some potential burdens, including errors and omissions if the work has design flaws. It can also be a challenge if the Contractor lacks certain expertise that is required for the completion of the work. For instance, a plug-and-go type installation might overload the circuits for the exiting panel, and the Contractor without electrical expertise might overlook the need to bring in an additional line for the equipment being installed. The turnkey Contractor might seek extra money but under a turnkey contract, the wording might be such that the Owner is not obligated to provide additional monies.

5.8 How to deliver

Individual commitment to a group effort – that is what makes a team work, a company work, a society work, a civilization work.

Vince Lombardi

There is no guaranteed perfect way to deliver a project. The two-centuries-old Design-Bid-Build method has had many successes, but it has also had many shortcomings. Looking to the past to resurrect the older Design-Build contract method of delivering a project can have advantages so long as the competence of the parties complements the goals of the project. Still, Design-Build is not a guarantee of perfection. Ultimately, the right team members may require more than just a Designer and Builder working with the Owner.

The more team members there are, the greater becomes the potential for effective contribution. The key to Integrated Project Delivery will remain a full team in the field and enough bench players to support and relieve them.

One sobering reality for the Contract Administrator is that by the time the Contract Administrator is involved in the process the form of project delivery and the contract itself have usually been determined and finalized. The challenge for the Contract Administrator, therefore, is to pull the personalities toward the common goal while recognizing that the details of the written contract have to be carefully observed and ultimately accomplished in a timely and efficient manner.

Based on the discussion in the previous chapter of types of compensation within different contracts, it is now possible to graph the basic contract delivery methods with typical compensation methods.

	Design-Bid-Build	Design-Build	IPD	PPP
Lump sum	Usual method	Often used	Can be used to establish budget	Will be established, but may have built-in escalation over the life of the project
Unit Cost	Mainly used in horizontal construction or for costing particular items within a lump sum	Rarely used	Rarely used	Not an effective means of delivering
Cost Plus	More likely used in a construction management Design-Bid-Build scenario, or for some change orders within an otherwise lump sum contract.	Could be used for part or all depending on the complexity of project and Owner's needs related to schedule	Will be used if incentives have been put in place to improve on Guaranteed Maximum Price lump sum.	Could be used within the PPP structure, but not necessarily for the overall contract pricing

Within all the cooperative processes, however, there always exists the need to "look out for #1." There is nothing wrong with such an approach. It is, in fact, a motivational approach recognized as far back as Aristotle around 350 BC, when he stated the true pursuit of all individuals is their own happiness.

Exercise 5.3

Take a moment
Recognize what motivates you

Assume that you are satisfied with what you are being paid to do your work. Rate from 1 to 8 (1 = most important, 8 = least important) what would motivate you to go above and beyond what you are being paid to do. Each motivating factor should be rated and no rating should be used twice.

Motivating factor	**Rating**
Job security	_____
Promotion and growth	_____
Good working conditions	_____
Interesting work	_____
Tactful discipline	_____
Appreciation by others of your work	_____
Help with personal problems	_____
Being in on things	_____

Discussion of Exercise 5.3

Such surveys have been done in the past and the results are often surprising until one really examines what does truly motivate each of us. As students we all can recognize that some assignments excite us far more than others. Those are the ones we put our "hearts" into, even though other assignments may require us to work as hard or harder in order to achieve a good grade.

The same is true in corporate America. Individuals come to work because they want (and need) a paycheck. They do what is their duty based on what they were hired to do, and at the end of a given period, they are compensated for their efforts. However, what makes someone work with their heart and not just their head? In all likelihood it is not the paycheck they are going to receive.

The survey you just took usually comes back with three motivating factors at the top:

1. interesting work
2. appreciation of work done
3. being in on things.

1. *Interesting work*. Looking at each of those factors individually, does it not seem that something that interests us would be a reason to put more of ourselves into it? Going back to that interesting school assignment, would it not be easier to approach that than something we have to do but do not like?
2. *Appreciation of work done*. Everyone enjoys being appreciated. As one person once told me, she could go for a week on one compliment. In fact, lack of appreciation is one of the main causes of dissatisfaction on projects for our crew members, who often say "I can do a hundred things right, but do one thing wrong, and that's what I hear about."
    a. I suggest to everyone that we need to sneak out on our projects, catch someone doing something right, and really let that person know about it!

3. *Being in on things* is a valuable way to build a team. If we examine IPD, that is what it is all about. The more we bring people into the planning of the project, the more we make our goals their goals.

Shortly, we will be examining many of the "killer clauses" that can actually pull us apart as a team, but for now, keep thinking about how we can make the team relationships stronger.

Look back at how you answered the eight motivating factors. It is possible that you thought about them differently than the three given as the most chosen by others. That is not wrong at all. You are unique, but so is everyone else. Circumstances change, and each of us is different. What is important is that we recognize motivation is important to the individuals on the team. As Vince Lombardi pointed out, individual commitment to the group effort is what will make the project a success. Finding the motivating factors for each team member might not be easy (unless of course you pass around a survey), but you can base your initial efforts to build team unity on either: (1) what truly motivates you will probably motivate others, or (2) what have been shown in the past as the three top motivators might be a good place to start with someone you have not worked with before.

* * * * * * * * * *

As a Contract Administrator, keep the list of top three motivators handy. They will help you build your team through the toughest of times.

* * * * * * * * *

INSTANT RECALL

- On the surface, Design-Bid-Build sets up the potential for adversarial relationships between the three main parties to the process – the Designer, the Owner, and the Constructor.
- The chief advantages of the Design-Bid-Build sequence are:
  1. The entire project is usually complete in design before the bidding phase, so the Owner has a clear understanding of final costs once the bid proposals are received.
  2. The Owner, through the bidding process, should get a very competitive price.
  3. The parties to the contracts have a clear understanding of their roles and the requirements of achieving project closeout.
- Multi-prime Contracting is most often used by public Owners, such as school boards, townships, or on state or Federal projects.
- Construction Management companies found a way to increase their volume of work, often without the risk they had previously assumed, when they did so much of the work as General Contractors.
- Since the Design-Builder is liable for all design ambiguities and defects, there is less arguing and fault finding and more solutions come forth more quickly.
- Contractors and often specialty Subcontractors are asked to install equipment to achieve a certain level of performance.
- Integrated Project Delivery or IPD has emerged as an effective method of pulling many diverse elements of the project team together while maintaining many of the established relationships.
- Just as there are "implied" commitments within a written contract, there are expectations for all the parties within the IPD project partnerships. These include:

1. full and open communication among all the partners
2. confidence that each individual brings expertise to share
3. trust that all the parties intend to work for the benefit of everyone else.

- Private Public Partnership (PPP) contracts are when a private entity "teams" with a public authority to produce a product that benefits the "community."

YOU BE THE JUDGE

A University in the Northeastern United States issued a request for Design-Build proposals for a new sports facility. A Contractor submitted a Design-Build proposal for $4,627,134. The University accepted the proposal and authorized the Contractor to proceed without a final price or scope being agreed. They began with the site work while the final drawings were being completed. Five months after the Contractor had been authorized to proceed, the final working drawings were finished. The University agreed to raise the price to a total of $7,157,051. The Contractor continued to work and completed the project on schedule, but claimed it had never agreed to a final price.

Upon completion, the Contractor asked for a final cost of $8,674,811. The University claimed the money above its revised $7,157,051 price should have been included in the original scope. The price was, after all, very accurate to the dollar, but there was no formal change order issued.

With no formal final price or scope set forth originally, will you consider the contract price based on the original notice to proceed, the revised price by the University, the costs sought by the Contractor? Will you allow some sort of parol evidence in the absence of fully documented pricing to establish the final contract sum?

Your ruling:

Find for the Contractor or the University

If for the Contractor, what amount:

Why:

IT'S A MATTER OF ETHICS

The student is encouraged to visit the Construction Management Association of America website (www.cmaanet.org). Included on the site is a code of ethics. Review for yourself or discuss among friends how good or effective you believe the code is and how well or effectively it is followed by others in the construction industry.

Going the extra mile

1. **Discussion exercise** (with friends or as part of a classroom discussion): The text referred to a form of project delivery known as PPP, or sometimes P3. In the discipline of decision making and life itself, some consider the 3 Ps to be Patience, Perspective, and Point of view. Consider how much of each of these 3 Ps you have, and would improvement in one or another help you as a Contract Administrator. How or why?

2. Do "You be the judge" and then go to the Companion Website to read how the actual case was resolved.
3. Effective use of the correct vocabulary improves communication and promotes respect among colleagues. The reader should become familiar with the key words listed below that have been used in the chapter. Definitions can be found in the Glossary. Additionally, the student is encouraged to use the flash card exercise in the chapter's Companion Website as practice for possible quiz or exam questions.

Key words

Building Information Modeling technology (BIM)

Construction Manager

CM at Risk

General Contractor

Guaranteed Maximum Price (GMP)

Integrated Project Delivery

Multi-prime Contracting

Performance Specification

Prime Contractor

Private Public Partnerships (PPP)

Special Purpose Vehicle (SPV)

CHAPTER 6

Interpreting the contract

There are often three ways a contract is interpreted – your way, their way, and the correct way. Finding the "correct way" without strained relations or even litigation is the challenge of all Contract Administrators. This chapter will focus on traditional rulings and interpretation of responsibility related to a construction contract, but it will also describe methods to understand and accommodate the needs of the whole project team.

Student learning outcomes

Upon completion the student should be able to. . .

- **know the importance of interpreting the contract as a whole**
- **distinguish the meaning and use of critical contract terms and, more importantly, know how they might affect the work and performance of the team**
- **analyze contract interpretation in relation to responsibilities of all parties involved in the project**
- **understand the relationship of contract documents and the resolution of ambiguities**
- **determine the importance of dealing with the entire project team in relation to the obligations each faces contractually**
- **appraise the legal implications of construing a document against the drafter in relation to the personal consequences of maintaining positive team spirit.**

6.1 Interpreting the contract

If you would win a man to your cause, first convince him that you are his sincere friend.

Abraham Lincoln

Certainly Lincoln's advice in a speech he made in 1842 is just as valid today. Of course, seated across the table from someone intent on getting his or her way at your expense can make Lincoln's suggestion a bit challenging. So let's start with a story, told to me by a good friend and a CEO of a major construction firm.

A CASE IN POINT

Rockie was just a young Project Manager, but he showed a lot of promise, so the Vice President (VP) of the company invited Rockie to come along to a special meeting. This was not Rockie's project, and there was good reason Rockie did not want to go along.

The Company had just gotten a major change order to fix a problem the Owner was not happy about, and probably resented having to pay extra. The Owner may even have thought the defect was borderline patent rather than a latent problem in design. So, to make sure there would be no more "handouts" to the Contractor, the Owner brought in "Big Bad John." Now, actually the names have been changed, but realize that "Big Bad John" (we'll call him John Jones) had a reputation far and wide of being the toughest Owner's Representative to have ever walked a project site.

No doubt things were taking a major turn for the worse if "Big Bad John" was doing the inspecting. Everyone was now concerned about getting through the nightmares of the day-to-day, and the looming titanic disaster if anyone made a mistake.

For sure, Rockie did not want to go, but he was reminded he was a young Project Manager (PM) and the rank of VP trumps PM, so Rockie was told coming along was not an option. Let's call the VP Joe. The two went to the first meeting with John Jones.

Gulp! It was just the three of them, and John Jones looked as mean as his reputation. Before John Jones started to speak, VP Joe did.

"Mr. Jones, I know your reputation. And I know exactly why you are here. But I also want you to know this. No matter what you say. No matter what you do. You cannot stop me from liking you."

By the fourth of July, Joe and John were barbecuing together.

Now that's the way the story was told to me, and I believe it, because when I have had the courage to say the same thing, it's worked for me.

As we start to look at the interpretation of contracts, we are going to understand one major factor. Contracts are almost always written by the party in control. In the case of Owners writing contracts for builders, there will also be builders who want to take the risk of building whatever will potentially make them money. Sometimes Constructors will take projects they never should have attempted, but the lure of building and the prospect of making money will have them tread where angels would otherwise fear to go.

So an Owner's contract is not written to favor the Constructor, just as a Contractor's is not written to favor a Subcontractor. Even with ConsensusDOCS, with its great attempt to be fair to all parties, we must recognize that all parties have needs, and fulfilling the need of one may be challenging to the needs of others.

As the Contract Administrator moves through the project, various opportunities will arise that relate directly to contract interpretation. Knowing the clauses (especially the "killer clauses," which will be discussed in the next chapter) is extremely important. Even more important is dealing with people you like, and who in return like you. It is easier to look for solutions among friends, and it is far better to find solutions than to argue faults among "used to be friends."

6.1.1 What constitutes the contract

Before we can interpret the contract, we must first be certain what is included in the contract. Let us look at some possible language related to what constitutes the entire contract:

> The contract consists of this signed contract agreement, the drawings and specifications, the General Conditions and Supplementary Conditions, and all Addenda, Modifications, and Amendments issued prior to this signed agreement. Unless referenced elsewhere within these enumerated documents, these documents constitute the entire contract.

At first that seems manageable, but the task quickly becomes daunting when one considers that specifications, including the General Conditions, can measure as much as a foot thick of pages, single spaced, with sometimes circuitous language.

In addition, the contract might include other documents "referenced elsewhere" within the General or Supplementary Conditions. These might include:

- the bid documents or proposals (often left out so as not to incorporate exclusions or items that might be in conflict with addenda, modifications or amendments)
- state or Federal regulations
- licensing requirements
- manufacturers' specifications
- special provisions.

Immediately, we see there are several different parts of the contract, and we even have to thoroughly read all the parts to make sure the phrase "unless referenced elsewhere" does not incorporate something else as part of the contract (e.g. payroll reporting documents for state or Federal work).

6.2 Ambiguities within the different contract parts

According to the law of nature it is only fair that no one should become richer through damages and injuries suffered by another.

Cicero

The contract is meant to record the complete understanding of the parties signing the contract. However, the contract is also meant to be read "as whole" and interpreted in its "entirety." Those two phrases can raise the most significant **ambiguities**. When going through all of the contract documents, it is rare that one or another does not somewhere have a conflicting detail with some other document in the contract.

In some cases, a Builder might wish to rely upon the bid documents or the original proposal, both being in writing, but unless they have been made a part of the contract documents by reference, they will not be considered part of the contract. As mentioned above, there is good reason why an Owner may choose not to include the bid documents in the final contract. Bid documents may have contained exceptions or clarifications, and the final contract should be void of exceptions. Additionally, all clarifications should have been cleared up between the parties prior to the signing. Therein, however, is a potential challenge.

6.2.1 Parol evidence

Often the first potential ambiguity is a different understanding based on something that might have been said. Do not rely upon verbal understandings made during or before the contract signing. Contracts will be read as though they represent the complete understanding of the parties, and any contrary thoughts testified to will probably not be admitted or enforced. Courts consider such testimony **parol evidence** – oral understandings that were not committed to writing and made a part of the contract.

A judge will want to read the contract and understand it in relation to what has been written.

However, important as is the written word, the interpretation of thoughts or oral understandings about the contract are as effective as the parties will allow them to be. This can happen only with trust and confidence, and so the Contract Administrator is the key player in establishing an atmosphere where the "talk can be walked" through the contract. This can only be done prior to any court actions, since the judge will try to place all weight on the clauses of the written contract.

The one area of special considerations for wording will be if judges recognize special meanings of words that might be specific to the construction industry.

Example: The **Critical Path Method (CPM)** of scheduling a project will initially be a little difficult to fully grasp for those who are not involved in the industry. Within the CPM vocabulary is an important word – **float**. Most Constructors will want to assert that they own the float, but if the Owner or Design Professionals use the float first, some courts might rule that they were entitled to use the float. This may be defined within the contract, but if it is not, it is important that the judge or arbitrator understands the implication of the word "float." For those who wish a more detailed description of CPM and float. consult the glossary and see "Going the extra mile."

* * * * * * * * * *

It is important to also emphasize that even after the contract is signed the Contract Administrator should not rely on what has been said. Make sure you get it in writing, particularly if it affects the contract in some way (e.g. change orders related to the schedule and/or money).

* * * * * * * * *

6.2.2 Patent and latent ambiguities

There are basically two types of ambiguities – patent and latent.

A **Patent Ambiguity** is a discrepancy that should have been recognized immediately by a knowledgeable and competent Constructor. These types of ambiguities can often be found during the design and estimating phase of the project.

Example: There is a discrepancy between the number of doors shown on the drawings and the door schedule. This is something that should have been caught by someone taking off the doors while putting the estimate together.

A **Latent Ambiguity** is one that no one expected or could have foreseen. Since these types of ambiguities often result in an increase in price they can be the beginning of difficulties between the Design Professional and the Constructor. One reason for this is that if the Owner sees the ambiguity as a result of the Architect or Engineer's error, then the Owner may charge the Architect for the cost of the change, deducting the money from the Designer's budget if the Owner–Designer Contract has such a clause, and they often do.

Some Contractors make it a habit to try to swap deductive changes requiring less work for design ambiguities in an attempt to lower construction costs without bringing significant backcharges upon the Designer. Although this can sometimes be helpful, it can also involve a lowering in quality or expectations for the Owner.

Examples:

1. The existing HVAC system was not able to handle the additional square footage added on to the building.
2. Specified laboratory equipment will not fit in the space as designed.

6.2.3 Scope clause

While researching the entire set of contract documents, the Contract Administrator should also read the **scope clause**. Although this may sound fundamental, many times items can be passed on through the scope clause to the Contractor, or particularly Subcontractors, which had not been expected.

Example: On some union projects jurisdictional disputes are sometimes resolved by the Contractor assigning certain work to a specific trade. Electricians may expect to install the solar panels, but the carpenters might claim it as roofing work, and only the actual electrical hook-up was in the electrical package, but the roofing Contractor, who employs only carpenters, did not see the panels were in the roofing scope. Now the roofer may have two problems: (1) A larger than expected scope, and (2) some angry union electricians he doesn't employ but who will be showing up for what they consider to be their work.

6.2.4 Precedence clauses

After the scope clause, the **Precedence Clause** is perhaps the most important clause for interpreting the contract as a whole.

All parts of the contract will be given the weight due to them. Whatever is part of the contract must have some meaning, even if it is subsidiary to other parts of the contract. Generally, something specific will govern over a general statement, but sometimes two or more statements might seem equally important.

Knowing there can be ambiguities between parts of the contract, the Designers and Owners usually state which documents will take "precedence" over other documents. Sometimes, for instance, specifications are put together on a "cut and paste" basis. This does not sound very professional, but imagine having to retype massive sections of specifications each time one designs a project. In the process of taking standard language from one project and incorporating it into another project, the specifications may also carry over some information that is not pertinent to the new project. This can cause some discrepancies or ambiguities between the specifications and the contract drawings.

The first thing to recognize about the Precedence Clause is that although it is meant to clarify the contract in regard to ambiguities, when it is combined with the Patent Ambiguity issue, then it can lead to further disharmony, unless handled early and wisely.

Example: The drawings state the block walls in the warehouse will be painted a high-gloss gray and the specifications state all walls throughout the project are to be flat white. This might even be considered by many as a Patent Ambiguity, but someone might have looked at the Precedence Clause, seen the details on the drawings are to take precedence over the specifications, and then the ambiguity might not have been raised. However, the prudent Contract Administrator will raise the issue just to be sure. Even though it seems obvious the warehouse walls should be a gloss gray, it is not worth creating the potential for disharmony when a clarification would resolve any possibility of a problem.

If raised during the bidding stage, the issue would be resolved by an addendum.

Ambiguities found in the bidding stage should be brought to the attention of the Designers for clarification. Even if a Precedence Clause does exist, any ambiguities found during the construction phase should also be brought to the attention of the Designers to make sure work is not done to the disappointment or dissatisfaction of anyone, even though it was done in accordance with the contract and the Precedence Clause.

In the gloss paint example above, it is possible, if it was not clarified during the bidding phase, that the Architect would have seen it as patent and it should have been brought up during the bidding stage, so there should be no change in cost. A painter who took it off gloss gray would have no problem; but the reason another bidder was low might be because the estimator had three other bids due the same day, and in putting the price together the painter had read only the specs, and had missed the detail on the drawing. That low bidder did not have the right paint. Now, either the painter or you or both could have a problem. Depending on the size of the warehouse, the difference in price for gloss gray paint could make a difference, particularly for a small painting Contractor, who we already know struggled to take off the project correctly. Caught in the middle, the Contract Administrator has a difficult challenge – one best avoided by a thorough review of scope at bid time or, at the least, before contract signing. If not, clear and open communication aimed at problem solving rather than finding fault will be the next best alternative.

Precedence clauses can become quite complex. Specifications might take precedence over the drawings, except for the details or schedules on the drawings. And addenda will usually take precedence over both the drawings and specifications, to the extent that they detail certain parts of the drawings or specifications for clarification.

6.3 Reasonable and logical

When it is not in our power to determine what is true, we ought to follow what is most probable.

Rene Descartes

The first level of understanding a contract is going to be to place the most reasonable and logical interpretation on the documents. For instance, precedence clauses will point toward resolving ambiguities between different parts of the contract. Important to a reasonable and logical interpretation is what the terms mean to those who wrote and signed the contract. Some terms may be defined within the contract, such as exactly who the Owner is, what is the scope of the project, and what will be considered the point of **Substantial Completion**. Specific terms will govern over more general terms, and sometimes this may create an ambiguity the Contract Administrator will want to have clarified.

Example: The specifications state the hardware on all doors will be brushed aluminum. The hardware schedule calls for all office doors to be polished brass. The Contract Administrator, or preferably the estimator, would want to clarify this ambiguity, but under the normal contract terms, the hardware schedule is more specific and would probably be the way the contract would be interpreted, unless a Precedence Clause stated differently.

6.3.1 The Spearin Doctrine

Since this text is not intended to make any of us lawyers, we have for the most part not cited actual case studies, but there is one important 1918 case that was eventually heard by the Supreme Court that has had tremendous effect upon the interpretation of responsibilities among the parties to a contract.

A CASE IN POINT

United States vs. *Spearin* is a landmark case in construction law.

Spearin had contracted to build a dry dock in the Brooklyn Navy Yard for a lump sum $757,658.70. As Contractor, Spearin had relied upon plans and specifications provided by the Federal government. Included in the scope was the reconstruction of a six-foot sewer that intersected the site. The Federal government had included in the plans the details for the relocated sewer, including the location and the dimensions. The work was completed and approved by the Federal government as per plans and specifications. The government knew that there were in fact drainage problems related to the sewer and the site, but it had not relayed any concerns to Spearin. Additionally, unknown to both parties, a dam in a connecting sewer beyond the limits of the contract but within the Navy Yard was about to cause a major problem. When a heavy rain came during a high tide, that unknown dam, which may have been the primary cause of the drainage issue all along, caused a heavy backwash flood that burst the newly relocated sewer built by Spearin, thus flooding the site and requiring extensive repair to the dry dock excavation that had already taken place. Spearin, placed in the position of doing extensive repair work, declined to go on with the contract, insisting the government pay for repair of the damage to the site, as well as repairs necessary to make the sewer system safe, or the contract was annulled. The cost of the repairs was estimated at $3,875. The government's position was that Spearin was responsible for fixing not only the damage but also the problems that had caused the damage. Fifteen months of back-and-forth negotiations, or lack thereof, finally brought the Secretary of the Navy to the position of annulling the contract and securing other Contractors to complete the work. That work, however, was completed under a much expanded scope and the original six-foot sewer was completely redesigned.

Up to the time of being removed from the project, Spearin had expended $210,939.18 of the original contract sum. Spearin had received $129,758.32 for the work completed.

The government's position was that Spearin should have proceeded with the contract, and, having refused to do so, was not entitled to anything except $7,907.98, which represented proceeds the government had received from selling part of the plant.

The Supreme Court had two issues to determine. First, was the sewer work part of the contract and done in accordance with the contract and the documents supplied by the Federal government? The Supreme Court recognized that "The provision for reconstructing the sewer was part of the dry-dock contract, and not collateral to it." (P. 248 U. S. 136.)

The more difficult decision was whether the Contractor had a right to refuse further work. Did this constitute a breach of contract on the part of the Federal government? Or, was the government Department of the Navy within its rights to remove and annul Spearin's contract?

The Supreme Court ruled "The contractor, upon breach of the warranty, was not obliged to reconstruct the sewer and proceed at his peril, but, upon the government's repudiation of responsibility, was justified in refusing to resume work on the dry-dock." (P. 248 U. S. 138.)

The government was found liable for all damages, including paying the difference between Spearin's expenditures and receipts $210,939.18 less the $129,758.32 originally paid, plus $60,000 that Spearin had shown he would have made in profit had he been allowed to complete the original contract.

This decision, then, has been a key in the interpretation of responsibility for design flaws in contract plans and specifications. Given plans provided by the Owner, Constructors cannot be held liable for defects related to the plans and specifications if the work has been done in accordance with those plans and specifications.

That ruling was made on December 9, 1918, and the construction industry, under what has become known as the **Spearin Doctrine**, has not been the same since.

6.3.2 Some initial effects of the Spearin Doctrine

First, the immediate contention will always be, "Was the work done in accordance with plans and specifications?" The Contract Administrator at this point must recognize the complication of providing design input through shop drawing submissions. That will immediately cloud the issue of who provided the plans.

This is also a reason why **means and methods** are usually not provided by the Design Professionals. When Engineers are Subcontractors to the Architect, the Architect does not want to pick up liability if the Engineer provides direction on how to do a particular part of the work, even if the Engineer is expert and knows a more effective means and method of doing something. To provide a means and/or method that might result in a failure would place responsibility back on the Design Professionals, who have probably passed these risks to the Constructor.

It is also important, therefore, that the Contract Administrator does not rely solely upon the idea that defects will always go against the Owner or Designer. If a prudent Constructor should have known there was a flaw or error, then the Constructor might be found liable, rather than the Owner or Designer.

Example: A Contractor recognizes a double door is to be cut into a load-bearing wall and no structural lintel is called for on the drawings. If the Contractor proceeds without asking for a clarification, then the Architect will most probably assert that a prudent Contractor should have brought the issue to the attention of the Designer and Owner before proceeding with work that could prove to be a problem, costly to fix, and possibly endanger people and property.

Recognizing the overall intent of the Spearin Doctrine to place design responsibility upon the Owner, and hence back to the Designer, changes have been made to the contract documents. Contracts have evolved to assist the Design Professionals and the Owner from assuming responsibility for what they deem is truly the Constructor's risk. To be fair, when one thinks of all the issues a Design Professional or an Owner might face, they will, as everyone does, look after protecting themselves first and foremost. This should not only be expected but it should not be condemned, since this is totally in keeping with being human.

6.4 Resolving ambiguities

A smile is the chosen vehicle of all ambiguities

Herman Melville

Not everyone will be smiling when a Contract Administrator points out an ambiguity in the plans and specifications. Usually ambiguities point to the fact that someone did something wrong, and no one likes to be told they are wrong.

As we had previously mentioned, there are two main types of ambiguity:

1. Patent Ambiguity: an ambiguity so obvious it should have been brought to the attention of the Designer earlier, probably in the bid stage, so it would have been accounted for in the original bid proposal. Designers might prefer to call all ambiguities patent, but they are not.
2. Latent Ambiguity: the second type, and actually far more common. With a short time to bid a project, many ambiguities go unnoticed. Often they are not found until work is actually begun and a **crash** is found in the field rather than on 3-D BIM drawings. Unfortunately, those ambiguities that are found in the field as latent often cause delays and inefficiencies in productivity. Even if the ambiguity results in a change order, the loss of productivity, and perhaps the extension of time to the job, can seldom be recouped through the change order. So, contrary to popular opinion, the Contractor often does not profit from change orders brought about by ambiguities. Owners, Designers, and Constructors do not need or want ambiguities.

Of course, the first course to resolving ambiguities is the Precedence Clause mentioned above. But, as we also mentioned, the order of precedence may not result in what the Designer really wanted. The Designer may have had something in mind that is not related or shown by the part of the contract that takes precedence.

6.4.1 Construed against the drafter

There is an important concept in construction documents, as in any contract law, and that is, that when there are opposing viewpoints, the documents will generally be "construed against the drafter." One can easily see how this is in keeping with the Spearin Doctrine. The person responsible for the mistake should be the person that bears the full consequences, but once again, there is always the possibility of adjustment for a Patent Ambiguity.

Worse, however, is the obvious issue that people don't like to be shown they are mistaken or they have done something wrong. This is not a way for a Contract Administrator to gain friends or influence with other team members.

So how, then, does one resolve ambiguities? The process in steps from quick to disastrous might be:

1. *Ignore at one's own risk*. Although this may sound ridiculous, ignoring at one's own risk happens more often than one might expect. There can be a couple reasons for this.
   a Someone on the project team thinks they understand what was really meant and interprets the ambiguity for what they believe will be either best or easiest. "Easiest," of course, can be a disaster in the making if the Designer never imagined easy was a possibility for aesthetically pleasing or for safety reasons.
   b The other, and often more challenging reason, is when someone from the project team does not see the ambiguity. Often the craft workers in the field see the drawings and do not see or refer to other parts of the contract, such as addenda or even a change order, which perhaps they were not told about because it did not arise out of a field condition. Looking at a detail on a drawing, the material is installed one way, while the Architect wanted something else entirely, as detailed on an addenda.
2. *Seek clarification through a Request for Information (RFI)*. Generally this is a very effective and safe approach. It should be considered the preferred method by the Contract Administrator. It can, however, cause difficulties if:
   a The resolution is costly and it is still within the original scope of work. A good strategy for the Constructor is to offer a solution when presenting an RFI, and make sure that it is a solution that will be what the Constructor would be able and willing to "live with" if accepted.
   b It is not within the original scope of work and the Designer is afraid it will come down as a cost under "errors and omissions" of the Designer.
   c Finally, for the Contract Administrator an RFI is not something sent and done. It is almost always something that needs to be followed up on until an answer is received. Create a tickler file for all RFIs so that they can be checked off when returned or followed up on if not answered.
3. *Meet separately with the Owner to discuss the defective work by the Designer*. This could be termed the "nuclear option," because it is surely going to explode all possibility of teamwork. Unfortunately this divide-and-conquer approach is the preferred technique for some Constructors. By taking such steps, the Constructor may be successful in the short term, but in all likelihood "lines in the sand" will have been drawn, and negative rather than positive approaches to challenges will be the way for the rest of the project. It is also possible the Contractor might decide to meet separately with the Design Professional(s) to discuss the imbalance in the Owner's approach. This too could be the beginning of more difficult relations on the project, since, in all likelihood, the Designers are under separate contract with the Owner, and their own success and perhaps future work depend on maintaining effective relations with their contractual partner more than on helping the Contractor through an ambiguity the Designers may see as an eventual threat to or attack upon themselves. **Machiavellian** tactics of playing one entity against another have been successful throughout history in the short term. In some circumstances a Contractor might resort to such an approach, but business is a long-term enterprise, and finding solutions that work for everyone is the best long-term solution.
4. *Offer swaps for ambiguities that might cost money, so the overall contract price remains the same or close to the same*. This technique is often employed, but the difficulty sometimes comes when one party, usually the Owner, does not feel it is getting a fair exchange. What it is giving up related to the contract is of more value than what the ambiguity would cost to fix. This leads to tensions on the project until the price is resolved. If it is never resolved to the satisfaction of the Owner, then future requests and other items may be more difficult to resolve.

5. *Prepare for litigation*. Of course the contract may call for another means of resolving the conflicts, such as arbitration, but when a part of a contract goes bad, the rest of the project is usually not going to go well. Detailed documentation will become a way of life for the Contract Administrator, and the role of accumulating evidence will absorb time that would be better spent completing project requirements on the way to a successful project closeout and final payment.

Knowing the above options and either creating more options or a blend of those possible approaches to resolution is a first and very important step related to ambiguities. The contract is not the project, it is merely the roadmap to reaching the Owner's ultimate goal. The physical completion of the goal is the Contract Administrator's best path to success. If we start with the goal in mind, then our greatest asset in achieving that goal is working with people. Although you did not create the ambiguity it can become your problem. However, a thorough understanding of the team can turn problems into the opportunities everyone talks about. By becoming a problem solver, we shift from why we have an ambiguity to how we are going to get past the challenge and turn vision into reality.

A CASE IN POINT

Walt Disney made a career out of telling stories, but this one is about him. They say that when he conceived the idea of Disneyland – an amusement park for the young and the young-at-heart – he realized this was not going to be like any other amusement park had ever been. Up until then, Ferris wheels, merry-go-rounds and wooden rollercoasters that went up and down but not in 360 degree loops were about all that existed as amusement parks.

Walt Disney wanted to create theme parks within his "Land." There would be Old Main Street the way grandparents remembered it, and an Adventureland for those who might never get to see the real animals in Africa. Plus Frontierland would let everyone step off into an adventure of the not-so-wild west, while others could journey through space in Tomorrowland.

Behind all of Walt Disney's success was just a little mouse, but sometimes success is found in paying attention to the smallest details. Disney knew the power of suspending disbelief even for just a while – of "staying young at heart" and believing dreams can come true. So in the center of it all would be a castle – Cinderella's castle. It has changed a bit over the years, but it still remains the center of Disneylands and the symbol of Disney Productions worldwide.

The point of the story for a builder, however, is this. When construction began, Disney insisted the castle was to be the first to be completed. In the center of the park, all the workers could always turn to see what they were building – a place for dreams to come true, if just for a while.

How often do we make sure all those working on the project really know what the vision is? When we build a hospital, are we just digging in the mud to pour foundations, or will someone's life someday be saved because we finished those footings on time?

Whether it's a dream castle, a hospital, a highway, a handicap ramp or any other construction, someone needs it to be done. We are those who make those visions – those dreams come true.

Of course, it helps if you either are young at heart or believe in the vision.

If we are going to read a contract, we had better recognize the power of the written word. Those who have written the contract documents have expectations based on the language that has been used, as well as the plans that show the final product in two dimensions.

There are, therefore, procedures that can help resolve ambiguities.

6.4.2 Request for Information

One of the most useful methods of resolving ambiguities when they or unexpected issues arise, is for the Constructor to submit a **Request for Information (RFI)**. This may ultimately lead to a change order request if the response entails actions that are outside the original scope of the contract.

One technique you might want to develop is to provide a solution whenever possible for the Request for Information. The reason for this is twofold:

1. A problem accompanied by a solution often gets returned quicker than one that does not have a solution and will require some research (usually put off to some other time).
2. You can propose a solution that will be better or more effective for you to perform.

Example: A Contractor was to suspend walkways in a hotel lobby from the roof, using a continuous rod. Supporting the walkways independently while getting them in place, one below the other on the same rod, seemed more difficult than another method the Contractor proposed. The second method was ultimately approved and saved considerable time.

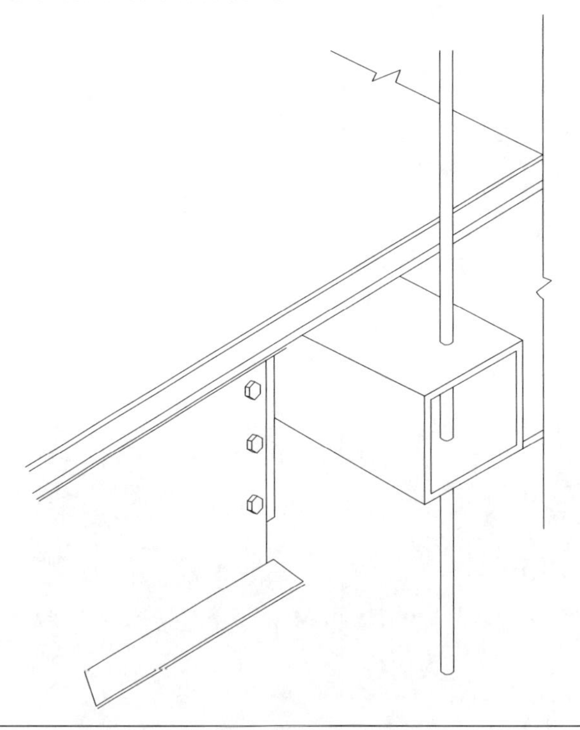

Figure 6.1 Suspension method? As originally designed

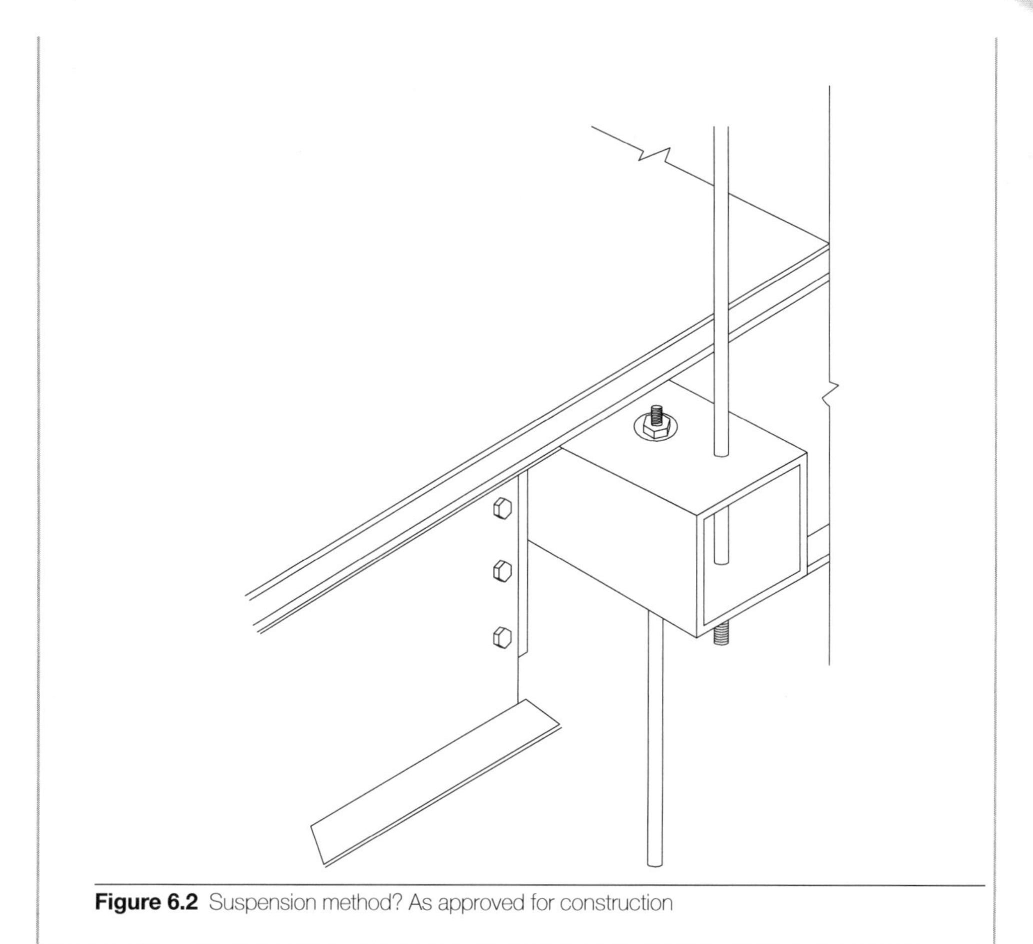

Figure 6.2 Suspension method? As approved for construction

6.4.3 Change order

One way ambiguities and other issues are resolved amicably is through the **change order** process. When actually constructing the project, it often becomes obvious that one way of interpreting what was required does not agree with what is actually needed. This usually results in the need to perform additional work for more money, although in some instances there may be a deduction in work and cost. Of course, change orders often are not the result of ambiguities but because the Owner has determined other needs than were originally discussed with the Designers. Most contracts contain a change order clause, detailing the process and methods for computing cost (and time extensions if appropriate) for alterations to the work from what is initially shown on the contract documents.

The need for a change order clause is rather interesting. Without a change order clause, the Contractor can actually refuse to do the change. It is difficult to believe that a Contractor would refuse extra work, which should mean more money, but it could happen, and in fact does happen, for reasons that will be discussed under Cardinal Changes in the next section.

For that reason, a change order clause is almost always included in the contract. Change orders are typical throughout a project. No one can get everything perfect from the conceptual stage. However, excessive change orders can be a burden on the Contractor, especially if they continuously delay the schedule without full compensation for the costs of delay, or if they impact on the smooth execution of the project and affect productivity.

They can also be a problem for the Designers, since many change orders have the potential of exposing something they did not do correctly from the beginning. This usually does not lead to harmonious teamwork if the Owner expects the Architect to compensate for errors or omissions.

A natural, selfish reaction to being hurt by one party is to try to hurt back. This can lead to changes in behavior, and one way to do that is to start to look at the terms of the contract for punishment rather than cooperation.

If the progress of the project has gone well, then changing how the parties interact can be detrimental to the future progress. The team needs to recognize that failures, mistakes, and ambiguities are part of the process. You cannot be Lincoln's sincere friend if all you do is find problems and want them resolved solely for your own benefit.

6.4.3.1 Cardinal Changes

There is a time, however, when "enough is too much." **Cardinal Changes** are the result of either too many changes or a change so large that it makes the scope totally different.

> **Example:** A project to build a five-store strip mall turns into a ten-store mini-mall, or an open air parking lot becomes a two-story parking garage, or a duplex becomes a row of condominiums.

There are some occasions when even if the Contractor would want to take the additional work – doubling the scope of work would seem a great boost to business, but sometimes other parties, such as banks or the bonding company, would prohibit the Constructor taking on a change that radically increased the scope of the work. The bonding company may be required to finish the work if the Contractor defaults on a project that is bigger than the company can handle at that time.

6.5 Administering the interpretation

All things are subject to interpretation. Whichever interpretation prevails at a given time is a function of power and not truth.

Friedrich Nietzsche

6.5.1 Flow down clauses

Prime Contractors will almost always want the Subcontractor to be subject to the same requirements as the Prime Contractor faces in fulfilling the contract. This means the burdens that are placed by contract on the Prime Contractor will also be placed on the Subcontractor. Rather than repeat all of the contract the Prime Contractor has signed with the Owner, the Prime Contractor will use **flow down clauses**. These will require the Subcontractor to perform in the same manner and with the same obligations that the Prime Contractor has to perform. Of course, the Subcontractor is going to do this in relation to its own work, but usually the flow down clause will address issues that are also indirectly related to specific trades, such as statutes and regulations in effect in a particular area, payment request and reporting forms, as well as special provisions based on unique needs of the Owner.

6.5.2 Backcharges

Backcharges are monies held back from a payment due a Contractor, Subcontractor, or Supplier because of presumed defective work or material. The Owner might also backcharge a Contractor for accrued **Liquidated Damages** (the agreed-upon cost that an Owner suffers for every day they are not able to use the structure after the contractual completion date) because of a late completion or projection of a late completion. Such charges might then be withheld by the Contractor from monies due Subcontractors or Suppliers deemed responsible for the delay.

When contract administration reaches this stage, there is usually something more than money being lost. Tempers can flare, if not in person, then behind the scenes in trailers and offices of all the parties, as each one blames the other. Finding fault reigns over finding solutions.

Based on flow down clauses, a backcharge by an Owner often affects one or more Subcontractors. Additionally, Subcontractors are subject to backcharges initiated by the Contractor. If a Contractor finds one of the Subcontractors not performing in accordance with the schedule or perhaps the quality expected, and this lack of performance has been addressed without satisfactory results, then the Contractor can decide to do the work or have another party do the work and the costs for that work will be "charged back" to the non-performing Subcontractor. This can be if the Subcontractor was officially notified of termination, or it can be that the Subcontractor is retained for the remaining work, but the work in question will be done by another party and the costs held against the non-performing Subcontractor.

This can also be true if the Subcontractor damaged another trade's work.

Example: The Roofer was not careful when hoisting the tar buckets from the heating unit below. The surface of the exterior wall needs to be repainted. Rather than have the Roofer become a painter, the Contractor subcontracts the work to a professional painter, and "backcharges" the cost of the repair to the Roofer. Of course, the Roofer might carry insurance for this, in which case, the solution might require a few more steps, but the prudent Contract Administrator will usually provide notice and withhold enough money to correct the work until it has been repaired properly.

Once again, the Contract Administrator must know the notice provision within the contract for backcharges. Although an Owner might backcharge the Contractor (or withhold payment from an invoice until work is repaired based on provisions in the payment clause), the Contractor may have other reasons to backcharge a Subcontractor. The notice provision within the subcontract, therefore, has to be followed to backcharge Subcontractors. In fact, giving notice will sometimes be enough to get the attention and change the effort of a poorly performing Subcontractor.

6.5.3 Estoppel

For all the written documentation of contractual requirements on how parties are to act during the project, there is one issue that will sometimes override and thus change the interpretation of some requirements. **Estoppel** is a legal term that prevents individuals from acting contrary to the way they have been acting or from denying what has already been established as true.

If, for instance, the Owner has continually paid for all change orders even though they have not been documented in writing in accordance with the contract requirements or have fallen outside the notice period, then suddenly to shift and fail to pay because they were not within the requirements can be challenged. Of course, the Contract Administrator should balance the cost of the challenge in time, money, and rapport with the Owner. The far better strategy is to remain within the contractual requirements, and set the standards for the team to always fulfill the *"quid"* so there is no delay in the *"quo."*

6.5.3.1 Promissory estoppel

There is at least one form of oral understanding that courts do allow, and that the industry accepts as standard without all the conditions of a contract having been fulfilled. During the submission of a bid proposal to an Owner, the Contractor receives many bids from Subcontractors and Suppliers. Almost always the Contractor has relied upon some of these prices from Subcontractors and Suppliers to submit the bid to the Owner. If it is successful, the Subcontractor whom the Contractor relied upon cannot withdraw its bid without just cause (a demonstrable bid mistake – not an error in judgment).

Promissory estoppel prevents one party from reversing itself from what it has agreed to do, such as a Subcontractor withdrawing a bid from a Contractor who legitimately relied upon the price when submitting a proposal to the Owner.

However, we must recognize that the Contractor also has obligations. One might consider these ethical considerations, but common law has also evolved to protect the Subcontractor in these relationships. If the Subcontractor's price was much lower than other prices "on the street," then the Contractor had an obligation before using the price of informing the Subcontractor that the price was very low. If the Subcontractor was so advised and still kept the price the same, then the Contractor had a right to rely upon the price, and the Subcontractor cannot subsequently withdraw the price.

Subcontractors, however, do have the right to withdraw their original bid under certain circumstances. For instance if the Contractor tries to negotiate a lower price or different terms, the Subcontractor is no longer required to honor the original offer. If the terms of the award are changed, or onerous provisions are placed on the Subcontractor that had not originally been part of the original agreement, then the Subcontractor can withdraw the bid.

Through all of the Precedence Clauses, parol understandings, estoppel practices, and promises kept and broken, the Contract Administrator has to first navigate the words of the contract, and then the actions of the parties to the contract.

Example: C&K Engineering (C&K) received a bid proposal from Amber Steel that was well below the next bidder's price. When told that it had an extremely low price, Amber Steel confirmed its price, and C&K used it to win the contract. Subsequently Amber Steel informed C&K it would not honor its original quote of $139,511. C&K was forced to find another Contractor and paid $242,171. C&K then entered a court action seeking recovery from Amber Steel for the difference in price.

This case took place in California in 1978. C&K was awarded $102,660. The reason was that C&K had fully warned Amber Steel that its price was low. C&K did not try to shop the price or place onerous conditions on Amber Steel. It merely wanted Amber Steel to comply under promissory estoppel with the "contract" that existed in the opinion of the court.

Exercise 6.1

Take one more moment
Whom can you trust?

This is a game economists like to play. It has been used worldwide, and it is a sort of variation of another game called "The Prisoner's Dilemma." We will be using worldwide averages related to the results.

What you have to imagine is that you are alone in a room with a button electronically connected to a bank account. Your task is rather simple. You cannot make money in this game, but you have been instructed by your boss not to lose money, just as we in construction, particularly during tough economic times, are given a project and told, "We do not have anything in this, but just make sure we don't lose money on the job. We took it to keep our people busy."

What you need to know, however, is that there are ninety-nine other people playing this game with you. Each of them is also alone in a room with a button tied to their own bank account. Each of you has an identical amount of money – $100,000 – in your individual account. You will each have the same instructions and choices, but you must realize that, with over six billion people in the world, the possibility of you knowing anyone else playing the game is extremely unlikely.

You (and everyone else) have two choices:

1. If you push the button, you cost anyone who does not push the button $2,000 to be deducted from each of their accounts, but you also protect yourself against anyone who pushes the button by having only $1,000 deducted from your account for each person who pushes the button.
2. If you do not push the button no money is deducted from anyone else's account because of your action, and, similarly, for everyone else who does not push the button, no money is deducted from your account. If you do not push the button, however, $2,000 times however many people did push the button will be deducted from your account.

Read the instructions as often as you wish. Then circle one or the other option.
I push the button I do not push the button

Discussion of Exercise 6.1

Worldwide, between 75% and 80% of the people push the button. The major reason given is that they do not trust anyone else to realize that not pushing the button is the best way to go, and they are sure most people will probably push the button.

If you did not push the button you would now be approximately fifty to sixty thousand dollars in debt. That alone is another good reason to push the button. If you did push the button you would still have between twenty and twenty-five thousand dollars in the bank account.

Basically, what we learn from this exercise is that there really is no win if there is no trust. We are all going to lose if we cannot establish an understanding that we are in this together. If the Owner, the Design Professionals, the Contractor, Subcontractors, and Suppliers do not trust each other, then the project is in trouble.

The challenge for the Contract Administrator, then, is to establish trust among all the parties. Communication is the catalyst that can make that happen, but trust is ultimately the social collateral that will make everything else possible among the team members. You can establish that through competence and managing the letter of the contract, or through leading others to fulfill the ultimate vision of everyone. Walt Disney built his castle first in the middle of his special kingdom. Others might build their castles in the sky.

As Henry David Thoreau said to those who did build their castles in the sky:

Now just build the foundations under them.

6.5.4 Substantial Completion

Substantial Completion is a very important time related to the project. A project is substantially complete when the Owner can use the structure for its intended purposes. Usually at substantial completion any Liquidated Damages cease. The Contract Administrator, however, should make sure from the beginning that the contract does not call for a two-tier Liquidated Damages approach. In a two-tier approach, at the time of Substantial Completion the cost of damages is lowered but not eliminated until final completion or acceptance.

Substantial Completion is nice, but if the **Pareto Principle** holds true for the closeout of projects, 80% of a Contract Administrator's efforts will be expended in the last 20% of the project. That might be an exaggeration, but everyone must recognize that getting to Substantial Completion is not the end of the project. Final payment is the goal, and that often requires a major push by the Contract Administrator. Working on project closeout from the start, however, can be a big help. Collecting operation manuals, for instance, is more easily done while the installer is on the project rather than after its crew has left and has gone on to other projects.

6.5.5 Punch list

A punch list is the document describing the work that needs to be completed or corrected before the project is considered finished.

No one likes a punch list. Design Professionals would much rather see a project that is fully ready to be termed complete, and Subcontractors hate to be called back for "nit-picky" items. So the pro-active Contract Administrator wants first to understand what the Design Professional and any other inspectors will be looking for, and then to alert the Superintendent, as well as other field personnel, as to what to be noting as they walk the project on a daily basis. A Superintendent walking the project with each Subcontractor a week before that Subcontractor leaves a project can be very valuable for all concerned. Design Professionals disappearing right after project meetings might be losing an opportunity to have a stitch in time save nine later on.

6.5.6 Warranty

Warranties are promises made by the Contractor for future performance of what has been accomplished. Most warranties in construction contracts run for a year, but the key is, when does the year start. The Contract Administrator must recognize a manufacturer's warranty may start upon delivery, or at least upon start-up of the equipment, but start-up may be months before the Owner actually accepts the structure as substantially complete, and that might still be months before final payment. So, it is important that the Contract Administrator coordinates the expectations of all parties regarding warranties.

Some "Green" construction contracts are also extending the life of warranties, so the Contract Administrator may want to carefully research the duration of all warranties.

In addition to understanding when a warranty actually begins and ends, it is also important to understand that a warranty is not a statute of limitations. If there is serious defective work that was not discovered until after the warranty, the Constructor might still be liable. Of course, such a defect may be due to a design error, and the whole warranty issue may become a major problem for more than the Contractor.

Compare contracts

The reader is encouraged to compare AIA A201 paragraph 3.5 with ConsensusDOCS 3.8. Initially the basic paragraphs of both documents match closely. However, ConsensuDOCS puts the start of the warranty upon the date of Substantial Completion. Although perhaps fair, since most equipment will be operational at that time, the AIA version is silent on that, knowing that this may not be in the best interest of the Owner if the Constructor takes a very long time to actually finish the rest of the contract and the total space is not fully turned over until late in the warranty. In subparagraphs 3.8.2 and 3.8.3 ConsensusDOCS further defines issues of warranty, relieving the Contractor from warranty of equipment supplied by the Owner, and details how special warranties will be handled regarding time extensions, as well as corrections of work and the effect on the warranty time frame.

A CASE IN POINT

Figure 6.3 Wolf
Photo credit: Charles W. Cook

We all love a story, but one might be wondering what a wolf is doing in a construction contract text. Stories, fables, nursery rhymes from our past are meant to teach us. Remember the story of the big bad wolf and the three little pigs? It is the original performance specification story of what happens when two of three pigs underdesign and take shortcuts in building their shelters. The big, bad wolf comes along as a *force majeure* and "huffs, and puffs, and blows them down." For most Contract Administrators, it is not the "wolf at the door" that will be the important issue. Fulfilling the contract conditions will be essential to a successful project. Cutting corners or taking shortcuts in the hope that no one will notice, or that the majority of the faults will escape the punch list and the really big problem that was "swept under the rug" will never actually matter, is not what being scared of the big bad wolf was supposed to have taught us long ago.

If we want to live happily ever after, we must start by making sure no wolf is ever tempted to knock on our door. To do that, we start at the beginning of the contract and pay attention to good construction practices right through the warranty. That's what we were supposed to have learned so many years ago. There are no shortcuts to doing things on time, within budget, meeting standards of quality, and from "once upon a time" to "happily ever after" always being safe.

In fact, related to safety, how do we approach the project? Do we see the possibility of an OSHA inspector in the same way the first two pigs experienced the wolf, or do we see ourselves throughout the project as the third pig, doing things correctly so that we are always prepared?

6.6 Dealing with the entire project team

If one advances confidently in the direction of his dreams, and endeavors to live the life which he has imagined, he will meet with success unexpected in common hours.

Henry David Thoreau

The contract is a very material form of commitment, and in the middle of the mud and minutiae of fulfilling the obligations of the contract it is sometimes difficult to pause and remember that we are ultimately fulfilling a vision, and probably not just one person's vision. The Design Professionals have a two-dimensional concept they want to see fulfilled, and in doing so they want to take pride in a "work of art." In fact, Constructors build for Owners the largest works of art made by humans.

Private Owners have a vision of what the structure can do for the growing success of their businesses, while public Owners want to realize the vision for the betterment of the community.

Beyond that, many Contractors, Subcontractors, and Suppliers want to be able to point to their own accomplishment, and every craft worker also wants to be remembered. How often does one hear a member of the construction community state with pride: "I helped build that!"

Beyond everyone's basic needs to feed themselves and their family, while keeping a roof over their heads, comes the desire in almost everyone to belong to a team and to be appreciated for what each person can contribute to the team. Social behavior has been the trait from the great apes through all the branches and development of the genus *Homo* to finally arrive at *sapiens*. We are social animals, and we need to be a part of the whole – a contributor to dreams coming true.

Two theories of motivation based on needs

Abraham Maslow established a theory of motivation based on levels of need, each higher level only becoming effective when the lower level had been achieved. For instance, if someone is starving today, they are not much concerned about the security of tomorrow; but, if fed well today, then knowing one will be able to eat tomorrow becomes a motivator related to the feeling of security. Once one feels secure, then belonging to a group becomes effective as a motivator, and once one belongs to a group, then being appreciated by the group is important; and finally, if you can fully enjoy doing what you are doing while helping others enjoy doing what they are doing, then you have become a very successful motivator under Maslow's hierarchy of needs.

Frederick Herzberg had a different view of motivation, but perhaps his most contrasting concept with Maslow's was his contention that there are factors in the workplace that cause satisfaction and other factors that cause dissatisfaction. A major challenge for the Contract Administrator wishing to motivate others on the project team is Herzberg's concept that unless one removes all the dissatisfiers, the satisfiers will not be effective. One can receive the respect and the sense of belonging to the team from everyone else, thus fulfilling the fourth level on Maslow's hierarchy, but if the project site is a dump and safety is not practiced, then dissatisfaction will certainly prevail, and, unless removed, the dissatisfiers will ultimately produce more negatives than positives in motivation.

Although seemingly at odds, the two theories can be compatible for the Contract Administrator who takes the time to learn what the true issues are for each member of the project team. The contract will generally be an excellent place to start. Knowing what is considered important will help focus the needs, and fulfilling the needs is an excellent step in promoting the positive. In the next chapter we will deal with "killer clauses," and successfully handling these will almost certainly eliminate many of the dissatisfiers.

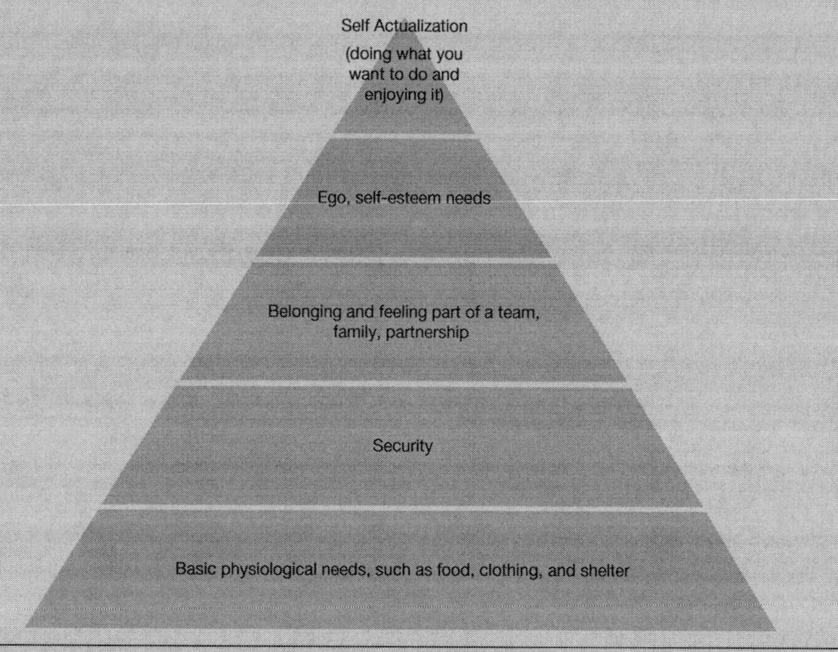

Figure 6.4 Pyramid based on Maslow's hierarchy of needs
Source: Adapted from Maslow, A Theory of Human Motivation, 1946

How, then, does the Contract Administrator become that visionary and the team member most involved in making a dream come true, when the details of contract terms occupy the immediate focal length of vision and perspective for that Contract Administrator?

In a previous chapter we looked at the "fight or flight" behavioral characteristics each of us have. We can see just how long we can avoid issues (flight), and, by effective use of a day planner, delegation, and self-discipline we will be able to bring ourselves to an effective level of management on the project (fight – for what is right).

It is worth repeating for everyone that having a certain behavioral trait is not a weakness unless we ignore the negative consequences of that behavior. Once we identify our tendencies, we are in a far better position to use all our traits to achieve success. If one trait or behavior detracts from certain situations, we can use other abilities to succeed.

Exercise 6.2

Take a moment
What "bugs" you?

How easy is it to get you in a bad mood? Knowing what demotivates us is just as important as knowing what motivates us to do a good job. Make a list of what prevents you from giving it your all. What turns you off?

Discussion of Exercise 6.2

It may seem strange to keep track of what annoys us, "bugs" us, demotivates us, but there are many who believe that if there is something that is bothering us – negatively motivating (or demotivating) us – then no amount of positive input or motivation will help us to do a better job.

There are many things that bother us. It could be lack of appreciation or recognition, long hours and poor pay, unpleasant or even unsafe working conditions, incompetent help, never being included in the decision making (or being a mushroom – kept in the dark, fed a lot of manure, and expected to produce).

What is important about this exercise for the Contract Administrator is not just to try to do something about it for oneself, but to recognize that whatever "bugs" *you* probably also annoys everyone else. The list you just created is the starting point for you to do something about the team atmosphere. As Mahatma Gandhi said:

Be the change you wish to see in the world.

6.6.1 Managing and leading the project team

We can be better managers simply by inventorying all our strengths and areas where we could improve and then working all aspects of our character to effectively achieve the tasks before us. But while a manager is entrusted to achieve the tasks, a leader is expected to work with people.

The Contract Administrator mired in the details of the contract may never get past being a manager, directing everyone to do their job on time, within budget, meeting standards of quality, and doing it safely. In the process, the Contract Administrator will never inspire anyone to go beyond their duty. No one will feel that they are making a dream come true. They will simply be collecting a paycheck. The "*quid*" for the "*quo*" of doing what they agreed to do.

Getting beyond managing the contract details to leading the whole team is the next step in successfully fulfilling the contract obligations. Can, in fact, the Contract Administrator lead others who do not have a contractual relationship with them? The answer might be better revealed by asking the question differently. Can the Contract Administrator effectively work by managing people with whom there is not a contractual agreement? That answer is a quick and resounding "No!"

You cannot manage people you cannot direct or, worse, intimidate in some "my way or the highway" fashion. That leaves a vast majority of the project team beyond the control of managing. This does not mean others within the team do not manage the people the Contract Administrator cannot manage. For example, a Subcontractor Foreman manages the craft workers of that trade. However, the direct management of most of the project team, and particularly the Owner and the Design Professionals cannot be done by the Contract Administrator.

Leading, then, is the next step in contract administration, and this has to be done from all positions to all positions on the project team. It is an old adage that in order to lead, one must have followers, but, based on contractual relationships, it is impossible to see the Contractor having the Owner or Design Professionals as followers.

The secret, then, is to put the castle in the center. Have everyone follow the vision. In meetings, in conversations, in resolution of ambiguities, have the vision be the center, and everyone will follow you to that.

Exercise 6.3

Take a moment
Add vision

Think of any building or structure you might like to work on. Write it down and then write what you see as the vision for that structure:

Discussion of Exercise 6.3

Certainly, the vision should be something that everyone can share. Although making money might be a motivating factor for everyone, you making money for yourself is not a shared vision. Before a price was even put on the work, the Owner had a need. The Design Professionals found a way to put that need into at least two dimensions. Now you and those who join the team through subcontracts and purchase orders are going to realize that vision in space and time.

Are you all working together to make life better for others, providing for the creation of better, faster, stronger products, making the world safer, more enjoyable, a friendlier place to do business, learn, relax, refresh, or call home?

It is through a joint vision you can all share that you will become the leader to that vision. As you move from detail to detail and challenge to challenge, ask yourself how you adjust your approach, your behavioral traits, particularly in relationship to others, so you all can more effectively reach the final vision – the goal – the castle in the center.

6.6.2 Managing the contract

Armed with the Spearin Doctrine and the knowledge that a contract will be construed against the drafter, a Contract Administrator might feel quite bold.

Unfortunately there are Constructors who make a practice of bidding a project low just to make back money on the change orders they feel are hidden in the ambiguities of the plans and specifications.

Owners and Design Professionals have therefore written specific clauses to protect themselves. These will be discussed in the next chapter.

It might be imagined, then, that contract administration begins already in a hole. If that is how one feels, that is how one will approach the project, and that is seldom good. Instead, one must recognize the challenges for what they are, and then find ways to ensure contract interpretation is just the beginning of under-promising and over-delivering.

6.6.3 Mediation and/or Arbitration

Sometimes nothing resolves the situation. The relationships have deteriorated and communication is no longer effective in finding solutions.

Generally, contracts will have some method for resolving conflicts that grow past the stage of finding a resolution amicably. Both mediation and arbitration are alternatives often written into the contract to settle disputes rather than going to court. They will be discussed in a later chapter, but mediation involves a neutral third party trying to find a mutually acceptable settlement between two parties. In arbitration, a neutral judge or panel of judges decides the dispute and assigns a settlement to the case.

These conflict resolution methods will be discussed in a later chapter, but each offers an alternative to court-resolved litigation. The potential of mediation is often not fully explored, but some contracts are now calling for project-site mediation as an alternative to post-project mediation. Ultimately, if a neutral party can get the parties back on track during the project rather than afterward, then collaboration may return between the parties and the success of the project once again becomes everyone's goal.

INSTANT RECALL

- Finding the "correct way" to interpret a contract without strained relations or, even, litigation is the challenge of every Contract Administrator.
- Contracts are almost always written by the party in control.
- It is easier to look for solutions among friends, and it is far better to find solutions than to argue faults among "used to be friends."
- Make sure you understand what documents constitute the entire contract.
- Be careful that the phrase "unless referenced elsewhere" does not incorporate something you did not expect to be in the contract.
- The contract is meant to record the complete understanding of the parties signing the contract.
- Testimony contrary to what is written in the contract will probably not be used to determine what constitutes the contract.
- Even after the contract is signed, the Contract Administrator should get future understandings committed to writing and not just rely upon what has been said.
- There are two types of ambiguity – patent and latent.

- After the scope clause, the Precedence Clause is perhaps the most important clause for interpreting the contract as a whole.
- Generally, something specific in the contract documents will govern over what is written in a general statement.
- The first level of understanding for a contract is what is the most reasonable and logical interpretation of the documents.
- Contracts have evolved to assist the Design Professionals and the Owner from assuming responsibility for what they deem or would prefer to be the Constructor's risk.
- A good strategy for the Constructor is to offer a solution when presenting an RFI.
- Create a tickler file for all RFIs so they can be checked off when returned or followed up on if not answered.
- Estoppel is a legal concept that prevents individuals from acting contrary to the way they have been acting or from denying what has already been established as true.
- Promissory estoppel prevents one party from reversing itself from what it has agreed to do.
- A project is substantially complete when the Owner can use the structure for its intended purposes.
- Most warranties in construction run for a year, but the key question is "When does the year start?"

YOU BE THE JUDGE

Go back to the example under 6.4.2, in which a design change was made to the support detail for two suspended walkways in the lobby of a hotel.

Those walkways were installed on the third- and fourth-floor levels in the atrium lobby of the Hyatt Regency Hotel in Kansas City. The walkways were about 120 feet long and weighed approximately thirty-two tons.

At 7:05 PM on the evening of July 17, 1981, 1,600 people had gathered in the atrium to watch and join in a dance competition. The two suspended walkways were occupied by twenty to forty people. Based on the change in design, the beams of the upper-floor walkway now supported both its own weight and live load along with the weight of the lower walkway and its live load. The upper walkway support system failed, collapsing onto the lower walkway, and then both walkways fell onto the lobby below. In the collapse, the water pipes supplied by tanks rather than public water were severed. There was no way to shut off the flow of water, and thus the lobby became flooded, making rescue efforts more difficult and increasing the potential of drowning for survivors who were crippled or disabled from escaping. The last survivor was rescued over nine hours after the collapse. There were 111 people killed immediately, and 216 suffered injuries – three of them later dying from those injuries, bringing the total death toll to 114, and making this tragedy the worst in toll of human lives related to construction design up to that time in United States history.

To go back through the design change process. Havens Steel Company, which was the steel Contractor responsible for manufacturing the rods, had asked for the change in design from the original plan since it would require the entire rod below the fourth-floor skywalk to be screw-threaded to accommodate the nuts that would hold both the fourth-floor and third-floor skywalk in place. It was assumed these nuts would be damaged as the skywalks were lifted into place.

The design change was made and approved as installed.

After the disaster, structural Engineers examined both the original design and the changed design and found both lacking. The original design might have supported 60% of the mandated minimum load, but the changed design, with all the weight being supported by the fourth-floor skywalk, could support only half or 30% of the mandated minimum load.

Compounding the potential for failure, investigators determined that there had been poor communication between Havens Steel and the structural Engineers, Jack D. Gillum and Associates. Gillum and Associates contended that the drawings were only preliminary sketches. Havens had interpreted them as finalized drawings. Calculations were not done by Gillum and Associates related to the proposed alternate plan.

Additionally, there were millions of dollars in claims for restitution to the victims and the families of the deceased.

Whom do you find against and in what percentage of amounts for damages to property and lives?

	Percent property damage	Percent lives
Havens:		
Gillum and Associates:		
Hotel:		

Go to the Companion Website to compare your ruling with the actual decision on the case, and its aftermath.

IT'S A MATTER OF ETHICS

Think to yourself or discuss with others: It was mentioned in the text that the "nuclear option" might be to discuss directly with the Owner the incompetence of the Design Professional. Is this ever the right thing to do? If so, how far must one be "pushed" before the "nuclear option" is exercised, and how would one go about it?

Going the extra mile

1. Complete the "You be the judge" exercise above and go to the Companion Website for a discussion of the case and to compare your ruling with the actual ruling.
2. For more information or advice on scheduling you might access www.whi-inc.com and search CPM or other scheduling descriptions contained on the website of these professionals experienced in construction projects and schedules.
3. **Discussion exercise** (in class or among friends about a team you know): Do a SWOT analysis of the team. It can be any type of team, but if you cannot think of one, do a local sports team. The key is to begin to use the SWOT tool to see how it can help you organize and position individuals to the best advantage of the project goals.
4. Effective use of the correct vocabulary improves communication and promotes respect among colleagues. The reader should become familiar with the key words listed below that

have been used in the chapter. Definitions can be found in the Glossary. Additionally, the student is encouraged to use the flash card exercise in the chapter's Companion Website as practice for possible quiz or exam questions.

Key words

Ambiguities
Cardinal Changes
Change order
Crash
Critical Path Method (CPM)
Estoppel
Float
Flow down clause
Latent Ambiguity
Liquidated Damages
Machiavellian
Means and methods
Mediation and/or Arbitration
Pareto Principle
Parol evidence
Patent Ambiguity
Precedence Clause
Promissory estoppel
Request for Information (RFI)
Scope clause
Spearin Doctrine
Substantial Completion
Warranty

CHAPTER 7

Killer clauses

Some elements of every contract deserve particular attention. How to spot these clauses and what to do about them during the construction process is vital to delivering the product and keeping the project team on task and out of difficult situations.

Student learning outcomes

Upon completion the student should be able to. . .

- **recognize the importance of reading the contract in its entirety**
- **know why specific clauses should be researched. These include:**
  - **scope clauses**
  - **notice clauses**
  - **termination clauses**
  - **exculpatory and risk shifting clauses**
  - **damages clauses**
- **understand the importance of payment and change order clauses and the role the Contract Administrator performs related to such clauses**
- **evaluate what is actual cost and what is not billable as a cost in accordance with the contract**
- **be prepared to track compliance with specific clauses.**

7.1 Finding the killer clauses

God is in the details.

Ludwig Mies van der Rohe, Architect

As our nation has become more and more litigious, Owners and Design Professionals recognize there are many possible exposures to lawsuits they might face from a long list of parties, the majority of whom they do not have a contract with and probably have never met:

- Contractor, Subcontractors, Suppliers
- workers of the Contractors
- staff or tenants or future tenants of the structure
- banks or lenders
- sureties and insurance companies

- patrons
- passersby or lawful visitors to the site.

So long as there are lawsuits, parties will attempt to prevent them before they exist. Acting in their own behalf, Owners and Design Professionals have included language in contracts to protect themselves against claims that they would prefer others to defend. The prudent Contract Administrator has to recognize and be prepared for the responsibilities passed on to the Constructor through what from the Constructor's perspective are considered "killer clauses."

From childhood we have been told that "Sticks and stones can break my bones, but words will never hurt me." Yet in business they can. We are using the term "killer clauses" because many in the industry have found that some specific parts of contracts can cause incredible harm to a project, to the parties working together on a project, and, in some instances, be the reason a company goes out of business. Still, we must remember they are just words. If we recognize them, understand their full meaning, know the intentions behind them, and commit to prosecuting the work in accordance with what we have agreed to do, then words will remain, just that – words to build by.

Still, in this chapter we will examine some very important clauses within most contracts that can be both a challenge to those who recognize them, and a definite danger to those who ignore them.

7.2 Notice clauses

When I was young I thought money was the most important thing in life. Now that I am old, I know it is.

Oscar Wilde

Of all the killer clauses, often the most innocent but potentially most damaging if not adhered to are the notice clauses. There can be many items or issues on a contract that require notice within a specific period of time. Just for a beginning list, here are some issues or events for which notice should be given in accordance with time restrictions within the contract:

- Differing Site Condition
- Request for Information related to an ambiguity
- change order for increased time and or money
- delay due to any circumstances beyond Contractor's control
- dispute notice
- request for payment.

Some of the most common notice clauses have to do with changes. The contract documents will usually state that a change condition from what is shown on the drawings must be brought to the attention of the Owner and/or the Designer within a certain number of days. Such a notice should be in writing. This time limit is often fourteen days, but it is not always fourteen days. It could be twenty-one days, or it could be seven days or any number of days.

Here is a further complication to notice clauses: although the general conditions might state fourteen days, the supplementary conditions (which the Precedence Clause states "takes precedence" over the general conditions) states, "due to the fast track nature of the project the Notice for all changes in contract price must be given within seven days of a changed condition to the plans and specifications. If such notice is not given, the Owner and Designers rely upon the understanding there was no change in price due to any defect or ambiguity in the plans and specifications."

This might seem to be unfairly punishing the Constructor, but the point of the notice clause is well founded. The Owner or Designer might react very differently to a situation if they know it is going to cause a problem or involve increased costs than either would have if they were not alerted to the situation. Getting the Owner and Designer involved in a change condition early enough will allow them to assist in the best possible solution for all parties. For that reason Contract Administrators will want to understand the length of time the notice clause permits, but the prudent Contract Administrator will want to alert all personnel on the project that notification should be as immediate as possible. "Bad news" is not always what one wants to deliver, but the sooner any changed condition is recognized, the more likely it can be turned from a fault-finding mission to a "teamworked" solution.

On the other end of the construction spectrum, bonding companies will rely upon notice clauses, some of them set by law. This will be discussed in a future chapter, but failure to provide notice within the time constraints of the bond will result in forfeiture of any rights to remedy or payment under the bond.

The same is true if a **Mechanics Lien** is filed late. A lien is sometimes used as a remedy against an Owner who has not paid for work that has improved the property. Again, liens will be discussed in a later chapter, but they, too, must be recognized for the constraints notice clauses place upon those who wish to file a lien.

7.2.1 Notice to proceed prior to pricing

It is also quite possible a change will be initiated by the Owner and/or the Designer. Often such changes are authorized to proceed without final pricing. Once again, this puts the Contract Administrator in a difficult position, one you do not want to turn into a "Monkey in the Middle" exercise.

Authorized to proceed for the sake of time, the authorization will usually state the Constructor has so many days to notify the Owner of any change in cost. Failure to do so will indicate there is no change in cost. The actual work may not only involve the Constructor's forces, but it may include several Subcontractors and Suppliers as well. The more parties involved, the more likelihood there will be one or more that do not get costs in to the Constructor within the time limit.

The Contract Administrator, therefore, must continue to follow up and communicate with all the parties involved in the change until the costs can be determined within the "notice provisions" of the authorization. At the very least, the Contract Administrator needs to advise the Owner and/or Designer that there is a change in cost and more time is needed to accurately price the change.

The second issue, however, can become, as stated in a previous chapter, that if the lump sum price as presented is not accepted, one of the ways to resolve the issue of a change order is through Time and Material pricing. Going back after the fact to determine Time and Material pricing for a change order can be very difficult and cumbersome unless the alert Contract Administrator has kept accurate records from the beginning. It may be easy to recognize additional material put in place, but separating actual hours for crews and equipment related to extra work while they were also working on contract work can be challenging, and it almost certainly cannot be done after the fact unless accurate records were kept from the beginning of work on the change.

Faced with an authorization to proceed without an approved cost for a change is a special and different kind of notice:

Contract Administrator beware!

Prudence will dictate a careful examination and monitoring of the work involved so the final costs, whether submitted lump sum, Time and Material, or under some other method, will be smooth and not cause disharmony among the parties. Owners and Designers do not want to be taken advantage of whenever they make a change, but at the same time Constructors and the Subcontractors and Suppliers that work for them need to receive proper compensation for what they did not originally intend to do.

7.2.1.1 What is a cost?

Even though the contract may call for resolution of some changes through the Time and Material basis, which is also considered a Cost Plus contract or contract change, it is important to recognize what constitutes a cost. The contract will usually clearly define what will be accepted as a Time or Material cost.

Sometimes recognizing true costs of labor can add to confusion in the field. Although a person's actual take-home pay might reflect one dollar value per hour of work, the billing for such an individual will normally be over twice the actual take-home. Not only are taxes taken out of the person's pay, but the actual hourly rate has several additional costs associated with it. In accounting terms these are called "payroll burdens," and they can include workers' compensation costs, unemployment taxes, additional state, Federal, and local taxes that are paid solely by the employer, pension and retirement benefit funds, union dues and contributions. Often a worker who sees a billing rate so much higher than what they know they get to take home believes the employer is getting rich very quickly off of Time and Material work, but the opposite might be the case if the actual costs of some of the payroll burdens have not been computed accurately.

Example: Each employer will have an experience modifier related to some insurances, particularly workers' compensation. Each trade, depending upon the history of accidents related to the trade and the cost of such accidents in human terms, will have a different rate from other trades. The greater the losses within a trade, the higher the rate of Workers' Compensation Insurance will be. The average experience modifier is 1.0, so whatever the rate of insurance for a particular trade on a project is, that worker's rate will be multiplied by 1, or the Contractor will pay exactly the rate set for a worker in that trade. However, if a Contractor has had several accidents, then that Contractor's experience modifier will go up. If the Contractor has had no or very limited accidents, the modifier will drop below 1.0. Based on the experience modifier, therefore, the payroll burden could be higher or lower than the actual universal workers' compensation rate for that trade.

One important note related to experience modifiers is that more and more pre-qualification submissions are requiring documentation that a Constructor's experience modifier is below 1.0. Contractors with poor safety records are being excluded from bidding work by some Owners.

The Contract Administrator needs to look carefully through the contract to determine what the Owner will accept as costs related to Time and Material work. It may be defined by the contract, but if it is not, you do not want it to become an issue of contention after you submit your first Time and Material invoice. One widely accepted principle is that no worker not in the field can be considered a cost. Anyone back in the home office is a cost to the Contractor's general and administrative costs (G&A, often referred to as overhead), and therefore is part of the overhead and profit markup and not a cost.

Finally, the Contract Administrator has to be aware of how disputes related to change order costs will be resolved. This has previously been discussed, but it is always wise for the Contract Administrator to review the possible scenarios to understand how to frame the change order request to the best possible advantage for being accepted without disagreement, since disagreements usually lead to delay in payment.

7.2.2 Payment clauses

Let us hope for all of us that money is not the most important thing in life, but for a corporation it undoubtedly is. It is the lifeblood of businesses. So, while we are on the subject of payment for changes, let us also look at payment for the entire contract. The payment clauses are not simply a matter of the "*quo*" for the "*quid*" a Contractor does. There are issues such as retainage that need to be taken into consideration. Sometimes the retainage will change at different stages of the project.

More importantly, there are implied issues related to the timing set forth in the contract related to payment. An invoice for the work accomplished to date is to be submitted by a certain date. Usually this date will be the end of the month. However, before the end of the month there will need to be some careful consideration of what can in fact be invoiced for the month.

The Contract Administrator will have to coordinate (usually through the subcontract or purchase order) with each Subcontractor and Supplier what costs will need to be included in the current month's invoice. This requires an accurate and honest appraisal of what progress will be made by the end of the month.

Although material supplied to a project is usually not invoiced until it is installed, there can be exceptions, particularly if the supplied material or equipment is proprietary to the project and of a costly nature. Where the material is stored, as well as access to that material, can be an issue if the Owner is expected to pay for it. Much of this will be detailed in the contract, but if it is not, it should be included as agreed to in writing in either project minutes or separate correspondence.

* * * * * * * * * *

Ownership of stored material is important. The Contract Administrator should realize that if a Supplier or Subcontractor goes bankrupt, and materials are stored at the Supplier or Subcontractor's location, it will become a legal labyrinth to get those materials released for use on the project.

* * * * * * * * *

Once the Contract Administrator has an understanding of what can be invoiced to the Owner for work that will be accomplished by the end of the month, it is often advisable to run through the billing with the Owner's representative (possibly the Architect) to make sure there is agreement on what can be invoiced. The next step is to create the invoice, with any required back-up forms. On public work these back-up forms can become quite voluminous since they may require payroll statements and information from each Contractor and Subcontractor supplying workers on the project during the month. This can also cause a "timing" issue, since some payroll ending dates will require submissions to post the scheduled date for submission of the invoice.

Coordinating the submission, therefore, is important in relation to the requirements of the payment clause. After submission, there are often a series of approval "checkoffs." Again, if it is public work, all the documents that were submitted as back-up have to be checked. The Owner and Designer might want to inspect the work again for both the actual quantity invoiced and, just as importantly, the quality of all that has been put in place.

If all goes well, the invoice will be approved and continued to be processed, but once again, the Contract Administrator in charge of payment issues will have a better time if good relations are maintained. Cash flow to the Prime Contractor and from the Prime Contractor to the various Subcontractors (usually through the Pay when Paid Clause, discussed in a previous chapter) will keep everyone happy. Failure to keep this vital flow timely, however, can be the beginning of trouble for everyone.

7.2.3 Notice of a bid mistake

Although not usually written into the contract, the Contract Administrator should be aware that notice of a bid mistake needs to be timely. In the pressure and tension to submit a bid, mistakes are often made. Some of those mistakes can produce a higher price submission, but often they can be responsible for a lower price than the Contractor would want to do the work for if awarded the contract. After the euphoria of winning the project, there is often the sobering reality that the project has to be built and there are many challenges even for the best Contractors. It is at this time that estimators often go back to make sure they actually included everything, and that all their numbers are correct.

If an error is found, there can be relief, but the error has to first be a mistake, and not an error in judgment.

Example: The Contractor discovered he had used the wrong price for the Electrical Contractor. Instead of $2,000,000, the price had been $200,000. That would be a mistake rather than an error in judgment. On the other hand, the Estimator might have told the Electrician to figure running the wiring as romex instead of through conduit to lower the price. Although that made the price lower it was not according to code or the Owner's wishes, so that would be an error of judgment and not cause to be relieved from the bid.

For the most part, a Contractor who is the low bidder will certainly try to look for ways to be awarded the work even if a minor mistake was made. If the mistake was major, there are three criteria that should be addressed as soon as possible if the Contractor expects to be relieved from the obligations of performing the contract at the wrong price:

1. Would performing the contract be a severe financial burden for the Contractor? If it would be, then the prudent Owner will not want to enter into the contract, for the very simple reason that the management of the project will almost certainly be a disaster, and the prospect of finishing on time is highly unlikely.
2. Did the Contractor actually exercise proper controls and checks when submitting the bid, and the mistake was still made without it being an error in judgment?
3. Was the mistake discovered in sufficient time to place the Owner in a *status quo* position? Basically, has no harm been done to the innocent party that did not make a mistake?

Point 3, therefore, is the key responsibility of timely notice of a bid mistake. The Contract Administrator must validate the contract price as early as possible if there is a possibility an error may have been committed. Waiting past a reasonable time may jeopardize the project and also put the Owner in a position where there cannot be a return to a *status quo* position – a position where they were prior to the award of the Contract, or not out any monies that cannot be recovered, or capable of continuing the project in some other way without severe financial consequences.

7.2.4 Delay notices

One of the killer clauses that can be very difficult for a Contractor is the **No Damages for Delay Clause**. This will be mentioned again under exculpatory clauses, but a No Damages for Delay Clause puts the Contractor on notice that there will be no payments made for additional costs if the Contractor is delayed during the project. Often this clause is written so that there will be no additional payment, even if the Owner or agents of the Owner (e.g. Architect) was the cause of the delay.

Immediately this seems unfair, and the Contract Administrator might decide there is no point in documenting delays if there is not a chance to be compensated.

However, there are actually two very good reasons to document a delay. Whether a notice clause for delay is contained in the contract or not, the Constructor should immediately notify the Owner, with copies to the Design Professional, of all delays.

First, although there may be no financial reimbursement for a delay, there very well may need to be a time extension. Time extensions without compensation are very common. Although the Contractor may indeed face costs, such as rental of equipment, that should increase the price, extending the time of a project may at least eliminate imposition of Liquidated Damages that could also be a financial burden.

Second, although the clause seems to unjustly favor the Owner, there are arguments a reasonable Owner will accept, and, if the Owner is not reasonable, then perhaps some arguments that will succeed in a court if the costs are significant enough to justify litigation (but try to avoid that). Such arguments include:

1. The implied elements of a Contract include cooperation. If one or another the parties to the project has not been cooperative or has intentionally done harm, then the courts will recognize the right of the Contractor.
2. The delay was totally unforeseeable by both parties, and was not something one would expect on a construction site.
3. The delay was caused by the Owner denying a fundamental obligation of the contract (e.g. Spearin left the project and claimed breach of contract for not paying for proper repairs to the sewer and the site, and then the Navy annulled the contract based on delay and abandonment. The Navy was found at fault).

Again, although a No Damages for Delay Clause may seem unreasonable, they exist because some Constructors in the past have used delays as a means to unwarranted financial gain. So, Owners and Designers have created clauses to eliminate unreasonable extras. Working with Owners and Design Professionals in open and honest communication is the best way to insure everyone will be treated fairly.

Exercise 7.1

Take a moment

Most contracts, such as the AIA and ConsensusDOCS, set time limits for notices. Go to Appendix I and read paragraph 8, "EXTENSION OF TIME AND SUBCONTRACTOR'S WAIVER OF DAMAGES FOR DELAY."

1. Does this subcontract contain a specific time limit for this notice?
2. What challenges might arise in relation to the contract with the Owner?
3. Is there a remedy for the Contract Administrator related to any differences between timely notice from the Subcontractor and the notice requirements with the Owner?

Discussion of Exercise 7.1

1. The word "promptly" is used, but it is sufficiently vague to cause varying interpretations, and, as we know, the Subcontract would most likely be construed against the drafter, the General Contractor.
2. The Contract Administrator faced with such a difference of notice provisions between the prime contract and subcontracts has to be aware of the delay and notice requirements and get prompt information from Subcontractors as well as Suppliers. Such an issue may seem unlikely if the subcontract would be worded differently, but the likelihood of such discrepancies in notice clauses or times is more common than one would think. Most pre-printed contracts for Subcontractors do not come from the same source as the Owner's prime contract. Vigilance and communication are the best assets for the Contract Administrator.
3. There is always the defense of the flow down clause for the Contract Administrator. Linking the Subcontractor to the terms of the prime contract is critical, and the flow down clause can do that effectively. However, for efficiency and rapport among the team members, the vigilance and communication mentioned above will be very important.

7.2.5 Waiver

Another way the No Damages for Delay Clause may appear is in the form of a waiver clause. Written into the contract, such a clause states that if notice is not given within the prescribed time period set forth in the contract, then the Contractor waives all future claims for increased compensation and/or extension of time related to the unexpected site condition or potential delay in the project.

Obviously, once again, the Contract Administrator has to be very careful to keep all parties informed of each potential delay, and then, within the notice period, make sure the effects of the change condition or delay are detailed. Even if a No Damages for Delay Clause exists, the Contract Administrator should make sure all delays are detailed and forwarded in writing, initially for negotiating purposes or later to provide documentation for supporting a future claim.

7.2.6 Notice of Differing Site Condition

Most differing site conditions occur underground, but not always. For example, sometimes they can be above the ceiling, such as hidden asbestos fireproofing. Once a condition has been uncovered that differs from what was anticipated or described in the contract documents, the Owner needs to be notified. In the case of exposing a hazardous material below or above ground, all work must stop until a remedy can be determined.

If hazardous materials are discovered as a **Differing Site Condition**, the prudent Contractor will pass the responsibility back to the Owner. Once involved in removal of a hazardous substance, the Contractor can become a **Potentially Responsible Party (PRP)**. That could put the Constructor in a difficult position related to responsibility for the safe removal and disposal of the hazardous substance.

The government looks for who put the hazardous substance in place. That is the PRP of record, but often that PRP cannot be found. There might also be a series of PRPs, but unless they can be found and prevailed upon, the last person involved might carry the entire burden of removal. If the Contractor that uncovered the hazardous material began removing it, or if more hazardous substances might have been removed during the "discovery" of the substances than should have been removed, the Contractor might then become the PRP responsible for the entire removal.

Of course, not all differing site conditions involve hazardous materials. Some are less harmful to the environment, but possibly quite expensive if not anticipated. Once again, the contract will probably require notice. There may again be clauses absolving the Owner from responsibility for differing site conditions, but notice should be given as soon as possible, and then discussion of solutions can commence. If it initially appears the Owner will not make financial remedies available, the solution sought should be the least expensive. Additionally, the Contract Administrator must continue to document costs for the potential of future negotiations or recovery.

7.2.7 Notice to request dispute resolution

Just as there is a notice time frame on advising earlier than later about a bid mistake, there is also the possibility of notifying in a timely fashion at the end of the job of a claim. No one wants to be blindsided months or years after a project with a claim. The contract may call for timely notice, and the courts may expect certain claims (e.g. a lien against a property) will be brought forth in a timely manner.

Constructive Acceleration is one issue that needs to be resolved before leaving the project behind. Constructive Acceleration is a term used to distinguish it from directed acceleration in which the Owner acknowledges the need to pay for the increased costs involved. In Constructive Acceleration the Contractor has had to increase costs, usually through overtime, to finish the project on the original contract scheduled date, even though there has been a delay not caused or controlled by the Contractor. The Owner, refusing to pay for any increased costs for the acceleration due to the delay (perhaps relying on a No Damages for Delay Clause), did not change the contract completion date. Now the Constructor is forced to pursue financial reimbursement through a claim.

Nonetheless, if timely notice for such claims had not been made, the Contractor may not be able to recover because the defense will be put forward by the Owner's attorney that failure to comply with timely notice negated the right to recover.

For a Contract Administrator wishing to pursue a Constructive Acceleration claim, documentation and foresight is essential. There are certain conditions you will need to prove and document:

1. The acceleration was caused by an excusable delay. You will need to fully document that the delay impacted the critical path and that the duration of the delay coincides with the additional effort expended. In other words, you cannot claim a one-day delay caused a week of acceleration.
2. You gave notice in a timely manner to both the cause of the delay and its potential impact.
3. Upon refusal of the Owner to authorize **directed acceleration**, you provide timely notice of intent to constructively accelerate the project and to file for compensation upon completion of the acceleration. Included in this notice and subsequent communications must be a thorough attempt to resolve all issues without the need to litigate or arbitrate the claim for Constructive Acceleration.
4. Full documentation of all cost strictly due to the Constructive Acceleration.
5. During 3 and 4 above, it will be wise to contact an attorney for advice and direction.

It is always important to recognize not all costs during an acceleration are the result of the delay or can be invoiced to the Owner.

Example: You decide it will be necessary to work around the clock on shift work to meet the original schedule. The rate for each of the two additional shifts is one and a half times the day rate. Only the additional half times the rate is an accelerated extra, since the hourly work would have been done on a day basis if the contract had been extended.

However, what you might be able to assert is the shifts were less efficient, perhaps due to artificial lighting, and so the cost of additional lighting along with a factor of inefficiency might be a legitimate cost.

7.2.8 Notice of termination

Termination on a project can be from the Owner to the Contractor or from a Contractor to a Subcontractor. Either way, such a step is very difficult. Usually, many attempts to avoid the termination have already taken place. Faced with no other alternative, termination is a difficult course and it might be best to have an attorney involved early in the process.

Contracts for good reason require specific time notices related to termination, and there is usually a specific requirement how the termination notice has to be delivered. Failure to comply with the time notice or elements of the notice requirements, such as sending the notice via e-mail rather than certified letter, may not change the fact a party has been terminated, but it may change the prospect of recovering costs related to the performance of the party that was terminated. Monies that were owed a terminated Contractor or Subcontractor might still be owed if the termination was done improperly (again refer to the Spearin case). Also, if a bond was involved and the termination was not done correctly, the bond may no longer apply.

There are basically two types of termination on construction projects: (1) for cause, and (2) for convenience.

7.2.8.1 Termination for Cause

Contracts will contain a clause that permits termination of the contract under specific circumstances. The termination procedure has to follow certain specific steps, including the notice has to be delivered within a specified period of time via a delivery method that can be demonstrated (e.g. certified mail).

If the contract does not contain a specific termination clause, there is the basic implied right to terminate upon notice of a grievous issue unresolved. In addition the contract might also contain rights of the Prime Contractor to perform alternative remedies such as backcharging for work done or material supplied by others, or directly paying other parties with a legitimate claim against the non-performing party.

It is important to make sure all the steps are carefully followed related to the contract provisions for **Termination for Cause**. Failure to do so may place the Prime Contractor in jeopardy for both the costs of taking over the work as well as payment due the removed Contractor.

In addition to contacting an attorney for advice and guidance, the Contract Administrator should be sure that:

1. All reasonable efforts have been made and documented to have the Contractor or Subcontractor perform in accordance with the contract.
2. Notice has been properly given.
3. Within the time frame allotted for compliance or rectification, the Contractor or Subcontractor has not made sufficient effort or progress to be considered performing in accordance with the contract.
4. The method of replacement (cure) for the problem is not excessive, but is a reasonable and fair attempt to resolve deficient issues in a timely manner.
5. Ultimately, although the cost may be greater than monies withheld, the accounting will treat the removed Contractor fairly, and monies owed will be paid in accordance with payment provisions within the original contract once the total extent of the costs are known.

In short, Termination for Cause is not a blank check for the Contract Administrator to repair damages or use another's problem to help one's own cause. The best way to avoid termination is to use the best Contractors and Subcontractors. The second-best way is to keep close and open communication with all the team members, recognizing the challenges, but also looking for solutions when they are small and first appear, rather than waiting until they become crises.

Exercise 7.2

Take a moment

In preparation for our chapter ending and website "You be the judge" cases, let us try one here. A $3.4 million renovation to the St. Bernard Port in Louisiana is awarded to a Contractor. The Contractor claimed they had completed the project, the Owner claimed punch list items were never completed. A court awarded the Port $216,000 in completion cost, and $459,000 to the Contractor for work performed but not paid for. The court then subtracted from the award to be paid the Contractor additional costs due to improper work and rework. The final amount owed the Contractor came to $101,000.

On appeal the Contractor sought payment for additional work it allegedly performed in the amount of $380,000. The Contractor pointed to the termination clause in the contract, allowing the Owner to terminate and complete the work as it saw fit. However, the Contractor also pointed out the work would be incorporated into a change order to be approved by the Engineer as to "reasonableness." Since the Engineer never approved a change order, the Contractor argued that the Owner was not entitled to the cost of the completion work after the Contractor had left the project, stating it had completed its work.

How do you rule?

In what amount?

Discussion of Exercise 7.2

The appeals court ruled in favor of the Port. The ruling confirmed that, reading the contract as a whole, the court could not find that recovery of costs for work not completed or unsatisfactorily completed was contingent on the Engineer's change order. The Substantial Completion clause clearly made the Contractor liable for absolute completion of the work. Nothing (including an Engineer's change order, a Certificate of Substantial Completion, payment by the Owner) would release the Contractor from that obligation.

It is always important to remember that clauses can be important, and even at times ambiguous, but the contract will be interpreted "as a whole" document. Sometimes that will favor the Contractor, and at other times it can favor the Owner.

A CASE IN POINT

Figure 7.1 Butterfly
Photo credit: Stanley Cook

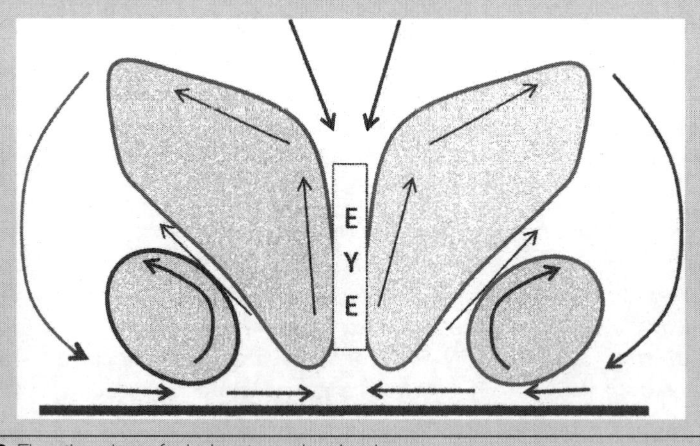

Figure 7.2 Elevation view of wind vortexes in a hurricane

This certainly can't be true, but I have heard it many times. They say that a butterfly's wings moving back and forth in China will start the slightest breath of air to move almost imperceptibly at first, but then it will become a breeze, and the air will shortly be carried up and over the Himalayan mountain range and come back down the other side much stronger as it blows across the deserts of Northern Africa and out into the Atlantic, where the sun will heat it until it becomes a major storm that will strike the coast of the United States two weeks later.

Well, that's the story I have been told, but it can't be true. Nonetheless it is a famous explanation of the Science of Chaos – established by mathematicians, physicists, and Engineers during the latter half of the twentieth century. The principle is that a small event can build over time and cause much greater harm, or "chaos" eventually.

Of course, why others have told it to me is not because it is a new science, but to remind me as a Contract Administrator or a Project Manager, it is easier to redirect the problem when it is as small as a butterfly than to suffer the consequences when it has become a storm.

7.2.8.2 Termination for Convenience

There may be circumstances under which a contract needs to be terminated without cause. For instance an Owner may have learned that the need for the structure has completely changed, or the Owner may have been bought by another company and the entire facility is being relocated. In such a case, the contract can be terminated. All of this assumes the parties are both acting in good faith. Courts will not look favorably upon a party that does terminate a contract in bad faith.

The Contract Administrator will most likely want to immediately review the **Termination for Convenience** clause as soon as the notice is given, but some points will almost be certain, including all progress work will have to cease immediately or within a very short and specific time frame set forth in the contract, and there will be directions on how to leave the site in a safe and environmentally appropriate condition. What the Owner will want to do is minimize additional costs, and what the Constructor will want to do is to be sure to be paid in full for all that has been done, and all that might be required to clean up and demobilize the site.

Here again, documentation by the Contract Administrator is important. Photographs of the entire project at the point of cessation will be very useful later if there is some dispute over how much had actually been accomplished. Additionally, all Subcontractors, and particularly Suppliers, need to be notified immediately, and any work still in progress needs to be documented. Some items may have been ordered that would be unique to the project, and if completed but still off site, the costs and determination of what to do with such items needs to be determined. In essence the Contract Administrator will want to produce invoicing and evidence of all the actual costs, overhead, and profit, and also any potential lost revenue that the termination caused. Depending on the circumstances and the rapport with all the parties involved, Termination for Convenience can work out well (with the prospect of future work for everyone with the same Owner) or be a "nickel and dime" negotiation over every single line item on the pay application.

Disappointment may run high for everyone on a Termination for Convenience. The Design Professionals will not get to see their vision fulfilled. The Constructor, whose pleasure is in building, not paperwork, misses the chance to finish a project and perhaps loses some significant profit while having to do a lot of general and administrative work just to get close to breaking even, and the Owner has spent a lot of hours and money moving forward for nothing. How bright the future will look for each of the parties depends on how well the present is handled, and that for the most part depends on how well the Contract Administrator can accurately and fairly document the past.

7.3 Exculpatory and risk shifting clauses

You cannot escape the responsibility of tomorrow by evading it today.

Abraham Lincoln

Many of the notice clauses relate to other clauses within the contract, such as delays, changes, and termination. The reason notice needs to be given is the Owner, the Owner's agents, and the Contractor, along with Subcontractors and Suppliers, need to know as soon as possible that there is a change from what was anticipated. There are other clauses, however, that may or may not require notice in accordance with the contract, but for which the Contract Administrator should still give notice and track closely both the financial and the schedule impact of the situations involved.

Some of the most egregious clauses in a contract from a Constructor's viewpoint are the **exculpatory clauses**. These are clauses that basically shift risk from one party to another, even though the party on the receiving end may have nothing to do with causing the risk or have any way to mitigate the effects of the risk. Some of the risk shifting might include:

- hidden conditions
- design deficiencies (particularly related to building codes)
- lack of payment
- delays
- performance criteria.

The evolution of exculpatory clauses has been varied and spread over time, but they usually have originated from one bad incident out of a million good ones. Unfortunately, it is the bad incident that gets the attention, and so Owners, for future protection based on what might happen, write exculpatory clauses to protect themselves. In essence, exculpatory clauses arise from a squeaky wheel, but the cure has not been just applying oil.

For the Contract Administrator the exculpatory clause that shifts someone else's risk (usually the Owner's) to the Contractor has many pitfalls and dangers.

We have already spoken about the No Damages for Delay exculpatory clause. In such a clause the cost of any delays that result from poor processing of information by the Designers or Owner, or any other delay, such as failure to inspect, coordinate access to the work site, or lack of response for changed conditions, are to be borne solely by the Constructor. We have mentioned that a prudent Contractor can get around such clauses in the right circumstances, but the problem with trying to nullify a No Damages for Delay Clause is that it costs money. Almost always an attorney will have to be involved if the costs cannot be settled amicably at the project site.

Another important issue related to exculpatory clauses is that although they shift risk, the risk is seldom insurable. During the bid stage, the risk has to be recognized, and sometimes the exculpatory clause is so potentially damaging that a factor will be included by some Contractors into their estimate. The unfortunate result for Contractors is that putting risk factors into the bid proposal probably means the proposal will not be the low bid. The unfortunate fact for the Owner, however, is any Contractor that took a risk and lost will undoubtedly have difficulties that will be reflected in time and quality of work on the project.

So there are good reasons for both the Owner and the Constructor not to rely upon exculpatory clauses to shift responsibility to other parties. Such clauses may be a great source of disharmony among a team, with great challenges for all of the risks each party always assumes based on their part in the Design-Bid-Build sequence. Nevertheless, the clauses exist, and the Contract Administrator needs to read the documents thoroughly and find any of the potential risk shifting clauses.

7.3.1 Differing Site Condition exculpatory clause

After the No Damages for Delay, perhaps the most common risk shifting clause is the Differing Site Condition exculpatory clause. In addition to giving notice regarding a Differing Site Condition, the Contract Administrator has to recognize if an exculpatory clause has shifted all responsibility onto the Contractor. This type of clause shifts liability for the conditions or accuracy of what the Designer has shown on the plans, specifications, and any additional site information supplied at bid meetings or by addenda to the Contractor. This usually is related to subsurface conditions, such as rock or underground utilities where no rock or utilities were indicated. However, it could also be related to conditions above ground, such as a bearing wall where a non-load-bearing partition was indicated for removal.

Caution: Although the Contract Administrator may think the drawings can be relied upon in regard to utilities, state laws require every Contractor that intends to dig below the surface of the ground to call and have all the utilities marked before any digging takes place. This can usually be done through a "One Call" service. In this way, when the Constructor digs it avoids disrupting the service of any buried utilities and having to shoulder the cost of dong so, and, in some cases, experiencing the danger of hitting unsuspected utilities. The issue becomes that the Constructor now knows there are utilities underground that had not been shown on the drawings, and that the excavation required will be more tedious and difficult in order to avoid disrupting service. If the exculpatory clause existed, shifting all risk of whatever is underground to the Contractor, then the cost for the additional work is expected to be borne by the Contractor, even though the Designers and Owner probably had much more time to have accurately discovered what was underground.

Again, Differing Site Condition exculpatory clauses can be beaten in court, but there is no guarantee. Some courts recognize that the broad interpretation of protection for the Owner by such clauses can go beyond the reasonable actions a Contractor should be expected to assume. However, some courts are not as lenient.

Example: (*Interstate Contracting Corp.* vs. *City of Dallas*) The City of Dallas withheld information it possessed regarding the subsurface conditions. The contract contained an exculpatory clause, but Interstate believed the bid process was too tight and broad to do its own subsurface investigations. The court upheld the City's exculpatory clause even though the City had information it had not shared prior to the bid with the Contractors.

It is important to recognize that one of the best defenses against a risk shifting clause is to be sure to look at the contract as a whole. Ambiguities may be a cause for negating the shifting of the risk.

A CASE IN POINT

Figures 7.3 and 7.4 Inserting a new water wheel into a restored grist mill would prove easier than constructing a new water wheel for a sawmill to be recreated where it had previously been
Photo credit: Charles W. Cook

Having successfully completed the restoration of an eighteenth-century grist mill on a Pennsylvania State Park, R. S. Cook and Associates were asked to bid on the re-creation of a water-powered saw mill that had existed nearby on the same State Park property. The saw mill was completely gone, but research had determined where it must have been located. Over two centuries had apparently changed the terrain significantly, but the State was convinced the saw mill in its original location would be an important addition to the entire Park experience, which also included many historic homes that had also been restored, including several by R. S. Cook and Associates, Inc. As a Contractor, everyone was comfortable with the team that had worked well together before.

Within the Owner's General Conditions there was a risk shifting clause termed a "No Compensation for Rock Excavation" clause. The Owner – the State of Pennsylvania – placed all burden of investigating the underground conditions upon each Contractor that wished to bid on the project.

R. S. Cook and Associates was the low bidder and awarded the project, but less than a week into the work, excavation of the proposed site revealed two-thirds of the saw mill footprint was going to require excavating bedrock under a thin surface of the topsoil. The company had not done any soil borings and had not included any contingency for rock excavation. Based on what was financially figured into the bid proposal for the project, the excavation of the rock would wipe out all the profit, and even represented a likelihood of a significant loss unless savings could be found somewhere else. With the grim prospect of rock before us, no one was figuring this was going to turn out well. The State insisted the "No Compensation for Rock Excavation Clause" would have to govern and no payment would be made for its removal.

Exercise 7.3

Before reading further, decide what you would have done in the above case to get relief from excavating the rock. Is there an ambiguity you would rely on?

How about a Termination for Convenience or Cause (circle one)?

Discussion of Exercise 7.3

As you read through the rest of the text, think of what you would have done or how you would have approached the Owner. Remember, the Owner was not trying to be difficult. The contract it was to enforce was not made by the personnel on the project site. They had superiors to answer to, also, so was there a legal argument or would R. S. Cook and Associates, Inc., have to continue with the work and accept a heavy loss on the entire project to remove rock that had not been included in the original estimate?

Closely related to the potential for underground differing site conditions are the concealed hazardous material site conditions. Some contracts deal with hazardous materials under an Indemnity Clause (discussed immediately below). Passing all responsibility of hazardous materials to the Contractor can be extremely burdensome if any hazardous material is discovered and all costs related to removal were supposed to have been included in the original lump sum price. There are basically four scenarios, and two are not good:

1. All the prices come in high to protect against the possibility of hazardous materials, but none actually exist, and the Owner paid more for the work than was necessary.
2. The hazardous material that was uncovered was included in the bid proposal of the successful Contractor and so the Owner paid what should be paid for the work.
3. Most of the bidders covered the risk for hazardous materials, but the low bidder did not. Finding no hazardous materials, the Owner did not pay for work not done.
4. Most of the bidders covered the risk for hazardous materials, but the low bidder did not. Hazardous materials were found and this caused considerable financial difficulties on the project for the Contractor. As a result teamwork on the project grew difficult, the schedule slipped, and the quality of the work was less than expected.

Today most Contractors are alerted to the potential of hazardous materials in buildings constructed before 1980. Such buildings often contain asbestos, including the tile flooring if it has not been replaced, and lead paint. Renovations to these buildings first require the safe removal of the hazardous materials. However, the Contract Administrator must recognize that contract wording may also require removal of materials that are found to be hazardous sometime in the future. The scare here is what materials might suddenly be considered hazardous? Of course, there will probably be much time before a new hazardous substance is identified, but if there is a topic in the news about a hazardous substance sometime in the future, the visionary Contract Administrator will be prepared.

The Contract Administrator handed a project containing any exculpatory clause needs again to practice prudent notification as detailed above and attempt to maintain effective and open communication with the Owner to find remedies that both parties can appreciate are in the best interest of the project as a whole.

7.4 Indemnity Clauses

Closely related to exculpatory clauses but quite different are what are known as **Indemnity Clauses**. Although a risk is being shifted from one party to another, this type of risk can usually be covered by insurance. For example, safety of workers is something that the Owner will expect the Contractor to be responsible for, and if an incident occurs, insurance should be in place to cover the incident.

Additionally, someone needs to cover the property against accidents. A form of insurance known as **Builder's Risk Insurance** might be purchased by the Contractor, but it is sometimes covered by the Owner. If this is the case, it should not be carried as a cost in the bid proposal, but the Contract Administrator should make sure after award that the coverage is adequate and the Contractor and the Owner are both covered.

Some people use bonds as a form of indemnification. Although not strictly an insurance policy, they do protect one party from the failure of another party to perform in accordance with certain provisions of a contract (e.g. payment).

The ultimate motive behind Indemnity Clauses is that the stronger party can dictate protection from the weaker party. Fortunately, there is protection from much of the indemnification requirements as opposed to the risk shifting exculpatory clauses. The most important habit a Contract Administrator can develop regarding Indemnity Clauses is to make sure an agent knowledgeable in both insurance and bonding has looked at the provisions related to indemnification. Even an insurance certificate does not guarantee effective coverage for all that might be detailed in the contract.

Example: Be careful when having insurance policies reviewed. Often generic exclusions for flood or tornadoes will be found in most policies, and you will need to purchase additional coverage if such natural acts could occur. Of course, there is always a choice to gamble, but if working next to a river, or in the Midwest during tornado season, you might want to bring any lack of coverage up before you find you are "not in Kansas anymore."

7.4.1 Broad Form Indemnification

Taking indemnification to the ultimate extent, Constructors are asked to indemnify others in the contract (Owner and/or Owner's agents) against all actions, including those caused solely by the Owner or the Owner's agent. This does appear to be excessive, and Constructors often balk at such indemnifications, but to get the work, acceptance of the clause is often required.

One possible relief for such clauses is coming in the form of state legislation. Some states no longer allow **Broad Form Indemnification**. The Contract Administrator should check with both the insurance agent and an attorney to determine whether such indemnification is permitted. If so, a defense of "unconscionable" (discussed under 7.6.1) may be in order.

7.5 Damage clauses

Most people are frightened of responsibility.

Sigmund Freud

Damage clauses can sometimes be interpreted as exculpatory clauses, but they deserve special attention, since they often have specific challenges and dangers.

7.5.1 Liquidated Damages

We have previously discussed **Liquidated Damages**, but it remains one of the most infamous and least popular of the damages. Although it is an agreed-upon amount that is "backcharged" to the Contractor for every day (or a specified period of days) the Contractor does not complete the work, there are several issues related to Liquidated Damages that should be recognized by the Contract Administrator.

7.5.1.1 Not a penalty

Courts have determined that Liquidated Damages cannot be a form of a penalty.
The damages have to be a reasonable and agreed-upon amount that as accurately as possible reflects the cost or loss to the Owner for every day that the property cannot be used.

Examples:

1. A cost to the Owner could be an extension of rental on a temporary property while the existing property is being renovated. Such a cost would probably be known based on conditions in the existing rental agreement.
2. A loss to the Owner could be the receipts on the sale of tickets to movies during the summer blockbuster season if the theater does not open as scheduled. Those costs could be estimated based on the number of seats and the cost per ticket, and a factor based on the history of previous summer movie attendance while subtracting the daily costs of operating the movie theater when it is open.

It is important, therefore, for the Contract Administrator to recognize the issues involved in whether the Owner can or cannot use the property for its intended purposes. Documentation of what the actual conditions are could be very important in removing Liquidated Damages assessed against the Contractor.

A CASE IN POINT

Sometimes the Contract Administrator has to recognize there is more at stake than just the cost of Liquidated Damages. During the early days of the Great Depression, the Circle Theater in Philadelphia was scheduled to open as the first air conditioned theater in the City. At the last minute the person in charge of the construction was informed of a major delivery problem. The chillers for the roof-top units would not be ready in time for the scheduled opening.

Figure 7.5 The Circle Theater, Philadelphia

Photo credit: Irvin R. Glazer collection of the Athenaeum of Philadelphia

As they say in show business, "The show must go on." In the first half of the twentieth century movie theaters were built for large audiences to attend. The people were going to fill the almost 3,000 seats of the theater to not only escape from the worries of the economic times for a bit, but also to escape the heat of the Philadelphia Summer. Not opening was not an option.

The solution: bring in huge blocks of ice, place them in the plenum, where the blowers could propel the cooler air through the ductwork. What the audience did not know did not hurt them, and they enjoyed the "refreshing" night out at the movies.

I know it happened that way, because my grandfather told us how he did it.

7.5.1.2 Mutually agreeable

If the amount of the damages appears to be inaccurate or actually being used in a threatening fashion as a penalty, then the Liquidated Damages provision is voided. Since they must be mutually agreeable, the best way to guarantee that they are agreed to is to make the Liquidated Damages part of the contract. In some instances, Owners may wish to impose a penalty, but courts have found that a penalty also requires a reward contained in the contract for finishing early.

7.5.1.3 Responsibility for delay

Even if a No Damages for Delay Clause exists in the contract, the Owner cannot delay the project and then assess Liquidated Damages. In fact, if there were mutual delays by both the Contractor and the Owner (or the Owner's agents, such as the Architect), then courts have usually not permitted Liquidated Damages to be assessed.

Once again, this is why it is imperative that delays be documented. Even though there may be no compensation from the Owner for the delay, the Contractor wants to also be sure there will be no Liquidated Damages assessed when the Owner is the cause of the delay.

7.5.1.4 Substantial Completion

As discussed in the previous chapter, there is a time in almost all projects when the Owner can use the project for its intended purposes even though there may still be some items to complete, repair, or "touch up." That moment is considered **Substantial Completion** of the project, and from that date onward Liquidated Damages should not apply. The Contract Administrator should make sure this is clear in the contract documents, or, if it is not, make a point to get it recorded early in the project (in a manner that does not raise concerns in the Owner that you anticipate being late on completing the project).

The Contract Administrator should check the contract carefully, however, since some contracts carry a reduced Liquidated Damages rate after Substantial Completion and up until final completion of all punch list items.

Contract comparison

The student is encouraged to compare AIA 201 General Conditions paragraph 9.8.1 regarding Substantial Completion with ConsensusDOCS paragraph 2.4.24, and the Engineers Joint Contract Documents Committee C 700 paragraph 14.04. In the AIA General Conditions, several paragraphs define specific issues that need to be fulfilled related to the Owner's initial and final occupancy of the space. The AIA document covers many contingencies, including partial occupancy in paragraph 9.9. Although still related to actual use of the space for its intended purpose, ConsensusDOCS has a much shorter version of what constitutes Substantial Completion, and further releases the Constructor from responsibility related to a Certificate of Occupancy if the failure to obtain such was not the fault of the Contractor. In the Engineer contract the Engineer assumes the responsibilities the Architects handled in the AIA documents, but the Engineers add the right of the Owner to exclude the Constructor from the premises except to remove equipment and materials or to finish or repair work not yet accepted.

7.5.1.4.1 Warranties

Although it is not usually considered a "killer clause," the Contract Administrator should not consider final completion and payment to be the end of all obligations. Warranties are in place in accordance with the contract documents, and defective work will need to be remedied effectively. As discussed in the previous chapter, how long the warranty period is and when it actually begins is often a point of contention among the parties, and the sooner such issues are defined, the better for all concerned.

7.6 Damages for defective construction

Truth never damages a cause that is just.

Mahatma Gandhi

Although there may not be a clause defining specific remedies for defective work, if the Owner is damaged in some way by the Contractor's improper work, there are remedies, whether defined or not.

The courts view two possible methods to make the Owner "whole":

1. the cost expended to correct the defective work
2. the loss in market value of the property because the defective work cannot or will not be repaired.

Perhaps courts are lenient on most Contractors in such cases, since they often award based on what is both expedient and less costly.

Example: The Owner discovered after moving into the project that the wrong color gray was painted on the walls, and now the mauve cubicle partitions do not seem as pleasing as the Architect's original aesthetic concept. The Owner wants a credit for the full amount of repainting the walls, even though it will not be done. The Contractor counters that, since it will not be repainted, the Owner should get only what would be a loss in value to the property, which in the Contractor's opinion is zero since no one really thinks the Architect's original design was that great of a color scheme. The Architect supports the Owner's position, but actually wants the walls repainted so the color scheme will vindicate his tastes. The Contractor counters that argument on the basis that it would be impractical to have to repaint the entire space now that everyone has moved in and access to the walls is greatly reduced.

Ultimately, the reader should ask, "Do I really want a court to decide such a case, or would it be better to studiously and carefully manage the project to avoid such a situation in the first place?"

Exercise 7.4

Take another moment
Back to that "cheat sheet"

At the beginning of this chapter we listed the following types of clauses we would be studying:

- scope clauses
- notice clauses
- termination clauses
- exculpatory and risk shifting clauses
- damages clauses.

Create a spreadsheet using these as major headings. Under each heading you will want to include any of the subcategories included in the text. That will mean notice clauses will have several additional lines under the main heading. Beside each heading you will want to have a column to identify the page or, even better, the paragraph where the subject clause can be found.

Another important column would be to list the contract documents in order of precedence. Make the spreadsheet work for you, but include everything you need.

Eventually, as you get further into the contract management process, you will want to pull out sections of the spreadsheet "cheat sheet." For instance, you will want to have a separate spreadsheet for the contract change order notification and follow-up process. There will need to be columns for the date the change was identified, the date the change was submitted for review, the date pricing or other means of compensation were submitted, the date of the approved change.

What is important is that you always have before you the condition time imposes on your communication.

Discussion of Exercise 7.4

The point of investing some sweat equity into the spreadsheet rather than just having one handed to you is not to see how perfect you can make it the first time or any time. The real value is in developing your own "cheat sheet" and, as you go forward, knowing you can change it for your own benefit. Store it, therefore, so it can be accessed and changed.

Knowing where you can find certain clauses in each contract you administer will ultimately save you a great deal of time, and making copies to help others on the team will also be a great benefit for everyone. Sometimes you will alert others to clauses they might not have paid enough attention to when they were getting into the project.

More important for you, however, is developing the discipline not to rely upon others for what you ultimately are responsible for related to the contract administration. You take the lead and help others through the process.

7.6.1 Damages to the Contractor from the Owner

Although the clauses in the contract have for the most part been written by the party that is in control, and almost always that is the Owner, the Contractor should still look to be treated fairly. Notice and documentation are important, and the Contract Administrator must keep current with both of those issues across a wide spectrum of project concerns.

One initial defense will be what courts consider a clause or action by one party to another to be "unconscionable." This, of course, can also be enforced against a Contractor acting in an egregious fashion toward a Subcontractor or a Supplier. In fact it is part of the case law associated with the **Uniform Commercial Code**, which helps to regulate commerce between states and parties purchasing and selling items. If ruled "unconscionable," a clause will not be enforced by a court. To get to that point, however, will require both time and money, and the Contract Administrator should be aware of the ramifications of such clauses in prosecuting the work, and the rapport that will be affected if one party begins to rely upon such a clause as a way of dealing with others on the project team.

It is also sometimes possible to seek damages from the Owner for delays caused by the Owner or the Owner's representative, but the presence of the above-discussed No Damages for Delay clause makes such a recovery difficult. Nonetheless, for a Contractor the three main areas of recoverable costs include:

1. rentals (e.g. the project trailers, idle equipment);
2. interest on loans used to pay payroll and materials if payment is delayed past the normal contractual payment clause cycle;
3. administrative burdens and lost profit if the delay prolongs the Contractor's ability to increase volume of work until the delayed project is completed.

The Contract Administrator has to be aware that there are many other clauses in the contract that can be quite damaging. Without careful attention and documentation, the opportunity to recover damages will be diminished. **Acceleration clauses**, for example, can require expending large sums of money to maintain a schedule that may have been delayed by the Owner or Owner's agent. Exculpatory clauses may initially seem to protect the Owner from any damages. Acceleration clauses and other clauses the Contract Administrator must carefully monitor will be discussed in later chapters.

Such clauses, and many others, are not in themselves "killer clauses." They become dangerous, however, when the Contract Administrator ignores them.

* * * * * * * * * *

As a final note for this chapter, the most dangerous clause in a contract is not usually the one you are prepared for. The one that will ultimately be the most dangerous is the one the Contract Administrator does not know about. A thorough reading of the contract is important. The Contract Administrator must resist the "you've seen one, you've seen them all," attitude. There could be a new clause somewhere that can be both dangerous if ignored and damaging if the obligation is not met.

* * * * * * * * *

INSTANT RECALL

- Of all the killer clauses, often the most innocent but potentially most damaging if not adhered to, are the notice clauses.
- Notice clauses are very important to adhere to. They might be required for:
  - Differing Site Conditions
  - Request for Information related to an ambiguity
  - change order for increased time and or money
  - delay due to any circumstances beyond Contractor's control
  - dispute notice
  - request for payment.
- The contract will usually clearly define what will be accepted as a Time and Material cost.
- Related to the payment clauses, the Contract Administrator will have to coordinate (usually through the subcontract or purchase order) with each Subcontractor and Supplier what costs will need to be included in the current month's invoice.
- The Contract Administrator should realize that if a Supplier or Subcontractor goes bankrupt, it will become a legal labyrinth to get materials stored at the Supplier or Subcontractor's location released for use on the project.
- Coordinating the submission of the invoice is important in relation to the requirements of the payment clause.
- Even if a No Damages for Delay Clause exists, the Contract Administrator should make sure all delays are detailed and forwarded in writing, initially for negotiating purposes or later to provide documentation for supporting a future claim.
- In the case of exposing a hazardous material below or above ground, all work must stop until a remedy can be determined.
- Termination on a contract can be from an Owner to a Contractor or a Contractor to a Subcontractor.
- When Termination for Cause is an issue, in addition to contacting an attorney for advice and guidance, the Contract Administrator should be sure that:
  1. reasonable efforts have been made and documented to have the Contractor or Subcontractor perform in accordance with the contract
  2. notice has been properly given
  3. within the time frame allotted for compliance or rectification, the Contractor or Subcontractor has not made sufficient effort or progress to be considered performing in accordance with the contract

4. the method of replacement (cure) for the problem is not excessive, but is a reasonable and fair attempt to resolve deficient issues in a timely manner
5. ultimately, although the cost may be greater than monies withheld, the accounting will treat the removed Contractor fairly, and monies owed will be paid in accordance with payment provisions within the original contract once the total extent of the costs are known.

- Some risk shifting clauses include:
  - hidden conditions
  - design deficiencies
  - lack of payment
  - delays
  - performance criteria.
- Related to exculpatory clauses, although they shift risk, the risk is seldom insurable.
- Closely related to exculpatory clauses, Indemnity Clauses also shift risk, but the risk can often be insured.
- The Contract Administrator handed a project containing any exculpatory clause needs to again practice prudent notification and attempt to maintain effective and open communication with the Owner to find remedies that both parties can appreciate are in the best interest of the project as a whole.
- Contractors should be alert to the potential of hazardous materials in buildings constructed before 1980.
- No Damages for Delay Clause puts the Contractor on notice that there will be no payments made for additional costs if the Contractor is delayed during the project, even if the Owner or agents of the Owner (e.g. Architect) were the cause of the delay.
- Whether a notice clause for delay is contained in the contract or not, the Constructor should immediately notify the Owner, with copies to the Design Professional of all delays.
- A waiver clause states that if notice is not given within the prescribed time period set forth in the contract, then the Contractor waives all future claims for increased time or money.
- The most dangerous clause in a contract is the one you do not know about.

YOU BE THE JUDGE

L&B Construction Co (LB) subcontracted to Ragan Enterprises, Inc. (REI) the electrical work on a State of Georgia project. As the General Contractor, LB signed a No Damages for Delay contract with the State of Georgia. Time extension would be allowed only if the Owner caused a delay. The contract read in part:

> The contractor shall not be entitled to payment or compensation of any kind from the Owner for direct, indirect, or impact damages.

LB then sent and REI signed a subcontract containing a flow down clause which read:

> Contractor shall have the same rights and privileges as against the subcontractor herein as the owner in the general contract has against the contractor. Subcontractor acknowledges that he

has read the general contract and all plans and specifications and is familiar therewith and agrees to comply with and perform all provisions thereof applicable to subcontractor.

REI eventually filed suit against LB when it was LB who delayed the project by two years. REI argued that the delay was not caused by the Owner, so LB could not invoke the flow down clause.

How would you rule (circle one)?

For Ragan Enterprise, Inc. For L&B Construction Co.

Why:

IT'S A MATTER OF ETHICS

At what point do Owners and Design Professionals protected by exculpatory clauses have an obligation to share information about a site that a Contractor would not be aware of based on a walk-through before bidding the project? Conversely, what is the dividing line for a Contractor who discovers an ambiguity in the contract documents that would give a competitive advantage to the Contractor during the bid submission process?

Going the extra mile

1. Complete the "You be the judge" exercise above and then go to the Companion Website to compare your decision with the actual ruling.
2. Also, go to the Companion Website to see whether your solution to the saw mill story was the same one that occurred during the actual project.
3. **Discussion exercise:** Which clause in the contract (other than the one you do not know about) has the potential to cause the most difficulty for the Contract Administrator?
4. Effective use of the correct vocabulary improves communication and promotes respect among colleagues. The reader should become familiar with the key words listed below that have been used in the chapter. Definitions can be found in the Glossary. Additionally, the student is encouraged to use the flash card exercise in the chapter's Companion Website as practice for possible quiz or exam questions.

Key words

Acceleration clauses
Broad Form Indemnification
Builder's Risk Insurance
Constructive Acceleration
Differing Site Condition
Liquidated Damages
Mechanics Lien
No Damages for Delay Clause
Potentially Responsible Party (PRP)
Substantial Completion

Directed acceleration

Exculpatory clauses

Indemnity Clauses

Termination for Cause

Termination for Convenience

Uniform Commercial Code

CHAPTER 8

Insurance

In these next three chapters we will discuss **insurance**, **bonds**, and **warranties**. Similar but different, these three aspects of construction need to be carefully monitored and protected, or incredible costs can occur both financially and in working reputation.

We will begin with the subject of insurance. For the Contract Administrator, who attempts to read policies and insurance requirements, this entire subject will seem rather daunting. Along with bonds, having a reliable and competent agent is essential for effective coverage at the best possible cost. Nonetheless there are certain fundamental principles of insurance that the Contract Administrator should know.

Student learning outcomes

Upon completion the student should be able to. . .

- **understand the challenges of assuming liability for events or issues on a project**
- **differentiate types of insurances and the purposes of each kind**
- **distinguish between an Employee and a Subcontractor**
- **understand the purpose of Workers' Compensation Insurance**
- **calculate the effect of an experience modifier on the cost of insurance**
- **understand the effect and consequences of misclassification of work on the costs of insurance.**

Exercise 8.1

Take a moment
What's in a word

Let's start this chapter with a bit of an exercise. Look up in a dictionary or online the meaning of three very similar words, and write a brief definition for each below that clearly distinguishes each word from the other two:

Insure:

Ensure:

Assure:

Discussion of Exercise 8.1

These words have become so interchangeable that in many dictionaries the definitions appear to be the same. The reason we start with these three words is they may help us in part to recognize the difference between **Bonds, Insurance** and **Warranties**. Just as "insure," "ensure" and "assure" are very similar but different, bonds, insurance, and warranty are very similar but different. If we do not truly understand the differences, we may not only overlook the purpose of each, but also misuse them to our own detriment.

What I want to propose is a analogy-type comparison:

Insure is to insurance as assure is to bonding as ensure is to warranty

Let us start with the word "insure."

For many, this is actually the only word that is used, and it replaces the use of assure and ensure. In conversation we might hear someone say "I want to insure everyone we will get the project done on time." Or, we might hear, "I will do everything it takes to insure we are finished on time."

Those are examples of when we should have used the words "assure" and "ensure," respectively.

What we should be using the word "insure" for is almost always related to an obligation of one party to make another party whole financially if certain events take place.

The key factor to recognize related to insurance in construction is that construction is a very dangerous industry. The potential always exists for severe loss to property or even personnel. The potential for tragedy or disaster is so real that parties that have little or no control over the possible consequences of an accident want to be protected against such an occurrence.

In construction, the financial obligation will be determined by the contract conditions, as well as the amount of exposure and potential loss that could take place. In most cases insurance of some sort will be carried by all the parties involved, but some insurance obligations will be set forth by the contract and will be the responsibility of one or another party to the contract.

So, for our purposes, to insure in contract administration terms means to carry a written policy or some set-aside sum of money that protects one or more parties against a financial loss, should damage or injury take place during the construction of the project.

Next let us look at the word "assure."

In the example above, it might have been better if the expression had been, "I want to assure everyone we will get the project done on time."

In this sense, we use the word "assure" to give everyone confidence. There is actually no financial obligation formally committed to, but the confidence we give to everyone will be determined by how much they can believe in our sincerity as well as our competence. For this reason, I want to suggest it is closely related to bonding capacity, which we will find is linked to both the character and competence of the principals and company that are being bonded.

Bonds are issued on the basis that they will never have to be used, since the company being bonded has provided enough information that they can assure everyone they will meet the obligations of the bid proposal, and, if awarded the contract, that they can both fully perform the work and pay all obligations related to the project. The bonding company that issues all the bonds (bid, performance, and payment) assures the Owner that the conditions of the bid proposal and the contract will be met.

If the bonding company has done its due diligence before issuing the bonds, the need for the Owner to file against any of the bonds should never happen. As we will see later in the next chapter, however, this is not always the final outcome. Just as anyone can assure, or attempt to bolster the confidence of someone else, that something will happen, things do not always work out the way we intended. And that is why Owners require bonds.

And the final of our three words is "ensure."

Again, going back to the example above, perhaps it would have been better to say, "I will do everything it takes to ensure we finish on time." Here we are injecting more than confidence into the expression. What we are saying is we will make certain it happens.

Now, once again, no one is perfect. Neither is everything we build going to be perfect. There are times we will have to come back and fix something. Occasionally it could be a serious latent defect. Other times it might be a piece of equipment that did not perform the way it was intended by the manufacturer. Whatever it is, we guarantee our work for a period of time. The Owner needs to know that, so there are warranties usually expressed in the contract.

When we sign the contract we acknowledge that we will do whatever it takes to ensure the structure functions in accordance with the obligations we have agreed to. In other words, we will *make certain* what we agree to happens.

8.1 Liability and insurance

Life is a shipwreck, but we must not forget to sing in the lifeboats.

Voltaire

Insurance rates are determined based on the potential loss due to occurrences that will happen over a period of exposure. Of course, the best way to avoid an insurance claim is to not have an "occurrence," but the rates of insurance are based on unfortunate events that will happen eventually.

Each party to the contract needs to recognize that it will have to provide some insurance to specifically cover its own potential loss. The Owner and the Design Professionals may pass some indemnification on to the Contractor, but each party needs to make sure it is fully covered for potential occurrences. The Design Professional as well as the Contractor may need insurance that remains in effect past the actual completion of the project, and such claims are often quite costly if they do occur. For the Contractor such a policy is considered Completed Operations Insurance, and for the Design Professional such coverage is usually part of the **Errors and Omissions Insurance**. Although the contract may call for specific insurance and levels of insurance, each party needs to be certain such insurance and the levels required are adequate.

Merely language in a contract, or even a certificate of insurance naming an additional party as being covered, is not adequate protection. The Contract Administrator needs a competent insurance agent to review the actual policy that is meant to protect the Contractor or the Design Professional. The amount of coverage may be inadequate, exclusions may be in place that should not be, and even the rights of an additional insured may be different than the coverage in the full policy.

It is also important to recognize that insurance is not meant to cover acts of incompetence by the Contractor or any of its agents. The Contractor is expected to perform in accordance with the terms of the contract, and if not, a performance bond rather than insurance would be the Owner's remedy.

Just as with bonding, the Contract Administrator needs to rely upon a trusted and experienced agent when it comes to insurance coverage. This person might be the same individual who acts as the agent for bonding, but in all likelihood the people handling the coverage for the bonding company and for the insurance company will be different individuals and quite possibly different companies.

The insurance coverage of most Constructors has been obtained prior to the project even being considered for bidding. However, the insurance agent should always review the bid and contract documents to make sure the coverage carried by the Contractor will be sufficient for the project that is being bid upon. It is possible that although one type of coverage the Contractor carries is more than sufficient, there may be other types of coverage that are below the contract conditions, or there can sometimes be coverage that is not included in the standard insurance the Contractor already carries.

The Contract Administrator should be familiar with the general types of insurance, but the "fine print" is still something a trained and experienced agent should review.

8.1.1 Risk management

As we discuss insurance, it is important to recognize that some coverage will be required by the contract. Other coverage is required by obligations set by government (including federal, state, and local). Beyond limits that are required by contract and law, however, there is also the amount of coverage a Constructor will want to insure a catastrophic event would not destroy the project or the company.

Generally these decisions are made above the level of contract administration. Principals of the Constructor have probably determined what the potential for financial disaster might be. In the industry, the decision about the limits of insurance is termed "risk management." The prudent Contract Administrator might always be mindful of any special or difficult issues that might require greater risk or higher costs if a failure or accident should occur, and so advise management.

8.2 Types of insurance

There are some important and basic types of insurance that Contractors should address in some way, and they are usually defined by the contract itself, including the dollar amount limits of coverage required:

- Builder's Risk Insurance
- Commercial General Liability (CGL)
- "Umbrella" Insurance
- payroll insurances.

One other form of insurance that may not be called out in the contract, but more and more Constructors are finding it necessary, is **Errors and Omissions Insurance**. In the past most Contractors have not carried Errors and Omissions Insurance, but as Constructors assume more and more input into design, the potential for exposure due to design errors or omissions becomes greater.

Examples:

1. A Subcontractor provides installation details on shop drawings that were not shown on the contract drawings. If a defect arises, the Subcontractor (and Contractor) might be held liable for design issues.
2. Constructors and essential Subcontractors (such as the MEP trades) give input during the design phase of Building Information Modeling. Failures in the design may come back to errors on who provided the input.

8.2.1 Builder's Risk Insurance

Builder's Risk Insurance is meant to cover accidental or unexpected damage to the project while it is under construction. Owners will often carry this insurance themselves, but that actually puts an additional burden on the Contract Administrator. Although the Contractor is relieved of the financial costs of the insurance, this is not really a savings for the Contractor, nor is it an added financial burden for the Owner who pays for it. If Builder's Risk Insurance is to be carried by the Contractor it would have been figured into the bid proposal when the estimate was being made, and it is sometimes carried by the Contractor as an additional line item on change orders, just as increases to the bond might also be included as an additional charge. So the Owner does not really save by having the Contractor provide Builder's Risk Insurance unless the Owner either self-insures (rather than pay a premium to an outside insurance company that has funds or assets sufficient to cover any loss) or has a lower rate.

If the Owner does elect to provide the Builder's Risk Insurance that does not mean a "get off free" card is presented to the Contract Administrator for the Contractor. In fact, making sure the coverage is adequate and the Contractor is fully covered for potential loss is a bit more difficult. Just seeing the coverage amounts in the contract does not mean the coverage extends to the Contractor, nor does it mean all potential losses or causes of loss are covered.

The Contract Administrator will want to know if there is loss from flood, tornado, and other natural occurrences. Also, is there coverage for items that have not been incorporated into the project, and for items that never will be (e.g. equipment or trailers)? Again, a good insurance agent is critical in discerning the extent of coverage and the potential for additional coverage.

Typically Builder's Risk Insurance covers losses or damage due to:

- fire
- lightning
- wind (but not necessarily tornado)
- theft and vandalism.

But Builder's Risk Insurance will often exclude damage due to:

- earthquake
- flood
- acts of war
- damage directly due to a tornado
- mold.

For many years, depending upon location, the threat of earthquake or flood has been recognized as a serious danger. In some parts of the country, tornadoes can be common enough to warrant weighing the risk to the project if such a disaster would strike.

Figure 8.1 Evidence that a tornado touched down and was the cause of damage to a project could negate coverage based on a "tornado exclusion"
Photo credit: U.S. National Oceanic and Atmospheric Administration

Fortunately for construction in the United States, acts of war do not seem to be a likely peril, but the threat of terrorist incidents has caused some concern in certain areas of the country. If the Contract Administrator is working outside the United States, the potential for coverage on all these threats by nature and humans should be analyzed.

The Contract Administrator needs to carefully examine the insurance policy to know what is covered, but just as importantly to understand what is excluded.

It is possible to buy coverage for an exclusion, or it is also possible to buy "special

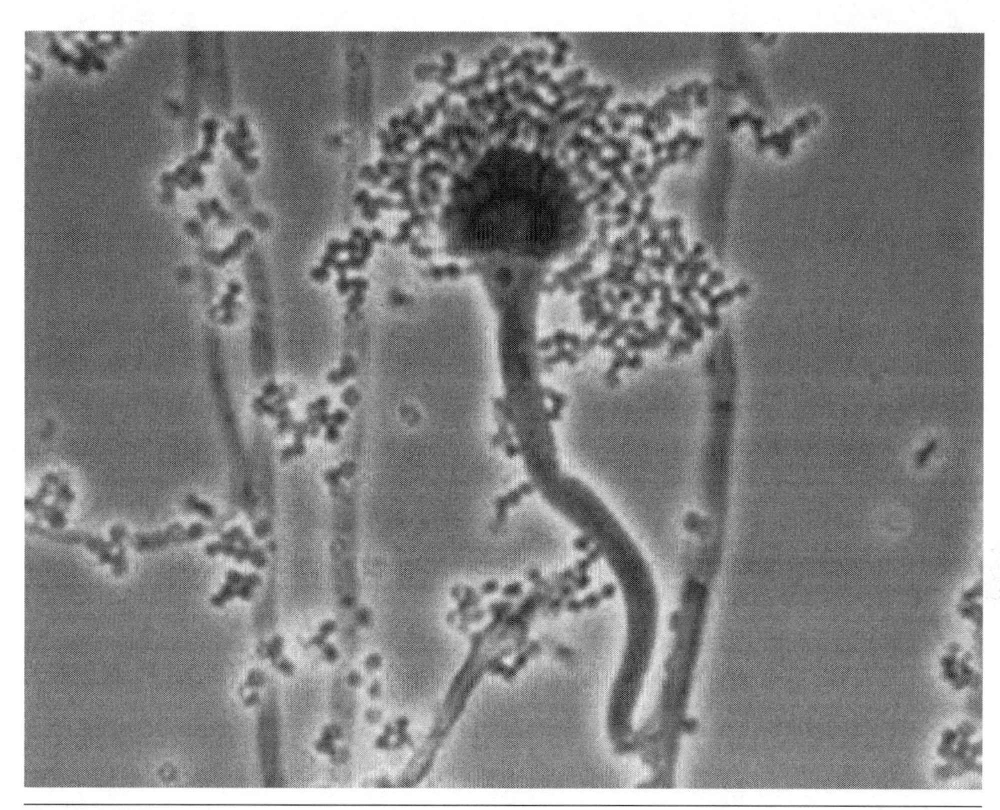

Figure 8.2 The mold spore *Aspergillus fumagatus* – microscopic and seemingly innocent – can attach itself to cells of humans with immune-compromised health conditions, such as asthma, and cause respiratory infections and even death
Photo credit: U.S. Department of Health and Human Services

perils" coverage for whatever is not covered by the original Builder's Risk Insurance policy. Coordination with the Owner is important in this regard, particularly in areas where flood, earthquakes, or tornados are a real threat.

Additionally, whether under Builder's Risk Insurance or "completed operations" insurance, the new environmental issue in many structures has become the presence of "mold" in many different but biologically unsafe forms. Although mold may not become an issue during construction, and therefore the exclusion to a Builder's Risk Insurance policy is not as important as it would be to "competed operations," nonetheless, the Contract Administrator should recognize the short- and long-term potential loss due to this "peril" that is increasingly excluded from the basic insurance policies.

The mold spore *Aspergillus fumagatus* – microscopic and seemingly innocent – can attach itself to cells of humans with immune compromised health conditions, such as asthma, and cause respiratory infections and even death.

Of course, the Design Professional must also be aware and recognize within the design process if the ventilation and flow of air throughout the project is adequate to prevent the build-up of moisture that might lead to mold in any areas. Courts may recognize that what caused the mold is where the ultimate responsibility will lie. If it is a design error or omission the Design Professional may be required to cover the loss, and the insurance of errors and omission may or may not apply. If it was caused by a latent defect in the construction, the

Contractor may be required to cover the loss. If, however, the mold was caused by an incident that is covered by insurance, then the parties may be able to apply against the insurance for the loss.

Examples:

1. The Design Professional might bear responsibility if the design failed to provide adequate ventilation in the showers of an army barracks. Mold build-up would eventually cause returning troops to require temporary barracks on another part of the post.
2. The Contractor could be responsible if substituted blowers in the duct system did not reach the proper cfm to ventilate the building and that was the reason why mold accumulated.
3. If a major storm damaged one of the rooftop units, and the drop in ventilation caused the mold to accumulate, then the insurance coverage for storm damage would probably eliminate the loss to the Designer or Contractor.

8.2.1.1 Time and amount

Builder's Risk Insurance is intended to continue through the duration of the project until its completion, when the Owner is able to occupy the building. Here again, it is important that the Constructor's Contract Administrator determines when exactly the coverage expires. Substantial Completion is usually defined as the point at which the Owner can use the structure for its intended purpose. If the Builder's Risk Insurance has expired at Substantial Completion, then the Contractor is not covered between the time of Substantial Completion and the final acceptance and closeout of the project. The 2007 edition of the AIA contract, for example, requires the Owner to obtain permission of the insurance company providing the Builder's Risk coverage to agree to the Owner occupying the structure. This is meant to protect the Architect as much as the Owner, since the voiding of coverage for the Owner could also void coverages protecting the Design Professionals.

It is also important for the Contract Administrator to recognize that some Owners doing renovation to an existing building may be covered for losses through the existing policy for the building. Once again, however, that policy may cover damage to the building, but does it cover the Contractor for all material, equipment, and any improvements made on the project at the time of the occurrence, particularly improvements made but not yet certified or paid for on pay applications?

In addition to making sure the Contractor is covered for all reasonable losses, the Contract Administrator needs to also determine the length of the coverage and the limits on the policy. Builder's Risk Insurance can be purchased for a period of time. If the project runs longer than expected (and most do), then the policy can be renewed, but it is important to make sure it has been renewed. This might be another block on the spreadsheet of contract issues the Contract Administrator makes when first reviewing the contract, or the Contract Administrator might want to put a note in a "tickler file" or on the calendar to make sure Builder's Risk Insurance is still in place after the initial expiration date.

Also, the amount of coverage is important. Would a total loss on the project be covered? Usually a loss due to a covered incident is more than just what it takes to put the material back in place. There are often removal and cleaning costs associated with a loss, and these exceed the original Unit Costs of construction. In other words $ 0.75 million of coverage on a $10 million project may initially seem to be a lot, but what would a total loss actually cost?

* * * * * * * * * *

The issue for Contractors and Subcontractors becomes: should additional Builder's Risk Insurance be purchased for any ordinary or special perils coverage? Builder's Risk insures the party that buys the insurance. The Owner will use Builder's Risk Insurance to make the Owner whole. The Contractor and Subcontractors may want to make sure they are covered to the extent they will need to be made whole.

* * * * * * * * *

8.2.2 Commercial General Liability (CGL) Insurance

Liability insurance in America first began simply as fire insurance. The protection of one's property against the catastrophic loss that a fire could cause was paramount. Before emigrating from England to found the colony of Pennsylvania, William Penn had experienced the devastating Fire of London and wanted his own city to be as "fireproof" as possible.

William Penn's original plan for a masonry city was his first defense against his beloved Philadelphia ever experiencing a fire so devastating as the 1666 fire that destroyed so much of London. In 1730, however, a huge fire began at one of Philadelphia's wharves and quickly spread across the street. Benjamin Franklin took that disaster seriously, and he began advising in his *Gazette* about how citizens should protect themselves and the city against another fire. He then brought forward the concept of creating a fire company.

Eventually, Franklin began discussing founding an insurance company for damage done by fire to houses. Specific construction would be required. He did not like the habit of carrying hot coals from room to room in an open shovel. He also thought it dangerous that some chimneys were constructed of wood and not stone or brick. He also suggested party walls should be fully bricked and plastered or mortared, without beams passing through, to prevent fires from spreading from one house to another as seen in one of the party walls of one of his properties in Figure 8.3.

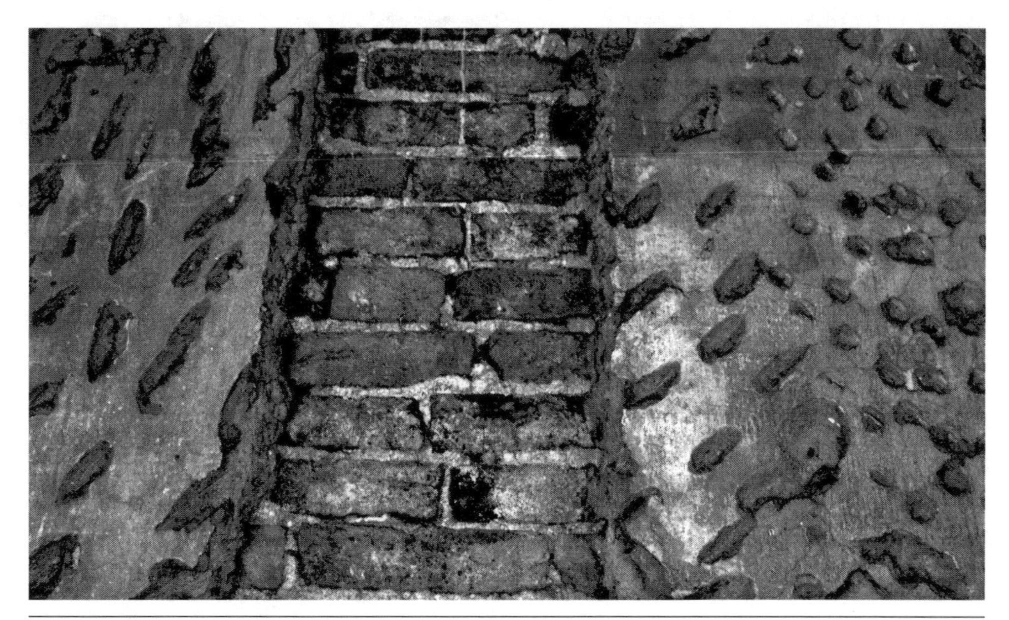

Figure 8.3 A party wall in one of Benjamin Franklin's properties
Photo credit: Charles W. Cook

Figure 8.4 Fire markers on exterior wall
Photo credit: Charles W. Cook

Fire companies with buckets and "pumpers" would be hired, or were attached to these insurance companies. The rumor spread that if they came to a house on fire that did not have one of their insurance markers outside, they would not fight the fire, but that is more fanciful legend than fact, since a fire in any part of the city represented a great danger to the rest of the city. In Figure 8.4 the tree insurance marker on the right was known as the "Green Tree," because unlike other companies, the company would insure a house that had a tree on its property.

Over time, it became obvious that more than fire could cause losses to buildings. Today, Contractors buy what is known as **Commercial General Liability Insurance (CGL)**. As mentioned above, included in their insurance coverage they may also want to have significant "completed operations" coverage. This is more and more important today, as many defects are not found until after the Constructor has completed the project and the Owner has paid for it in total. Perfection does not last forever, but if some tragic event has taken place well after the structure is completed, and if it is determined to be due to a latent defect unknowingly put in place by the Contractor, the loss could be catastrophic without insurance coverage. Basically "completed operations" insurance is an extension of CGL insurance.

The purpose of CGL is to protect against unexpected loss, but not to protect against improper workmanship. Generally CGL can be considered protection against third-party claims. Contractual requirements are not considered liability that is covered. The Contractor cannot protect against what the Contractor has agreed to do in accordance with the plans and specifications. CGL insurance is not intended to insure against poor workmanship or incompetence in managing the progress of the project. On the other hand, if during the installation or after the material is in place something happens that damages property or people, then that is when the insurance would be called upon.

If an insurance claim is made, the insurance company has the duty to defend the insured as well as indemnify that party against the claim. If there is the potential to settle the claim, then the insurance company will probably do so as expeditiously as possible. No further costs should be borne by the Contractor for an insurance claim, although it is quite possible insurance rates will go up in the future based on the "experience" of the loss by the insurance company.

Of course, when rates go up, Contractors often go shopping, and this should be a caution for the Contract Administrator. A good insurance agent will again help in regard to terms of coverage.

8.2.2.1 Terms of coverage

An insurance company may issue an **occurrence policy**, and so long as the policy was in place when the incident causing the loss occurred, the Contractor will be insured, regardless of when the claim is made. A "**claims made**" policy will cover incidents that occurred during or before the policy so long as they are reported during the term of the policy. Occurrence policies tend to be more expensive because the Contractor is paying for what is called the "tail" cost up front. The Contractor might leave the insurance company with unknown liability for events that will become losses in the future. The "claims made" policy, however, often has escalating costs with a built-in required commitment. The insurance company may require the Contractor to sign on for a period of years, and raise the rates over that specified period depending on the actual losses experienced.

By now it should be obvious that all Contract Administrators, whether working for the Design Professionals, Owners, Contractors or Subcontractors need to have proof of insurance from the other parties involved. Such proof is usually required by contract to be submitted prior to the start of work on a project. This adds an additional burden for the Contractor's Contract Administrator, since the schedule will often be so tight that Subcontractors suddenly appear to perform and there is no proof of insurance. Most Contractors, therefore require such proof to be submitted along with the signed contract prior to start of work, but all this entails another tracking chart or tickler file.

8.2.2.2 Defending claims

Whether for a Design Professional or a Constructor, the Contract Administrator faced with a claim will want to know who pays for the defense under CGL coverage. Usually the policy will be written that the insurance company will pay for the defense. In such a case, be certain the cost of the defense does not decrease the total limit of coverage. Sometimes the cost of defense will be an additional burden of the insured. The attorneys will be provided by the insurance company, but the actual cost will be outside the cost of the premium.

Regardless of who pays for the defense, the Contract Administrator should do as much as possible to provide whatever is needed to defend the claim successfully. What should also be recognized is that the cost of defense is probably the lesser of the overall costs. The insurance company, in lieu of defending a claim, may be more interested from a financial standpoint in settling the claim. As a Contract Administrator, you may find yourself in the difficult position of knowing the insurance company is talking a settlement to mitigate its own exposure as much as possible, while perhaps you believe you are not even responsible for any part of the loss.

8.2.2.3 Deductibles and policy limits

Insurance policies are written with deductibles, which means the insured will pay the first costs of the loss up to a certain amount. Sometimes, in order to negotiate a low premium, the deductible is quite high. This could mean some significant cash flow up front if a justified and significant claim is made. On the other end of coverage could be limits that were very low. A significant occurrence could put the loss over the policy limits and, once again, the Contractor is faced with a significant loss.

8.2.2.4 Additional insureds

In most contracts between the Owner and Contractor, there will be a requirement to name the Owner as an **additional insured**. Also there might be a requirement to name the Design Professionals, which would include the Architect and any consultants such as Engineers as additional insureds. AIA contracts require the Owner, the Architect, and the Architect's consultants to be named as additional insureds. The Contractor might then also require Subcontractors to name the Contractor as an additional insured. Generally an insurance policy is to cover the person paying for the policy against claims that might be made against the insured, but when another party is in a position to demand inclusion as an additional insured, they expect the same coverage as the person buying the policy.

This has become so common in construction that it usually does not increase the cost of the policy, since it has been factored in already over several decades of similar practice. However, a Contract Administrator for any party in the construction process must recognize that a simple certificate naming a company as an additional insured does not guarantee the coverage over time or to the extent one might expect. Policies can be cancelled and notice might not be given to an additional insured. In addition, it is possible that exclusions not listed on the certificate but contained in the policy will void protection for a claim not covered. Reading the policy itself is the only way to be sure coverage is adequate and conforms with what is intended. That again should become the responsibility of a good insurance agent.

Some Owners have also required resubmission with each payment application on a monthly basis as proof that insurance coverage is still in place. The Contract Administrator faced with such a requirement will need to alert the insurance company and also Subcontractors so that this additional burden for submission becomes routine and does not hold up applications. It becomes particularly burdensome when a new Subcontractor submits an application for the first time and is unaware of the requirement, thus prolonging the submission to the Owner.

8.2.2.5 Waiver of subrogation

An insurance company would generally like to retain the right to pursue compensation for a loss from the party that actually caused the loss, assuming it is not the party that the company has insured in the first place. On a construction project this can be quite important. Owners who have passed indemnification for their actions on to the Contractor, and Contractors that have passed such liabilities on to Subcontractors through flow down clauses, do not want to be defending claims after the fact. Both the AIA and ConsensusDOCS require waiver of **subrogation** for the parties, which means that once a claim is paid by an insurance company, no actions can be taken against a third party that may have been the original cause of the claim.

8.2.2.6 Timely notice

If a Contract Administrator receives notice of a claim, the insurance company should be notified immediately, even if the claim on the surface seems to be without merit. Failure to notify the insurance company on a timely basis might negate the coverage in some instances. In addition to notice, the Contract Administrator will want to provide as soon as possible any evidence or information in support of negating the claim or reducing the potential loss.

The Contract Administrator in gathering information must remember the insurance company is going to be looking in two directions: (1) Was the Contractor actually covered for such an occurrence by the insurance, and (2) does the claim have merit? In providing notice and assembling information, to the best of one's ability, the Contract Administrator will want to answer the first in the affirmative and the second, if possible, in the negative.

If the insurance company is not certain of the coverage or some portion of it, it might initially accept the claim with reservation. What it is basically doing is preserving its rights to drop out of defense of the claim if for some reason it determines the Contractor actually was not entitled to coverage under the policy. Although this may be rare, the Contract Administrator must again be careful. Close communication throughout the claim process, along with a strong and constant perception of any negative signals the insurance company might be sending, is essential should it attempt to drop the claim part way through. The more an insurance company remains involved in a claim, the more likely it is to take the defense to the end if it cannot make some settlement, but if the Contractor believes there might be an eventual withdrawal of the insurance carrier from the coverage, it is a good idea for the Contract Administrator to seek independent counsel from the company's own attorney. All of this can become quite expensive, but the potential loss can also be catastrophic if the Contractor or, in the case of an Errors and Omissions claim, perhaps a Design Professional, is suddenly left hanging without the representation of the insurance company's attorney.

8.2.3 "Umbrella" Insurance

Within the construction industry, parties seldom complain they have too much insurance. They might claim it is costly, but basically insurance costs are part of doing business, and a way is ultimately found to pass the costs on to the consumer. However, there are times when the coverage on a CGL or Builder's Risk Insurance policy is inadequate for the overall costs of the claim. Almost all Contractors carry what is known as **"Umbrella" Insurance** or sometimes **excess insurance**. The Contract Administrator should be aware if the company carries excess insurance rather than "Umbrella" Insurance, since excess insurance differs from "Umbrella" Insurance in that it extends coverage only on a specific part of the insurance or underlying policy, and excess insurance sticks to the exact terms of the original policy it augments. "Umbrella" policies will sometimes add additional coverage above and beyond the underlying policies. With an "Umbrella" policy, what the insurance company is doing for any policy it holds is stating that in addition to the limits of coverage as defined in the policy, the insurance company will simultaneously provide an additional amount up to the limits of the "umbrella" which protects all the subsidiary policies. For instance, if there were a terrible accident involving the company's truck and driver, and the automobile coverage for the company was only $500,000, but there was a $3 million "umbrella," if the claim was greater than half a million dollars, the "umbrella" would be used for the additional amount, up to the additional $3 million limit.

A CASE IN POINT

No amount of insurance can cover some things, and for that reason I have not used names in this story.

A very good Constructor had a contract to add a significant addition on to a store for a major national chain. The site was hundreds of miles from the Constructor's hometown, and his employees spent the week at a nearby hotel, and they would travel home for the weekends.

Close to the completion of the project, a long holiday weekend arrived, and the employees went home to enjoy the extra time off. The Building Manager had asked prior to their leaving if they could let him have the keys to their storage and staging area. That space would shortly become the store's own storage space and he wanted to move some pallets into the area and out of the store so they would not be in the traffic pattern of the customers.

The Contractor's Superintendent was of an accommodating nature, and so he agreed.

That weekend, after the store's pallets had been moved into the space, the door was not relocked. With the door open, a couple of children found their way into the space. Some open buckets containing caustic liquids seemed of interest and a game of who could splash whom the most soon ensued, and the children quickly suffered major burns on their hands and faces.

That Contractor is a very good friend of mine, and I know that finding who was at fault and covering costs through insurance for the disfigured children has never healed the memory for that Contractor of what happened.

8.2.4 Payroll insurances

States will collect a form of tax known as **unemployment insurance** or UI. This is a Federal requirement based on the Federal Unemployment Tax Act and rates are set by the United States Department of Labor. Each state is required to collect sufficient taxes to insure unemployment benefits for those out of work can be paid. The percentage of tax collected from each employer is determined by the experience of all employers within that industry. The higher the historical unemployment within an industry, the higher the rate will be to fund the unemployment insurance for those making jobless claims.

In addition, there is another insurance collected based on payrolls known as **Workers' Compensation Insurance**. Although perhaps the collection seems like a tax, Workers' Compensation Insurance is clearly an insurance against the potential cost of injuries to employees.

Workers Comp, as it is usually referred to, is also based on the historical experience of workers within an industry, but it is further based on the actual work each person performs. Carpentry, for example, is considered a more dangerous form of employment than accounting work. For that reason the Workers' Comp rate for a carpenter is higher than the Workers' Comp rate for an accountant.

Workers' Comp covers medical expenses as well as the cost to recuperate from the injury, or disability costs if there is a permanent loss of some faculty or function. Lost wages during recovery will also be paid for. In addition, if there was a tragic loss of life, there are death benefits for the survivors.

Workers' Compensation Insurance has an excellent purpose behind it. Many Contract Administrators faced with a claim against the CGL or Builder's Risk might want to challenge or defend the claim. Workers' Comp is intended to get money immediately to the individual in need without the delay of court litigation or negotiation of the claim. If the claim is false, relief can eventually be sought, but when employees are in desperate need of care and recovery, denying them the financial means to do so would be tragic and certainly a miscarriage of what American workers injured on the job, as well as their families, should expect.

There is, however, an important financial benefit to the employer. The injured party cannot collect additional monies in a lawsuit against the employer. The loss will be paid for in accordance with the set Workers' Compensation rates. However, there are two ways in which an employer can be held liable beyond the protection of the Workers' Compensation Insurance:

1. The employer failed to carry or pay for the Workers' Compensation Insurance.
2. The employer willfully disobeyed standard practices and thus intentionally put the worker in harm's way.

The Contract Administrator will need to be familiar with the specific laws in the state in which the employer is providing work. Note that each state's laws for Workers' Comp vary, and Constructors who work in multiple states will need to be careful that they follow the appropriate regulations.

One potential variant is that the injured worker can go outside the immediate employment and sue another party for loss and injuries. This is not the same in all states, but it is another reason Owners and Design Professionals demand indemnification from such actions against them. It is possible an employee working for the General who is injured by the Subcontractor's crane that fell on him might attempt to sue, if not the Subcontractor, then perhaps the crane Supplier. That would be a second-tier party from the original protection of the Workers' Comp for the Contractor.

8.3 Experience modifier

Experience keeps a dear school, but fools will learn in no other.

Benjamin Franklin

Statistics are maintained for all industries and for each trade within the industries. Being kept over a large number of employees throughout the country, the rates for each trade are based on the average claims made by employees that are injured. The higher and more expensive the losses due to claims for a particular trade, the higher the Workers' Compensation rate for that trade will be.

That process provides the industry average by trade. However, some employers recognize that safety is the crucial fourth leg of any project – the first three being (1) schedule, (2) cost, and (3) quality. In fact, safety should be considered the *sine qua non* ("without which nothing"), since without safety you do not do the other three. Unfortunately, some Constructors disregard or pay little attention to safety on their projects. It would not be fair to treat Constructors at the two extremes the same. Therefore insurance companies are allowed to calculate an **experience modifier** into the calculations for the costs of Workers' Compensation Insurance. The average experience modifier will be 1.00, which would mean the Contractor or Subcontractor with such an experience modifier would be charged the exact industry rate for each trade employed for the work they do.

The employer that has a very good safety record will have an experience modifier below 1.00. That modifier will then be used to adjust the rate of the insurance costs for that Contractor. And the employer that has a poor safety record will have a modifier greater than 1.00, and the cost for that employer will be above the average.

Examples: Let us say the Workers' Compensation rate for a mason is $9.77 per $100.00 of payroll the individual is paid.

1. The average employer with a modifier of 1.00 will pay $9.77 in Workers' Compensation Insurance every time he pays a $100.00 in masonry payroll. Note this is not take-home pay, but the amount of pay before taxes and other deductions.
2. The employer with a good safety record might have a .95 modifier: .95 x $9.77 = $9.28, so that employer will pay only $9.28 for every $100.00 of masonry payroll.
3. The employer with a less favorable safety record might have an experience modifier of 1.05. That employer will pay 1.05 x $9.77 = $10.26 for every $100.00 of masonry payroll.

Based on the above example you can see how financially important a good safety record can be. Many private Owners will not permit Contractors with a greater modifier than 1.00 to bid on their work. They realize that on a significant project involving a great deal of labor an elevated experience modifier means significantly higher costs.

On competitively bid work, the employer with the lower experience modifier in the example above has a distinct labor advantage over the employer with the above-average modifier.

Additionally, not only Owners, but also insurance companies and bonding companies will monitor the experience modifier rate (EMR) of Contractors. A poor EMR might preclude a company from wanting to insure or bond a Constructor. Bonding will be discussed in the next chapter, but without insurance or bonding, safety is indeed the *sine qua non* for the Contractor wishing to bid or perform work for many clients.

Exercise 8.2

Take a moment
You do the math

On a project that will have $1 million of masonry labor payroll, how much higher will the employer with a 1.05 experience modifier be over an employer with a .95 modifier just based on the Workers' Compensation costs? Base your calculation on the rate being the same as in the example above ($9.77/$100 of payroll).

Employer with 1.05 experience modifier: __________

Employer with .95 experience modifier: __________

Difference: __________

Discussion of Exercise 8.2

Your answer should have come out as an advantage just under $10,000 dollars the more safety-conscious employer already has over the competitor with a poorer safety record. That may not seem like much compared to the $1 million of overall payroll, but every advantage counts. That could mean an additional profit of $10,000. Also, this example was based on a contract that had $1 million of masonry payroll. That would be an exceptional masonry project, but there are many projects with $1 million of payroll among several trades on the project and several of them may easily have higher rates than the masonry trade.

Owners, therefore, are looking for the most cost-effective Contractors to do their work, knowing payroll insurances can save them significant amounts of money.

Also, a Contractor with an excellent safety record might easily have a better than .95 modifier. And Contractors with good safety records spread the attitude of a safe jobsite to the Subcontractors they attract to work with them. Their safety modifiers fall also.

In short:

Safety pays!

The Contract Administrator will want to recognize safety is paramount when managing the project. It is an important financial factor. It can be the reason an Owner does or does not invite a Contractor to bid on a project. And, most importantly, contrary to Cain's attitude, we are our brother's (and sister's) keeper. It will always be our day to "watch them." We want to watch them go home just the same as they came to work.

8.4 Fraudulent record keeping

Things gained through fraud are never secure.

Sophocles

With hundreds and thousands of dollars involved, it is unfortunate that some Contractors find ways to cheat the system. If they are caught, the failure to obey the law can lead to severe fines and imprisonment, but the chance of getting caught has not always been as great as the desire for some to not pay the proper payroll insurances. This can also be a serious problem for the worker who is not covered and sometimes cannot then claim for costs incurred during hospitalization and recovery.

8.4.1 Misclassification

The easiest method of fraud in reporting payroll insurances related to Workers' Compensation is misclassifying the worker for what is being done. There are several trades involved on every major construction project. Some have much higher Workers' Comp rates than others. It is very easy after the fact to classify an individual as having done the lower classified work for hours that they were actually doing the higher rated work.

Example: Working on a concrete structure forming and pouring walls in one state has a Workers' Comp rate of $11.88 per $100.00 of payroll. Working on sidewalks and floors in that state is only $5.72/$100.00. If no one from the government is there to monitor (and they are not), then by the time reporting is due for payroll records, the person who worked more time on the walls has, remarkably, spent almost all the time on the slab.

In the example above, the Contractor is saving over $6.00 for every hour such an employee can be misclassified, and if you multiply that out over an entire workforce, that can be significant.

Additionally, the rates of some classifications are even higher than those in the example. Workers on structural steel erection might have a rate of $22.00 +/- per $100.00 of payroll. Finding ways to avoid that cost can lead to major savings for the Contractor that is not "playing by the rules."

8.4.2 Subcontract vs. employee

Similar to Workers' Comp misclassification is the payment of an employee as a Subcontractor. The purpose of this strategy is to remove the necessity of paying any payroll taxes whatsoever. The amount paid can be handed to the individual as a lump sum check for work done on the project. The individual receiving that check is supposed to properly report the income on his or her own tax forms. That seldom is the case. The Contractor working in that manner is able to escape scrutiny with one exception. By law the Contractor is expected to send a 1099 form to each "Subcontractor" who earned more than a certain amount set by the IRS, such as $650.00. This form is strictly intended for Subcontractors that are not registered with the IRS as a corporation and thus reporting income in that manner.

Again, legitimate Contractors would do this already, but it is often ignored by those who are acting outside the rules.

8.4.3 "Pay under the table"

The expression to "**pay under the table**" refers to the act of paying a person directly in cash so there is no record or trace of the payment. Paid in cash, the person(s) does not have taxes removed from his or her pay, but his or her pay is usually much less than the standard pay for someone doing the work legally. In addition, however, the Contractor paying in such a method does not have to pay any of the payroll taxes or insurances that would be necessary for such a worker. Paying "under the table" can lead to great savings, but also very severe consequences if one is caught. Some Contractors working in this manner have even told their workers that if they are injured, that is the day they will be put on the payroll so they will be covered for Workers' Comp. However, the actuality of that happening has not been 100%.

Such practices, of course, also raise the cost of Workers' Comp for those who are doing what is right by their workers, and failing to pay payroll taxes not only affects all taxpaying citizens, but it deprives local, state, and Federal governments of needed revenue to improve services, including repairs to our infrastructure, and to provide education for all our children.

The subject of ethics will be discussed in our final chapter. Defrauding the government may be a means to financial pleasure, but it will ultimately have other consequences even if one is not caught. There is for the Contract Administrator attempting to work in such an environment one additional word of advice. Local, state, and Federal governments are now recognizing the incredible loss that misclassification and other fraudulent reporting is costing. More and more rules and monitoring will be taking place. Even if not caught, the stress on those that have to constantly look for who might be watching will be increasing. If health is a part of happiness, the short-term gain may no longer be worth the long-term penalty. If you live each day as though it is the last day of your life, then probably someday sooner than later you will be right.

INSTANT RECALL

- Insurance rates are determined based on the potential loss due to occurrences that will happen over a period of exposure.
- There are some important and basic types of insurance that Contractors should address in some way, and they are usually defined by the contract itself, including the dollar amount limits of coverage required:
  - Builder's Risk Insurance
  - Commercial General Liability (CGL)
  - "Umbrella" Insurance
  - payroll insurances.
- One other form of insurance that may not be called out in the contract, but more and more Constructors are finding it necessary is Errors and Omissions Insurance.
- Unless the insurance policy, itself, is reviewed, the amount of coverage may be inadequate, exclusions may be in place that should not be, and even the rights of an additional insured may be different than what is expected or a contractual obligation.
- It is also important to recognize that insurance is not meant to cover acts of incompetence by the Contractor or any of its agents.
- Builder's Risk Insurance is meant to cover accidental or unexpected damage to the project while it is under construction.
- Typically Builder's Risk Insurance covers losses or damage due to:
  - fire
  - lightning
  - wind (but not necessarily tornado)
  - theft and vandalism.
- But Builder's Risk Insurance will often exclude damage due to:
  - earthquake
  - flood
  - acts of war
  - damage directly due to a tornado
  - mold.
- Builder's Risk Insurance is intended to continue through the duration of the project until its completion, when the Owner is able to occupy the building.
- Basically "completed operations" insurance is an extension of CGL insurance.
- The purpose of CGL is to protect against unexpected loss, but not to protect against improper workmanship.
- Sometimes the cost of defense will be an additional burden of the insured.
- Regardless of who pays for the defense, the Contract Administrator should do as much as possible to provide whatever is needed to defend the claim successfully.
- Reading the policy itself is the only way to be sure coverage is adequate and conforms with what is intended.

- The Contract Administrator, in gathering information, must remember the insurance company is going to be looking in two directions: (1) Was the Contractor actually covered for such an occurrence by the insurance, and (2) does the claim have merit?
- If the insurance company accepts the claim with reservation, it is basically preserving its rights to drop out of defense of the claim if for some reason it determines the Contractor actually was not entitled to coverage under the policy.
- If the Contractor believes there might be an eventual withdrawal of the insurance carrier from the coverage, it is a good idea for the Contract Administrator to seek independent counsel from the company's own attorney.
- Workers' Comp covers medical expenses as well as the cost to recuperate from the injury, or disability costs if there is a permanent loss of some faculty or function. Lost wages during recovery will also be paid for. In addition, if there was a tragic loss of life, there are death benefits for the survivors.
- There are two ways in which an employer can be held liable beyond the protection of the Workers' Compensation Insurance:
  - The employer failed to carry or pay for the Workers' Compensation Insurance.
  - The employer willfully disobeyed standard practices and thus intentionally put the worker in harm's way.
- The higher and more expensive the losses due to claims for a particular trade, the higher the Workers' Compensation rate for that trade will be.
- The average experience modifier will be 1.00, which would mean the Contractor or Subcontractor with such an experience modifier would be charged the exact industry rate for each trade employed for the work they do.
- Owners are looking for the most cost-effective Contractors to do their work, knowing payroll insurances can save them significant amounts of money.

YOU BE THE JUDGE

An Insurance Company (USAA) denied coverage to Homeowner (Davis) citing an exclusion in the "all risk" policy of any damage "caused by, resulting from, contributed to or aggravated by any earth movement, including but not limited to earthquake, volcanic eruption, landslide, mud flow, earth sinking, rising or shifting. . . " Davis maintained he had bought an "all risk" policy, and he showed that the damage was caused by a Contractor that failed to correctly reinforce structural elements of the foundation and had not graded the subsurface soil properly. The resulting settling damaged the house.

How do you find:

Against the Insurance Company

Against the Homeowner

Against the Contractor

IT'S A MATTER OF ETHICS

Some Contractors save money by not paying insurances and taxes on employees they treat as Subcontractors. It is estimated that thousands of dollars are lost to townships and up to billions of dollars are lost to states because employees are treated as Subcontractors and Workers' Compensation Insurances and payroll taxes are neither reported nor paid. What would you do if you believe a competitor is treating some or all of its employees in this way and not paying taxes? What if you believe a Subcontractor on your project is doing the same? What if you think someone in your own company is being paid in this manner? Do your actions change and, if so, why?

Going the extra mile

1. Study the "You be the judge" exercise above, write your opinion, and then go to the Companion Website to determine how your judgment agrees (or not) with the actual judgment by the court.
2. **Discussion exercise** (to yourself or among classmates): Brainstorm and discuss how many ways effective safety practices can help a Constructor.
3. Effective use of the correct vocabulary improves communication and promotes respect among colleagues. The reader should become familiar with the key words listed below that have been used in the chapter. Definitions can be found in the Glossary. Additionally, the student is encouraged to use the flash card exercise in the chapter's Companion Website as practice for possible quiz or exam questions.

Key words

Additional insured	Insurance
Bond	Occurrence policy
Builder's Risk Insurance	"Pay under the table"
"Claims made"	Subrogation
Commercial General Liability (CGL)	"Umbrella" Insurance
Errors and Omissions Insurance	Unemployment insurance (UI)
Excess insurance	Warranty
Experience modifier	Workers' Compensation Insurance

CHAPTER 9

Bonds

An Owner might look upon a Contractor's bond as a form of insurance, but a bond really is not insurance, and the Contract Administrator needs to recognize the differences and the consequences associated with defaulting on a bond.

Student learning outcomes

Upon completion the student should be able to. . .

- **recognize the different types of bonds**
- **distinguish between Responsible and Responsive bidding**
- **recognize the importance of the three Cs of bonding – character, competence and capital.**
- **understand issues related to bonding capacity**
- **know the effects of the Miller Act and Little Miller Acts**
- **understand issues and obligation in relation to defending a bond claim.**

Please note: Although this chapter deals with bonds and bonded projects, many of the points discussed regarding the rights and obligations of Owners and Contractors are also relevant to non-bonded projects. For example, causes for termination and defenses for improper termination can be just as valid in non-bonded work. The Contract Administrator will need to recognize that provisions within the Contract will govern in non-bonded work just the same as if the project was bonded. The bond does not change the contract. It merely adds an additional layer of protection or assurance for the Owner in regard to how the contract might ultimately be executed and completed.

9.1 Bid Bonds

In all our contacts it is probably the sense of being really needed and wanted which gives us the greatest satisfaction and creates the most lasting bond.

Eleanor Roosevelt.

Bonds are going to be issued on behalf of a Contractor with two levels in mind: (1) a single project limit, and (2) the total amount of work (bonded and non-bonded) the Contractor has currently under contract.

Examples: A Contractor might have a limit of $1 million for a single project, and an overall limit of $5 million for total work being performed:

1. If the Contractor has $4.5 million worth of work currently underway, then the bonding company would most likely refuse offering a Bid Bond for a $1 million project.
2. If the Contractor could show the projects underway were so far along that the actual value of the work remaining was well under $4 million dollars, then the bonding company might agree to a $1 million Bid Bond.
3. If the Contractor had well below the total bonding capacity of all the projects (e.g. $2 million of work on hand), but wanted to bid a $1.5 million work, the bonding company might refuse providing a Bid Bond based on the project being 50% more than the Contractor is approved for a single project. The bonding company may appear to be arbitrary, but it knows from experience that jumping ahead in capacity too quickly can lead to significant difficulties, one of the most crucial being cash flow.

Perhaps because **Bid Bonds** usually cost nothing, the Contract Administrator often takes these bonds for granted. However, they are where everything begins, and in that sense they are the most important in securing.

For public works and for some private work, the Owner will require a Bid Bond to be submitted with the original proposal. The Bid Bond will be the assurance from a financially reputable company that if the Contractor that submitted the proposal is not willing or capable to sign a contract based on the bid proposal, the bonding company will pay the difference between the submitted bid of the Contractor withdrawing from the bid and the next lowest Contractor. There is usually a limit of exposure, such as 10% of the defaulting Contractor's submitted bid.

In some instances, Owners require certified checks or cash deposits in lieu of a Bid Bond. This practice has become less popular since it can tie up precious funds of a successful Contractor for a period of time (and often the certified check is kept on the second and third Contractors, who are usually without hope of eventually being awarded the work). If a Contractor does not fulfill the bid proposal the certified check is used to reach an agreement with the next lowest Contractor. Fortunately, most Owners have seen that this does put some good Contractors on the sidelines for bidding other projects, since significant cash in a certified check prevents them from bidding another project while that money remains unreturned. Such Owners have turned to Bid Bonds instead.

For the Constructor a Bid Bond represents a major commitment and investment. Bonds are issued on the basis they will not be used or called upon. A major mistake during the bidding phase of a project, however, is always a possibility, and so an Owner might call upon the bonding company to comply with the terms of the Bid Bond if the Constructor believes the mistake is too great to actually complete the project.

If a bond were insurance, that would be the end of it, but a bond is not insurance, and so if a bond is called, that is just the beginning. The bonding company is going to seek repayment from the Contractor for any monies lost.

In essence, the difference between insurance and a bond lies in the fact that the cost of insurance is based on actuarial tables that take into account that in the world of construction accidents and events will occur at a rate that is tied to historic statistics of such incidents. The rate of insurance, therefore, is based on such probabilities over a given period of time for the policy.

In a bond, the failure of the Contractor to perform properly is not expected to happen, and therefore there are no historic statistics on which to base the cost of the Bid Bond (or other bonds).

Although it is not supposed to happen, it does, so how do bonding companies function differently than insurance companies?

9.1.1 Three Cs of bonding – character, capital, competence (or capacity)

As mentioned, the Bid Bond is where it all begins. Before issuing even a Bid Bond, the bonding company is going to carefully analyze the ability of the Contractor to perform the work for which they are seeking bonding. Bonding companies are not going to bond a housing Contractor to suddenly build hospitals. They do not expect a hospital Contractor to be qualified to build a bridge. They would not expect a bridge builder to start erecting wind farms. Although it might be possible such Contractors could make the transition, the bonding company will justifiably be initially skeptical and more research about the company would be required.

The research on each Contractor seeking bonding is based on the **three Cs of bonding** – character, capital and competence. Competence is sometimes also referred to as capacity. At any given time in the world of bonding one or another of the three Cs might have a greater value. Also, it is possible, based on certain Contractor characteristics, that one or another of the three Cs will have more influence on the bonding company's decision.

Examples:

1. When bonding companies have had a particularly difficult period financially, they may place greater emphasis on the capital available to the Contractor they are bonding than they would do when the economy has been very good.
2. A Contractor with an exceptional track record in a particular kind of work might be given a larger bond than usual for a project that is within that Contractor's expertise.
3. The character of a Contractor (honesty, integrity, reputation) can sometimes convince bonding companies to stretch bonding capacity a little more than would normally be the case related to the other two Cs.

In some respects, it should be mentioned that public Owners and others who require a Bid Bond from a Contractor's bonding company are actually having the bonding company pre-qualify the Contractor before the bid submission. The bonding company is required to research the Contractor, and if it believes the Contractor capable of submitting a bid on a particular project, the Owner has the assurance that some research has been done and they should expect the contract obligations can be performed by the Contractor.

9.1.1.1 Character

It should be stressed that everything begins with character. A reputable bonding company is not going to deal with a Contractor that cannot be trusted. The bonding company is going to ask for financial statements related to the company, and these have to be relied upon as truthful. A principal in a construction firm that might "doctor" the books or falsify something as important as percentage complete project reports can be the reason a bonding company would lose significant money if it made the mistake of bonding such a Contractor.

Moreover, in order for the Contractor to obtain bonding, the bonding company is going to want to secure the assets of the principals of the company. If the principals of the company are not willing to risk their own personal wealth, then the bonding company is going to reduce the amount of bonding the company can have. In addition to the principals of the company, the bonding company will also expect the spouses of the principals to sign their assets over to the bonding company. In the event a bond claim is made and the bonding company is required to pay for what it has assured the Owner that the Contractor could perform, then the bonding company will go against the company's assets first, then the principals' and their spouses' assets.

Of course, if the Contractor has defaulted, the bonding company cannot act recklessly while trying to complete a bonded project, but generally the cost for completing a project is going to run higher than it would have done, had the Contractor been able to continue and finish without defaulting.

Needless to say, this form of personal indemnification gets the attention of the principals and their spouses. It is also the reason many Constructors are seen as either extraordinary entrepreneurs, placing their own "fortunes" on the line, or gamblers willing to risk it all for another shot at something never done before.

Either way, the bonding company is going to want to know that the character of the Contractor can be trusted, and that the information the bonding company obtains from the Contractor is truthful and complete. The bonding company in the end does not want to gamble. It wants no risk on its part. In addition to the Contractor's personal qualifications, it also wants to have confidence in the ability of the Contractor to put an estimate together.

That leads to another C – competence.

A CASE IN POINT: CHARACTER

Hank Neiman had built a strong and prosperous renovation and repair contracting company for himself. When his step-son died, he had no one to leave it to, so he began negotiations with the son of his old family friend.

Robert Cook and Hank Neiman eventually came to terms, with Hank staying on for a period of time during the company's transition. Robert Cook's background was more heavy and industrial, having built power plants and other industrial structures around the country. The low-volume renovation and repair work did not hold his interest or ambition for long.

Soon Robert was seeking larger projects, and eventually these would be in the public sector, something Hank Neiman had never done. Once the "specter" of personal indemnity with bonding arose, Hank Neiman decided it was time for an earlier than originally agreed-upon exit.

Under the leadership of Robert, the company continued to serve clients well, but the economy was not favorable for construction at the time, and growth and prosperity was being hampered. Larger contracts were just out of reach. The Reliance Insurance Co. served as the bonding company for R. S. Cook and Associates. (Reliance had played an important role in bonding history, as you will discover during the "You be the judge" exercise at the end of this chapter.) Without sufficient capitai, the main criteria for bonding R. S. Cook and Associates was always the character of Robert Cook, himself.

Then, one day, something would happen that would shake the confidence of the agents at Reliance, and they would be concerned the Bid Bond they had issued was in jeopardy.

(To be continued)

9.1.1.2 Competence

Competence is often also termed capacity, usually referring to the equipment and personnel available to do the work that is being considered for bonding. A Contractor bidding on highway work would need far different equipment than someone intending to erect a skyscraper. A Subcontractor that is required to provide bonding for a concrete slab project would need much different personnel and equipment than a finish carpentry Contractor who intends to install fine casework and wainscot on a floor of executive offices.

Competence to do work begins with knowledgeable and experienced people to put the estimate together. The bonding company will want to know that the estimator or estimating team has had experience in the type of work that is being bonded.

Prior to bidding the work, if the cost increases over the expected amount, the bonding company may withdraw its Bid Bond, so it is important for the estimator to keep everyone informed if the project seems to be going over what has been anticipated.

If, after bidding the project, there seems to be a discrepancy in pricing between the low bidder and the rest of the bids, the bonding company may even require some justification before issuing a performance or payment bond.

A CASE IN POINT (*CONT'D*): COMPETENCE

Robert Cook had hired Lee, an individual with a strong background in public transportation work. Lee knew how to estimate and construct for light rail, railroad, and other transportation projects. Recognizing his competence, Robert almost immediately agreed to let Lee bid on a project for the South Eastern Pennsylvania Transportation Authority (SEPTA). That project was won, it was completed on time and within schedule, and Lee proceeded to estimate more work for SEPTA. The renovations and repairs Contractor was definitely branching out into new areas – something bonding companies often question, but both the background of Robert and Lee's own history were satisfying the bonding company that the competence existed on the team to do the extremely different work from what the company had historically done.

Shortly, however, Lee submitted a bid on a very important project for SEPTA. It would be the subway station that connected to the Gallery – America's first urban mall. This had been the dream of world-renowned city planner Ed Bacon. The importance of completing the project on time was paramount, but when the bids were opened, another concern suddenly arose.

R. S. Cook and Associates was not just the low bidder, but they were *extremely* low. All the other bidders were much higher, and everyone was wondering if Lee would be able to prove the major mistake he had made was a mistake in calculation and not an error in judgment. If he could, he would be able to withdraw the bid. What SEPTA would be allowed to do with the Bid Bond might be another matter.

At this point the competence of R. S. Cook and Associates, Inc. to perform the work for the submitted price was in question.

(To be continued)

9.1.1.3 Capital

Of course, money is a very critical element in order to figure bonding capacity. With the indemnification of the assets of the company to the bonding company, along with the psychological indemnification of the assets of the principals and spouses of the principals, the bonding company will be able to calculate just how much would be at stake on any given project that could easily be reimbursed by seizing the assigned assets. Along with a single project, the bonding company will also look at the entire amount of work the Contractor is doing, and most importantly, the current balance sheet showing the profit or loss the company is experiencing on all its projects.

Basically, the bonding company does not intend to have a loss based on the first two Cs – character and competence – but it will make certain that it has enough capital in reserve that if there is a loss, it can recover all the monies it has to expend. This is the theory. It does not always work that way, and that is why there are good years and bad years in the bonding industry.

If there is a loss during the Bid Bond stage, the bonding company will not expect to lose the entire bid proposal amount. It is bonding just the amount necessary for the Owner to sign the next lowest Constructor to a contract. It is possible that the next lowest bid could be incomplete in some way, and, therefore, unresponsive. In such a case the next qualifying bidder would be the one awarded the contract and the bonding company would be expected to make that difference up between its bonded client's bid and the next qualifying bid.

A CASE IN POINT (*CONT'D*): CAPITAL

Lee denied he had made a mistake.

SEPTA hesitated awarding the contract to such a low bidder.

Based on the difference in the low bid and all the other bids, the capital necessary to complete the project if Lee had made a mistake seemed greater than defaulting on the bid proposal.

The bonding company questioned whether just defaulting on the bond and seeking compensation from the personal indemnification of Robert Cook would be less costly than writing a performance bond for the project and then expecting and thus waiting for R. S. Cook and Associates, Inc. to go bankrupt and thus require the bonding company to assume the cost of completing the project.

Although the agents for the bonding company were still confident of the character and competence of Robert Cook, there were grave doubts.

However, in just two weeks the estimate was validated by a private outside firm.

Budget considerations are always a concern for a public agency, and SEPTA was no exception, so it could not ignore awarding the low bid if it had the performance and payment bonds in place.

Capital became the final issue for the bonding company, and Robert Cook's assets were stretched to the limit. It was then that Lee came forward and pledged his entire savings, including what he had put away for his retirement (no IRAs or 401s existed at that time).

The amount was not enough to cover the shortfall, but the gesture was more than enough to convince the bonding company (armed with the independent validation of the estimate) that R. S. Cook and Associates, Inc. could do the work.

And we did.

Sometimes the key to estimating is seeing a way to do a project that no one else imagines. The project was completed on time, within budget (along with a very nice profit), meeting standards of quality and doing it all safely.

9.1.2 Responsive and Responsible bid

There are basically two ways a bid must qualify. It must be **Responsive** and the Contractor must be **Responsible**.

9.1.2.1 Responsive bid

A Responsive bid is one that has included all the responses required in the bid proposal. These can include acknowledging all addenda (even if an addenda that was not acknowledged would not have changed the price), or properly filling out all the blanks and signing the proposal, or including the Bid Bond, itself, or meeting necessary Minority Business Enterprise (MBE) and Women Business Enterprise (WBE) percentages of involvement in the project, or acknowledging any statutory and regulatory requirements.

If a Contractor includes exclusions not requested by the Owner, this would also be a reason to find the Contractor "nonresponsive." A Contractor often feels the need to do this, and there may be good reasons. The Contractor has found that a specified piece of equipment will not work as intended, or the schedule of construction needs to be modified, or the Contractor will try to do its best to meet MBE/WBE requirements but was not able to do so at bid time. Or, a Contractor might be pricing the project during times of uncertain or volatile conditions for certain critical items, such as fluctuations in steel pricing that occurred earlier in this century (and, based on the rising costs of iron ore, that may happen again). Then Contractors feel the need to protect themselves against such unknown costs. Unfortunately, these are all causes for losing the bid for being nonresponsive. So it is best, if possible, to have those conditions addressed by addenda prior to bid submission.

This may seem unfair toward the Contractor who attempts to qualify the bid in an important way, but the Owner also has a responsibility to all the other Contractors. If a Contractor has the chance to qualify the bid, and then knows what all the other prices are in a public opening of the bids, then the Contractor would have the opportunity to fine-tune the price based on what has been submitted by everyone else. Basically, no change – omission or addition – to a proposal should be allowed that would give any one Contractor a competitive advantage over anyone else submitting a proposal.

9.1.2.2 Responsible bid

A Responsible bid is one by a Contractor that can in fact do the work the project will require. By being bonded, the bonding company is basically verifying the Contractor is responsible. This again goes back to the act of assuring the Owner that the Contractor can be counted upon to perform the obligations of the contract if it is awarded based on the proposal submitted. The Owner relies upon the bonding company to have made certain the Contractor has the knowledge and integrity to commit to the terms of the contract and abide by them. The Owner also wants to be assured the company has the experience, personnel, equipment, and, most importantly, financial resources to see the project through to an on-schedule conclusion.

That said, the Owner will ultimately look to the performance and payment bond as a sort of insurance, even though they really are not. Still, at the Bid Bond stage, provided there is no difficulty with the low bidder, the Bid Bond is really just a form of assurance.

However, there are two cautions the Contract Administrator should be aware of at the Bid Bond stage:

1. It is at the consent of the surety that a performance bond and a payment bond are issued after the Bid Bond has been submitted by the low bidder. The surety seldom charges for

a Bid Bond, in part because unless the bid is the lowest qualified bid and accepted for a contract there is no income flowing to the Contractor for putting together the proposal – usually an expensive proposition without the added cost of a Bid Bond. Language in the agreement between the Contractor and the bond provider will usually allow the bonding company to back out of providing the performance and payment bonds if it feels there is an irregularity, or some issues have arisen to suggest the Contractor is at risk. If this occurs, the Contractor may have a promissory estoppel argument. Just as with a Subcontractor's proposal, the Contractor relied upon the bonding company when it provided a Bid Bond to also provide the performance and payment bonds if they were successful. Such an argument may work, but close relations and clear communication from beginning to hopefully "never ending" with the bonding company is a far better strategy for the Contract Administrator.

2. One caution related to "responsible," however, is that the bonding company must also be a responsible bonding company. Public Owners often publish a list of approved bonding companies. There have been some instances when fraudulent or unacceptable bonding companies have been the cause for rejection of a bid as being "nonresponsive" due to a bonding company being considered "irresponsible."

Figures 9.1 and 9.2 Subway station, under construction and completed
Photo credit: Charles W. Cook

From a large hole in the ground to a finished subway station, sometimes effective contract administration begins after the bid and before the contract is signed. Particularly in public work, it is sometimes necessary to convince the Owner that the Contractor can do the project. In this case, once the low bidder was deemed capable of performing the work, a rather interesting irony developed. Although the actual opening of the Gallery – a revolutionary downtown shopping mall – was delayed due to the stores' not being ready, the subway station connecting to the mall was finished on schedule. Originally believed to have made a bid that was too low, the Contractor excavated and built the new station on time, within budget, meeting standards of quality, and without a single lost-time accident.

Exercise 9.1

Take a moment
Minding the three Cs

Sometimes the best way to avoid a bad situation is to anticipate it and then take corrective action before it becomes a problem. To do that you need to know what could be a problem in the future. Once again, this is catching a challenge when it is a butterfly and not a full-blown hurricane. Below are listed the three Cs. Write one potential issue for each why a bonding company might deny a Contractor a performance bond after they have already issued a Bid Bond:

Character:

Competence:

Capital:

Discussion of Exercise 9.1

We must keep the communication lines open and that becomes a part of the character bonding companies are looking for when they provide bonding to a Contractor in the first place. Below are partial lists of reasons why a bonding company might hesitate to issue or even deny a Contractor a bond related to each of the three Cs:

Character

- The Contractor exceeded the approved limit set by the bonding company for any one project when submitting the bid. (*Solution*: Whenever a project appears close to the approved limit, keep checking and keep the bonding company advised of the potential increase. Unfortunately, some Contractors will have to not bid a project if the amount is not approved by the bonding company on a timely basis, but that would be better than bidding it, being low, and then not being able to furnish a performance and payment bond. Also, if you know soon enough, forming a joint venture can often increase the bonding capacity of the combined entity over what the Contractors individually could achieve.)
- A project or projects that were reported by the Contractor to be doing well are now known to be in trouble for schedule and/or costs, based on letters from concerned Owners or Subcontractors to the bonding company. (*Solution*: Credibility is essential. If a bonding company hears the good, the bad, and the ugly from you, it will be more inclined to accept your explanation than if it learns of problems from others. If a Contractor is having difficulties, it is better to have the bonding company as an ally than a reluctant partner,

looking for how it can best keep itself whole. Even worse, you do not want the bonding company as an adversary intent on securing its position against your assets. Some bonding companies can actually bring expertise to a situation that might be new to you, and working together, they can help eliminate difficulties down the road. They have a history of working through difficult circumstances and tough economies, so keep them as a partner through mutual trust.)

Competence

- The bid proposal submitted was exceptionally lower than the estimates of everyone else who bid the project. (*Solution*: Immediately check for a bid mistake. Remember there are circumstances under which a bid can be withdrawn without fault. It has to be a genuine mistake and not an error in judgment. If you are certain of your numbers and that you can do the work for what you presented [perhaps you have a technique or method no one else had considered], then you may need to have the estimate independently verified for the bonding company, detailing how you intend to perform the work and why it can be done for what you estimate.)
- You lose a critical member of your company. (This can happen to the best of companies. It can be tragic, through illness or accident, or a rival Contractor may have made an offer the employee should have refused but could not. The larger the company, the easier it is to replace individuals, but all Contract Administrators should be thinking of succession and growing the team. Good mentoring programs should replace the silos approach to project management. The more confidence a bonding company has in the team, the less likely it will consider the season over if one player is no longer in the lineup. For that very reason a one-person operation will always have more difficulty getting significant bonding than a larger corporation. As part of character, the bonding company will naturally be concerned about health of the single Constructor.)

Capital

- The Company is suddenly advised of an IRS or other government audit. (*Solution*: Audits in themselves are not reason to panic. Keeping two sets of books and not letting the bonding company know that someone has been "cooking the books" is a reason to be concerned. The bonding company that discovers your profit is actually a loss on your real balance sheet is going to immediately take positions directly behind the Federal government. Cheating on your bonding company is second only to cheating the government. The best way not to do that is not to do that.)
- Mechanics Liens or notices against payment bonds are being served. (*Solution*: Sometimes there will be problems that have legitimate explanations. If you have been having difficulty with a particular Subcontractor and you have reason to withhold payment, that needs to be part of your conversation with the bonding company before it comes to you for an explanation based on what the Subcontractor has already done. Not all liens are valid or legitimately filed. Not all notices against a payment bond are correct or accurate. It is easier to anticipate the issue than to have to explain it after the fact. Remember, your agent has someone he or she has to report to also. When that boss hears of a problem, the immediate response is, "What's wrong with your client?" We cannot fault the boss for reacting as humans usually do. If your agent is already apprised of the possibility of such a situation, then the boss might already be on your side, or your agent can quickly quell the problem for you.)

- The bank says you have maxed out your line of credit. You do not have the funds to either start a new project or complete your work on hand. (*Solution*: Cash flow is essential. As the Contract Administrator, make sure all the projects are current on payment. Then, before bidding on any project, take a careful analysis of what is on hand for cash flow, should you be awarded the project. In this case a close relationship with the bank is also critical. It needs to know the true profit and loss on your balance sheet and for the projects you have currently underway. Some Contractors have been known to think "I need to get another project so I will have enough money to finish the one I already have." Such a solution only digs the hole deeper. The bank is certainly looking after its own interests, but in doing so it is taking a careful analysis of what you actually can perform effectively. A long-term relationship with a good banker is the equivalent of a valuable member of your "board of directors." Take the banker's counsel wisely, and the company will grow – gradually but successfully.)

9.2 Performance bonds

I know the price of success: dedication, hard work, and unremitting devotion to the things you want to see happen.

Frank Lloyd Wright

A performance bond is the Owner's safety net in case something drastic goes wrong with the Constructor who signed the contract. On Multi-prime Contracts each Contractor will be required to furnish a bond, and this can definitely complicate the schedule and other matters on the project if one of the other primes does go into default on the bond. The Contract Administrators for all the other parties involved, including the Owner and the Design Professionals, will have a difficult time coordinating the schedule, and teamwork will be essential. We will discuss some of the issues related to how bonding companies might respond to a claim against their bond shortly, but first let us consider exactly what the performance bond entails.

Once again, a performance bond is not an insurance policy. Insurance is intended to indemnify against a potential but unknown event, while the bond is for a default of the Contractor on its contractual obligations, should that occur. The risk of insurance is based on an unforeseen but always potential accident in a dangerous construction environment, and it does not insure against the improper workmanship of the Contractor. The bond does require the surety to step in and complete what the Contractor has not done correctly or in relation to the contract requirements, or if the Contractor is no longer able to perform its contractual obligations.

A performance bond involves three parties:

1. the Principal, usually the General or another Prime Contractor in a multi-prime project, but, of course, it could also be a Subcontractor;
2. the obligee for whom the Principal is required to perform the conditions of the contract, and since this is almost always the Owner, that is how we will refer to this party;
3. and the Surety, itself. The bonding company is sometimes referred to as the Obligor, but we will refrain from that in this text, in favor of plain English.

What is important at every step of the way is to realize the Surety is there for financial support, should it become necessary. The people who work for the bonding company understand the fundamentals of construction, but they are not Constructors themselves. The surety will not step in and complete the project with their own forces. They will need to acquire expertise and Contractors to do whatever is required to fulfill the contract if the performance bond is called upon by the Owner. Often, this takes a considerable amount of time, and in the process the non-defaulting Subcontractors and other Multi-prime Contractors (if the project involved other primes) can be delayed. This is a time for a Contract Administrator to pay attention to notices for delay and start documenting the delay as soon as it occurs.

9.2.1 Miller Act and Little Miller Acts

Performance and payment bonds are often required based on the size of the contract. Through what is known as the **Miller Act**, Federal work requires bonds once a contract exceeds $100,000, and states contain the same provisions in what are known as **Little Miller Acts**.

9.3 Removing a Contractor

I do the very best I know how – the very best I can; and I mean to keep on doing so until the end.

Abraham Lincoln

Whether on a bonded project or non-bonded work, terminating a Contractor or Subcontractor is usually a very difficult decision to make, not based on the non-performance of the Contractor or Subcontractor to be removed, but based on the difficulty of replacing any entity that has had some impact on the schedule, cost, and quality of the work to date. If you are a Design Professional advising the Owner, or a Contractor being terminated or involved in the termination of another party, the prospect of staying on schedule is slim, the chance of coming in under budget unlikely, and everyone will probably be chasing patent and latent flaws for quite a while.

There are several reasons an Owner will consider terminating a Contractor and requiring the bonding company to find a way to complete the work.

9.3.1 Termination for default

Most contracts will have a provision within them allowing the Owner to terminate the contract based on default – the failure of the Contractor to complete one or more portions of the contracted work. If a Contractor becomes bankrupt and/or incapable of continuing, removal for default is obvious.

Sometimes the issue may not be actual bankruptcy, but the Constructor is having difficulty with cash flow. If the Owner cannot improve the cash flow (and perhaps does not want to, based on the possibility of a Contractor in default), then the bonding company might step in and help the struggling Contractor in order to avoid a default. This, however, requires quick and honest discussion with the bonding company.

9.3.2 Termination for failure to pass on payments made by the Owner

Owners expect that the payments they make to a General or another Prime Contractor for work done by Subcontractors or for material supplied to the project by vendors should be passed on to the respective parties in a timely fashion. Obviously most Contractors have specific provisions regarding payment to Subcontractors and Suppliers, including the Pay when Paid and Pay if Paid Clauses, but Owners do not want exposure to having to pay twice, so they will certainly consider terminating any Contractor that is not paying those responsible for the work the Owner has already paid for. When we discuss payment bonds later in this chapter, we will recognize another defense the Owner has regarding failed payment, but any Contractor not paying for work done and paid for is a troubling sign for an Owner, and cause to consider termination before matters get worse.

9.3.3 Termination for lack of progress

In a previous chapter we mentioned how the schedule is the primary focus of an Owner once the project contract is signed. Getting the structure done so it can be used for its intended purpose is paramount. In fact, most contracts contain the term "time is of the essence." For the Owner, it almost always is. If a Contractor is in trouble and does not have the ability to supply enough personnel, equipment, material, or Subcontractors to prosecute the work, then an Owner may have to resort to removing the Contractor. The likelihood of getting back on schedule if a major Contractor or Subcontractor is removed is remote, but the opportunity to not lose more time might be a critical factor in the decision to terminate.

9.3.4 Termination for Contractor's refusal to perform

Of course a Contractor has the right to refuse to continue working under certain circumstances. The Spearin case is an example. Another possibility would be a **Cardinal Change** or changes to the original contract, making the new project beyond what the Contractor could reasonably perform. The Owner ultimately has the right to terminate if the Contractor refuses to do the work in a timely manner. Once terminated, of course, the Contractor can prosecute the case to recover lost revenue and profits if the Contractor was wrongfully terminated. Here again, the documentation by the Contract Administrator will be extremely important in determining who acted properly and what the consequences, particularly financial, for both parties should be.

Counsel from the Design Professionals on such a termination should be very carefully given. In fact, it hopefully will not have to be given.

One major exception to total stoppage of the work is if the Owner fails to pay in accordance with the contract. The Contract Administrator should research the notice clause (hopefully this would already have been done and noted on the spreadsheet developed earlier). In most cases the failure to pay on a timely basis will be considered just cause to stop work. Of course, late by a few days is different than not paid at all. So the Contract Administrator will want to recognize intent and pattern related to payment.

9.3.5 Termination for failure to comply with specific or significant provisions of the contract

There are sometimes paragraphs within a contract that both parties might ignore and still work through to a successful conclusion. Although that happens, it should not be considered that anything within a contract should be ignored. There are, however, some requirements that are not always seen as significant to the Constructor, but if they are ignored, they can be cause for termination. For example, with the submission of pay applications, the Constructor along with all the current Subcontractors might be required to submit payroll forms, documenting wages and percentages of employment by race or gender. Failure to submit these forms on a timely basis can hold up payment. If the submission or lack thereof demonstrates a failure to comply with the contract requirements, then that is a breach of contract and termination might follow.

Similarly, the contract might call for specific safety standards and requirements. If a Contractor or Subcontractor continually or knowingly violates those standards, the Contractor could be terminated.

Every Contract Administrator must recognize the Owner has authority to remove or terminate a Contractor and/or request that a Subcontractor be removed from a project for failure to comply with the requirements set forth in the contract, which includes issues most closely related to schedule and quality of the work.

9.3.6 Termination for convenience

As discussed in Chapter 7, courts will usually uphold the right of an Owner to terminate work on a contract simply for their own convenience if it has been expressed in the contract. An Owner may in the middle of a project recognize that the structure is no longer needed. It is possible a company may be bought by another company and consolidation of the facilities to another location might be the direction the new Owner is taking. There are many reasons a project might be stopped, and so every Constructor has to recognize that possibility.

For a Constructor, losing work to do is not good news, but looking for a silver lining, termination for convenience does not involve the bonding company needing to take over the work. In fact, depending on the size of the project, the Contractor's bonding capacity may increase by having the amount of bonded work on hand reduced by the amount of work remaining on the terminated project.

9.4 Contractor's rights during termination

I am extraordinarily patient, provided I get my own way in the end.

Margaret Thatcher

It is highly unlikely most Contractors will sit idly while they are terminated for any of the causes mentioned above other than for convenience. If a Contractor is found in default by the Owner, most Contractors are not going to agree. Assuming there has been ample dialogue between the Owner and the Contractor, and if the agents of the Owner, such as the Design Professionals and the Construction Manager if one is involved, have also entered their opinions on the side of the Owner, the Constructor will have a difficult time refuting the default.

If the work is bonded, the Contract Administrator will want to work closely with agents of the bonding company to mitigate damage. Whether bonded or not, there are definite steps to take.

Certainly the Contract Administrator's documentation will be critical, but it will also have to be accurate. If, for example, the Constructor was delayed for no fault related to the prosecution of the work, then the Contract Administrator must be prepared to at least show:

1. the specific cause of the delay;
2. the duration of the delay itself (and any other consequential impacts of delays on other events);
3. the actual and specific impact of the delay on the progress of the work;
4. possibly the cost of correcting the delay if any acceleration occurred because of the delay.

Of course, if the Contract Administrator has done this all along, then for most Owners acting in good faith there would be no termination. Still, the Contract Administrator needs to be ready for such a situation.

9.4.1 Contract Administrator's efforts in relation to termination

Assuming the Contractor has not been terminated for convenience, the Contract Administrator has several steps to take:

1. If it is a bonded project, you should contact the bonding company immediately. Some dialogue with the bonding company should already have taken place if there had been any previous discussions about termination between the Owner or any of the Owner's agents and the Contractor. Being first with the bad news is sometimes considered dangerous if "they shoot the messenger," but being second with bad news is even worse, since you may already be considered guilty before your side of the case is heard.
2. Contact your attorney. Termination for Cause is going to be a legal labyrinth and the sooner a competent attorney in such cases knows about the issues and how to direct you, the better.
3. Do not accept the termination without response, particularly if you feel there are any errors or discrepancies, based on the way the Owner handled the notice of termination or on what the Owner is basing the reasons for termination.
4. By contract there might be a notice period for you to correct the items or actions that are bothering the Owner. Carefully analyze what that would mean and whether the action to correct would be more prudent than continuing the fight over termination.
5. See if you can negotiate an alternative approach with the Owner rather than termination. Termination is not going to be easy for the Owner. If you can show a better way, it might be accepted. Sometimes you will need to determine exactly what the issue is that is causing the Owner to act so drastically by pursuing termination.
6. Document fully all you have done to date on the project and off site if there are deliveries or materials in the process of being manufactured. Take photographs of everything you can. You need to know what material and equipment you have on site that might be used by others if you are forced to leave. You might also want to record the impressions of everyone who had a significant part in the construction. Obviously, construction has many members who migrate to other projects and other companies. Discussing the project while

they are still available and the memories are vivid can be very important. At the very least, check the daily work reports. Make sure they are up to date and fully tell the story.

All this also equates with what you must do if you are terminated for convenience. You must prepare a complete and accurate total of all costs to date in order to recover later. The Owner (and bonding company if a bonded project) do not have a blank check to make things right. If you are wrongfully terminated you will be entitled to all the monies you expended and additional costs that might include lost revenues, general administrative costs, and lost profits. If the Contract Administrator, however, does not keep very accurate records, much of those dollars may be lost.

7. Contact others who might be affected by the termination. Subcontractors and Suppliers will need to know their status and what their options are, going forward. The payment to Subcontractors through the Contractor will have to be carefully accounted for going forward, based on what has been paid and what has been retained by the Owner. A looming difficulty for the Contractor and the Contract Administrator is the potential for claims by Subcontractors and Suppliers against the Contractor. As much as possible the Constructor does not want to fight a two-front war, with the Owner coming from one side and the Subcontractors coming from the other. Close and honest communication, along with accurate records of fulfilled payments and other obligations, will be the Contract Administrator's greatest opportunity to create allies rather than enemies.

* * * * * * * * * *

An important habit a Contract Administrator should develop is checking daily work reports on a weekly basis. Incomplete daily work reports are useless and a waste of time for whoever is filling them out. Rather than make work that winds up being of no value to anyone, protect the accuracy of the reports for when you need them by making sure the person filling out such reports knows why they are important and just how they should be written.

* * * * * * * * *

9.5 Surety's obligations

A promise made is a debt unpaid.

Robert Service

If a Contractor does default on a bond, the Owner has a few options. The surety that bonded the project may not be in the best position to bring someone in and complete the work. In such a case the surety will be obligated to pay the Owner what it costs the Owner to have another Contractor complete the work. The surety's involvement will be strictly financial. There is usually a limit stated in the bond of how much the bond is worth, and therefore, the Owner and its agents need to recognize they do not have a blank check for any and all work they might intend to do. In fact, the bonding company is required financially to pay only for the contract work. Changes to the contract have to be carefully documented and in most cases (unless a patent defect) they are the responsibility of the Owner and not the surety.

If you are a Contract Administrator on a project where another Prime has defaulted on a bond, this is another reason to keep careful records of what additional work was done due to the default. In some instances, the bonding company will assert a defense against third-party claims, but if the additional work is carefully documented and shown to be the direct result of the default, some courts are allowing the expanded costs through the Owner's original claim.

If there are Liquidated Damages to be assessed on the project, the surety will also assume responsibility for them. The surety cannot rely upon the good will of the Owner to ignore the delay damages that might have been exacerbated by the need to remove one Contractor and bring in another.

In addition, the surety may be responsible for attorney's fees, penalties, or interest on amounts that might have been due Subcontractors or others but were withheld improperly.

9.5.1 Surety's defenses

The surety, however, is not without rights and defenses. In fact, any right the original Contractor that defaulted might have asserted, the surety is also entitled to assert. Again, if at all possible, the Contract Administrator of a defaulting Contractor wants to work closely with the bonding company (if it will allow it) to help mitigate the potential loss. In some instances, however, the relationship with the bonding company has deteriorated so much that further communication and work is out of the question. The atmosphere of cooperation usually depends on how early and often the defaulting Contractor informed the bonding company of the impending situation.

In addition, the performance bond may include several defenses, the most common being a notice clause. The surety needs to be informed within the time set forth in the bond. There might also be restrictions on how the bonding company is to be informed. A conversation on the phone will not likely be acceptable, based on the language of the bond.

The counsel of the Design Professionals in a bond default situation can be helpful to the Owner, and ultimately the whole atmosphere during a difficult situation can be positively or negatively affected by other parties on the project. Architects consider it good practice to:

1. notify both the Contractor and the bonding company of the potential for default in a timely manner, so that a potential resolution might be found before default is necessary;
2. meet with the surety, the Contractor, and any interested parties to try to find a potential solution;
3. advise the surety based on the terms of the contract, if no solution can be found. It is important to be sure no actions or conversations made in points 1 and 2 above will negatively impact the ultimate removal of the Contractor for default.

Once a Contractor has been removed, the surety is not the sole entity with financial responsibility. The Owner is still obligated to pay (now usually to the surety) any monies currently due, and eventually any remaining funds in the contract that would have been paid to the Contractor, had the Contractor completed the work within the terms of the contract.

Of course, the surety is not first in line. Often when a Contractor defaults, there are monies unpaid to the IRS and some state wage taxes that are often not paid.

This can actually evolve into another defense for the surety, which might assert that the Owner had improperly overpaid the Contractor and that the remaining funds were inadequate for all that remained to be done. Again, the Design Professional who usually has the Expressed and/or Apparent Authority to approve invoicing needs to recognize the potential difficulty that overpayments that were certified by the Architect can create in the final accounting for a defaulted Contractor. This is not a reason to cut a Contractor's legitimate pay request, especially since we saw the ramifications of cash flow on a Contractor, but it is to alert Design

Professionals that the responsibility of payment applications is a double-edged sword. Not paying enough can create a cash flow disadvantage for a Constructor that could be the cause of a bankruptcy, and paying too much might leave too little in the Owner's account to complete a project under a defaulted performance bond situation. The key will always be how cognizant were the Owner and the Design Professional that there was an error in payment.

9.5.2 Good faith

Performance bonds can be viewed as another layer of risk for the Constructor. By law, Owners may be required to include them. In other instances, having a bond may give some additional comfort that the project will be completed as intended, but it is certain that an Owner never really wants to call upon a performance bond to complete a project.

Fulfilling the contract is the path to success. Not requiring a performance bond to be activated is part of that success, but it is still a challenge. Construction is hard work. It requires dedication, from the first time an Owner speaks with a Design Professional through all the challenges of construction and up to the final sweep of the broom or dust brush into the pan before the Contractor walks off the site. There are many reasons for failure, but there are many more stories of success. The bond assures the Owner, but nothing can take the place of what Frank Lloyd Wright termed dedication, hard work, and unremitting devotion to the things one wants to see happen. The Contract Administrator is at the center of that hard work.

9.6 The new Contractor's *caveat constructor*

Beware you do not lose the substance by grasping at its shadow.

Aesop

The common Latin term of *caveat emptor*, meaning "let the buyer beware," should be also a warning for the Contractor who assumes the work of a defaulted Contractor. More work may seem like a good thing, and the Owner and others will certainly welcome someone competent to step in and finish the project, but it is not an easy thing to pick up part way through a project. Commitments to existing Subcontractors and Suppliers will have to be analyzed and sometimes renegotiated. Initial productivity will probably be much lower than would be typical if the project were smoothly progressing from the beginning, and so estimates might need to reflect that. And that can lead to others questioning your reasoning, but it needs to be considered. It will almost certainly require more staff to get the project restarted and finally flowing smoothly than had been the case with the defaulted Contractor. These are burdens the new Contractor will need to be compensated for, or there will be a significant loss.

Most importantly, the schedule has to be realistically adjusted. Hidden in the schedule might be Liquidated Damages, and the new Contractor should be very careful that neither the Owner nor the bonding company, if one is involved, places those Liquidated Damages back on it, particularly if there is no adjustment for the completion of the project. If faced with Liquidated Damages, the new Contractor will have to consider what acceleration costs will be necessary to meet such a date, but the preferable manner of handling Liquidated Damages is not to agree to them, or not to agree that they are its responsibility, but that they should be placed on the previous Contractor. The other alternative would be to significantly change the completion date for Liquidated Damages to accrue.

This is all part of the negotiations, but the Contract Administrator should recognize that the power to negotiate will disappear once the contract is signed.

9.7 Payment bonds

No complaint is more common than that of a scarcity of money.

Adam Smith

Payment bonds are furnished to an Owner as an assurance that the Contractor will pay monies to Subcontractors and Suppliers as they are due them. This can vary, since the terms of the contract with the Subcontractor usually reflect conditions dependent upon payment from the Owner (Pay when Paid and Pay if Paid Clauses). Payments to Suppliers are often based on a thirty-day period from time of delivery. In some instances, depending on the size or uniqueness of the order, as well as the lack of a "track record" or experience the Supplier has with the Contractor, terms such as COD. (**Cash on Delivery**, which means the Supplier will be paid upon delivery of the material) may be used. Some Suppliers also offer discounts for prompt payments, such as a 2% discount if paid within ten days, but the drawback to this is: (1) During times of low interest rates, a discount by the Supplier can be a tremendous benefit to the Contractor but a severe drain on profit to the Supplier, or (2) Contractors often stretch the payment to thirty days but still take the discount.

In essence, the payment bond is again assurance to the Owner that the Owner will not be faced with having to pay twice for material, equipment, or workmanship that has been put into the structure. The Contractor furnishing such a bond can also ask Subcontractors to furnish bonds. These would be furnished to the Contractor and not the Owner, but it becomes a layer of protection that extends payment bond protection for the Contractor who has paid Subcontractors and those Subcontractors have failed to pass on payments that are owed to their Subcontractors (termed **Sub-subcontractors**) or their Suppliers.

9.7.1 Tiered protection

Not everyone that may have contributed to the project is protected by a payment bond the Contractor supplies to the Owner. The expectation of coverage is linked to how closely the entity is to actually supplying labor and/or material or equipment to the Contractor and the project. The parties covered under the bond are known as **first or second tier subcontractors or suppliers**. Usually any entity that is farther removed than having directly furnished labor or material to the project is not covered by the payment bond. Also, generally, only the first tier Supplier is recognized as a potential claimant under a payment bond.

Example: The hardware Supplier that provided all the locks, handles, hinges, and closers for the doors on the project would be covered, but the metal Supplier that furnished the brass and steel to the manufacturer of the hardware would not be covered, nor would the manufacturer of the hardware be covered if the hardware was bought from a Supplier, as it typically is, rather than directly from the manufacturer. Basically, a Supplier to a Supplier is not covered by the payment bond.

For a Supplier it is very important to establish whether you are supplying to the Contractor or a Subcontractor.

The intent of covering just the first and second tier Subcontractors and only the first tier Suppliers is that monitoring all the potential claimants that are so far removed from the project would require extraordinary and unusual effort to monitor effectively. It is also worth recognizing, however, that a Sub-subcontractor would be second tier on the Owner's payment bond, but would be first tier to the Contractor if the Contractor had required the Subcontractor to supply a bond, and a Supplier to the Sub-subcontractor could then be considered first tier on the payment bond protecting the Contractor.

All of this gets very confusing and complicated even for the courts that might be required to decide a case related to who was and was not covered. Essential to such a process is to first determine who is a Subcontractor and who is a Supplier. Subcontractors are expected to supply labor in addition to any material they install on the project. Suppliers are those parties that deliver or make available material or equipment that is directly incorporated into the project or essential to the Subcontractor's construction of the project, such as essential equipment rented to erect material. The Supplier, however, does not actually provide labor in the process except perhaps for the delivery of the product itself.

One exception courts have sometimes recognized is that a manufacturer might be covered if it supplies a unique or custom object such as equipment or furnishings that had to be specifically manufactured for the project off site before it was delivered and installed without significant modification.

Every case may prove different from the one before it. One can never be sure how one court or another will interpret the "closeness" to the Contractor relative to who is actually covered. Therefore, the potential of someone claiming against the bond for payment is always present, and it can often come as a complete surprise, because the claimant seems far removed from what one would expect.

The lesson for the Contract Administrator ultimately is that the best defense against claims on the payment bond is to make sure all parties are paid in an appropriate fashion in the first place. For years the AIA has provided payment applications that help to do this as much as possible with those directly involved with the project.

* * * * * * * * * *

If a claim against a payment bond is made, no doubt the bonding company will notify the Contractor. At the very least they will want an explanation. This is time for complete honesty and accuracy. The claim can be defended, but the Contract Administrator will need the facts in order to do so. An important action that might be overlooked, however, is to contact one's own attorney separately from the attorneys the bonding company might provide. The bonding company will mount its own defense. To the extent that it will protect the Contractor, that is fine, but recognize that indemnification by the Company and probably principals of the Company will cover the costs that the bonding company might incur. If the bonding company does have to pay the claimant, the bonding company will come after the Contractor and its principals for reimbursement. And that can make the bonding company a less than perfect ally, relative to the Contractor's total interests.

* * * * * * * * *

9.7.2 Beware middle corporations

Reversing roles, if the Contract Administrator is a Subcontractor or Supplier wishing to be protected by a bond, then be certain you are not being pushed down the tiers by a middle corporation set up by the actual Contractor. Although such a procedure might not be looked upon favorably by a court, a Contractor might set up a middle corporation that will deal with the Suppliers. If a problem arises, the Supplier has already been pushed down to a second tier by the middle corporation, and if this is upheld, the second tier Supplier cannot file against the bond of the actual Contractor.

9.7.3 Making a claim against a performance bond

Timing is essential related to bond claims. Failure to make the claim notice or to actually file suit against the Contractor's bond can serve to negate any rights to recovery under the bond.

Therefore, in addition to the notice clauses within a contract, the Contract Administrator needs to recognize the notice requirements for making a claim against the bond. Fortunately the time to file an actual lawsuit is fairly liberal. The suit must be brought within one year of the last day labor for the claimant was on the project. Unfortunately, the time for actual notice is rather tight, considering the usual payment flow relative to applications and work. Usually a claim notice must be delivered within ninety days of the last day labor or materials were supplied to the project site.

For the Contract Administrator working for a Subcontractor with a potential claim, this can be quite difficult. There could be many different scenarios, but let us imagine a Subcontractor that has been slowly but completely paid through all the work done on the project. The chronology of the final payment might run in the following way:

1. The Subcontractor finishes all items, including the punch list the first week of one month.
2. The Subcontractor then submits an application for payment, which cannot be submitted to the Owner by the Contractor until the end of the month.
3. At the end of the month (twenty-three days after the Subcontractor left the project), the Contractor requests on a pay application the final payment due the Subcontractor. *Twenty-three days have expired without reason for notice.*
4. The payment by the Owner is not due the Contractor for another thirty days (at best). *Another thirty days expired without reason for notice.*
5. And then the Contractor, based on a Pay when Paid Clause, will process the payment to the Subcontractor, which could take perhaps another fifteen days. *Perhaps up to another fifteen days expired without reason for notice.*

So there could remain only twenty-two days left to determine if the slow payment is not the Contractor's fault (e.g. because the Owner has typically taken more than thirty days to pay the Contractor). If it is not the Contractor's fault, making a claim against the bond will certainly damage relations and jeopardize future work with that Contractor. It is at this time that any Contract Administrator's rapport with the whole project team, and the open communication between all parties, can be very important in determining what course is best.

If the choice becomes obvious to claim against the bond, the Contract Administrator needs to research to whom the notice needs to be sent and in what manner. Oral claims will not be acceptable. Usually there is a requirement to provide proof to the court that the claim was sent and received. The claim might need to be sent to the bonding company, but not always. The

Contractor needs to be advised of the claim, and the prudent Contractor must then advise the bonding company and their own attorney as noted above. The notice needs to also include the amount of the claim. Of course there will almost surely be some counter-arguments to why the claim is invalid, inflated, or bogus, but being as accurate as possible at this stage will prove very helpful if the claim does evolve into an actual lawsuit.

The issue becomes even more difficult for a Sub-subcontractor. Payment flow to a second tier entity is even slower. A Contractor might not know who is supplying labor or material to a project through a lower-tiered entity. For that reason some states have instituted a policy requiring notice to the Owner and/or Contractor as soon as a purchase is made or work is started on a project. In this way parties are informed of who is doing work, and payments can be tracked or documented in various ways to make sure no claims "come out of nowhere."

9.7.4 What can be claimed

The Contract Administrator will need to accurately demonstrate the amount sought from the payment bond, particularly if it is to include work not documented through a payment application. For instance, if the Contractor has not written a change order approving additional work, then that additional work will have to be justified to the bonding company and, if necessary, to the court. The better the documentation, the easier it will be for those not involved in the project, or perhaps anyone in a court not acquainted with the construction industry, to understand the justification for paying for the extra work.

In addition, the Contract Administrator might want to collect for costs associated with delays or acceleration. These will be very difficult to prove without substantial and accurate documentation. Bonding companies may attempt to negotiate such settlements, whereas courts generally recognize actual costs without profit as the correct compensation under the bond. Interest on the money may be allowed, depending on the laws of the state, but generally the cost of attorneys to prosecute the claim will be denied. This makes it even more imperative that the Contract Administrator do as much work as possible to document accurately the exact claim, showing the details and facts clearly.

Of course, avoiding the bond claim in the first place would be the best course of action, but that is not always possible, and if the amount warrants the claim, the Contract Administrator will have to focus on justifying all the costs, including extra work and even interest if possible that were the result of failure to pay. At some point the Contract Administrator will have to recognize what the advantages might be to accepting the bonding company's view of the facts or going forward with a claim through the courts. Taking the claim to court may become necessary, because the bonding company does have defenses related to a bond claim.

9.7.5 Defending a bond claim

The first defense a bond company will assert, if it is available, is late notice. If the claim against the bond has not been given in sufficient time, that defense alone will be applied.

If the claim has been properly and timely made, however, the bonding company is required to make a response within another period of time, usually between thirty and sixty days, regarding what amounts are considered correct and what amounts are considered disputed relative to the claim. If there are any amounts that are undisputed, they should be paid without further delay, and the bonding company should not appear to be withholding them as a method of negotiating the remaining amounts.

The surety, however, may raise any defenses against the claim that the Contractor could have raised regarding the performance or inaccuracies of the billing requests by the claimant against the payment bond. One of the most dangerous clauses used against the claimant for filing against the bond is the Pay if Paid Clause. Courts do not always look favorably upon the Pay if Paid Clause, but the drawback to it is that it will cost money to determine whether the court will recognize it or not, and usually the cost of attorneys' fees to prevail or find out might not be allowed. Again, the Contract Administrator will have to determine the cost of litigation versus what can reasonably be achieved in a judgment.

9.7.6 Liens versus bond claims

Contract Administrators will need to consult an attorney regarding the potential effects of filing a **Mechanics Lien or construction lien** rather than a claim against a bond. In some states Owners prefer to have a payment bond in place because it eliminates the right or necessity to lien the property. In other states, there are different rulings, and the bond is just another way of filing for unpaid work or material on a project. The Contract Administrator should also check before the situation arises about the actual timing of liens. The notice period is again essential, or the claimant's rights can be denied for late filing.

9.7.6.1 Waiver of lien rights

Liens against the Owner's property are not appreciated by the Owner, and the payment application practice often includes a waiver of all lien rights by everyone seeking payment. In addition, there is usually a required statement that anyone seeking payment, particularly final payment, has paid all parties all monies that are due them. In addition, if the project is bonded, the Contractor may ask for a waiver of claims against the bond up to a specific date of the payment request.

Working with reputable Owners and Contractors, such waivers are executed without exception or worry, but the Contract Administrator working with parties for the first time needs to be careful and develop good relations and understanding as quickly as possible to ensure options for recovery of legitimate costs are not in jeopardy, or the exposure is kept to a minimum.

Liens will be discussed more extensively in a later chapter.

9.8 Subguard insurance in lieu of a bond

Avoiding danger is no safer in the long run than outright exposure. The fearful are caught as often as the bold.

Helen Keller

Some Subcontractors cannot be bonded or do not want to be bonded, and in such a case the Contractor has the choice of not requiring a bond or using what is known in the industry as Subguard, which is actually a form of insurance. During the construction of the Cardinals Stadium in Arizona, Subguard was used in lieu of performance or payment bonds for the Subcontractors.

Subguard is strictly an insurance policy in which the General Contractor will be compensated by an insurance company rather than a bonding company if a Subcontractor cannot complete the contractual obligations.

Also, as an insurance policy, the fees are based on actuarial tables, and the Subcontractors are not scrutinized to the extent they would be by a bonding company for character, competence, and capital. The policy is not intended to be monitored on a Subcontractor by Subcontractor basis, but it is based on insuring the entire project, or it can also be issued to cover all Subcontractors on all projects for a specific period, usually annually or for a longer period of time such as five or ten years. Again, the insurance company, unlike a bonding company, wants to spread its exposure so as to meet the overall averages of the industry for failure. It is possible, the first year a company uses Subguard would be the worst year of a longer period. Spreading the duration of the insurance coverage allows the insurance company an extended period to collect fees if the first year was the worst for loss.

Another major drawback for Contractors using Subguard is that the policy can be cancelled and coverage lost, but once a bond is written it is in place and cannot be cancelled. The bonding company still has defenses, but the bond remains in place for the duration of the project.

Finally, the primary market for Subguard is private rather than public construction projects, since it cannot take the place of government required bonds, and it is really intended as a protection for the General Contractor or another Prime Contractor, and not the Owner.

Of course, as discussed in the previous chapter, insurance of any type is one way to mitigate the costs of doing business in a dangerous industry. Helen Keller may be right that we cannot avoid danger in our careers, but we have to protect ourselves against it. The Contract Administrator's best defense is neither being fearful nor bold, but being alert and prepared. Agents of bonding companies fully understand the hazards of construction. Character, competence, and capital are the keys to successful contracting – whether bonded or not bonded.

INSTANT RECALL

- Bonds are going to be issued on behalf of Constructor with two levels in mind: (1) a single project limit, and (2) the total amount of work (bonded and non-bonded) the Contractor has currently under contract.
- The bonding company is going to seek repayment from the Contractor for any monies lost.
- The research on each Contractor seeking bonding is based on the three Cs of bonding – character, capital and competence.
- There are basically two ways a bid must qualify. It must be Responsive and the Contractor must be Responsible.
- A Responsive bid is one that has included all the responses required in the bid proposal.
- A Responsible bid is one by a Contractor that can in fact do the work the project will require.
- It is at the consent of the surety that a performance bond and a payment bond are issued after the Bid Bond has been submitted by the low bidder.
- Whenever a project appears close to the approved limit, keep the bonding company advised of the potential increase.
- If you know soon enough, forming a joint venture can often increase the bonding capacity of the combined entity over what the Contractors individually could achieve.
- If a Contractor is having difficulties, it is better to have the bonding company as an ally than a reluctant partner looking for how it can best keep itself whole.

- There are circumstances under which a bid can be withdrawn without fault. It has to be a genuine mistake and not an error in judgment.
- As part of character, the bonding company will naturally be concerned about the health of the single Constructor.
- Not all notices against a payment bond are correct or accurate. Keep your bonding agent advised of potential difficulties. It is easier to anticipate the issue than to have to explain it after the fact.
- A Contract Administrator must pay attention to notices for delay and start documenting any delay as soon as it occurs.
- Through what is known as the Miller Act, Federal work requires bonds once a contract exceeds $100,000, and states contain the same provisions in what are known as Little Miller Acts.
- Most contracts will have a provision within them allowing the Owner to terminate the contract based on default – the failure of the Contractor to complete one or more portions of the contracted work.
- Usually expressed in the contract, courts will uphold the right of an Owner to terminate work on a contract simply for their own convenience.
- Payment bonds are furnished to an Owner as an assurance that the Contractor will pay monies to Subcontractors and Suppliers as they are due them.
- Not everyone that may have contributed to the project is protected by a payment bond the Contractor supplies to the Owner.
- Timing is essential related to bond claims. Failure to make the claim notice or to actually file suit against the Contractor's bond can serve to negate any rights to recovery under the bond.
- The defense of a bonding claim is going to rest with the bonding company, but if the Contract Administrator for the Contractor can clearly demonstrate the claim is invalid, that will be very important.
- Some Subcontractors cannot be bonded or do not want to be bonded, and in such a case the Contractor has the choice of not requiring a bond or using what is known in the industry as Subguard.

YOU BE THE JUDGE

Appearing before the Supreme Court:

Pearlman, Trustee in Bankruptcy for the Dutcher Construction Company

Reliance Insurance Co., which provided surety bonds under the Miller Act for work Dutcher in bankruptcy failed to perform on the government's St. Lawrence Seaway Project

Claim:

Pearlman seeks to reverse a lower court's ruling to turn over funds to the Reliance Insurance Co. as the surety all monies the government made available to the bankruptcy trustee that had been withheld from payment to Dutcher until after another Contractor had completed all work originally in the Dutcher Contract.

The government had terminated Dutcher's contract by agreement after the Contractor had financial difficulties and could not finish the work. In accordance with the payment terms and flow of payment applications, the government was holding considerable sums not yet paid to Dutcher. Upon completion of the work by another Contractor, the government still had $87,737.35 remaining from the withheld amount, and this sum was turned over to the bankruptcy trustee. Reliance Insurance Co. claimed a prior position over the bankruptcy trustee, but the referee in the bankruptcy court denied Reliance Insurance Co.'s claim. Upon appeal, the District Court reversed the bankruptcy judge's ruling, and subsequently Pearlman appealed to the Supreme Court of the United States, which granted certiorari, because various courts throughout the United States have made rulings both ways – i.e. some have ruled the bankruptcy trustee takes precedence and thus the surety is only in line the same as all creditors, while others have ruled the surety comes before any monies should be turned over to the bankruptcy trustee.

Pearlman argued that under the Bankruptcy Act that governs all bankruptcies, the surety is not given special consideration, and therefore should not expect any from the court.

Reliance Insurance Co. countered that its rights preceded the rights of bankruptcy since its obligation and the obligation of Dutcher in return was before the bankruptcy took place.

Pearlman responded that such a position was not valid in light of the requirement of the Miller Act, which placed the surety in the same position as any other creditor to the Contractor.

Make your ruling below, including a filing why you find for either Pearlman or Reliance, and then check your ruling with the actual case on the Companion Website.
Find for:

Reason:

IT'S A MATTER OF ETHICS

Bonding companies almost always insist on the principals of the company signing over their assets to the bonding company in the event of a default. In addition, the bonding company will want spouses to sign over their portion of jointly held assets, and usually their own individually held assets. Once done, that is usually the end of communication on the subject unless a difficulty arises and the assets need to be taken by the bonding company as payment for a debt. How fair is it that spouses are probably not informed on a regular basis of how each bonded project is doing? Is it fair for the principal to move assets once pledged? If a principal needs cash to continue working, is it appropriate that some of the monies in bank accounts be used, even though they may have been used to secure the bonding in the first place?

Going the extra mile

1. Study the "You be the judge" exercise above, write your opinion, and then go to the Companion Website to determine how your judgment agrees (or not) with the actual judgment by the court.
2. **Discussion exercise** (with your classmates): Discuss how you would approach asking your spouse to sign over all the family's assets to the bonding company so the bonding company can recover any loss it suffers in the event of a default.

3. Go to Appendix II and check the sample Daily Work Report. Are there areas that should be changed, items that are missing? When on a project, you want to make sure the Daily Work Report is filled out each day and accurately reflects the work done or not done (including the reasons for not accomplishing a task).
4. Effective use of the correct vocabulary improves communication and promotes respect among colleagues. The reader should become familiar with the key words listed below that have been used in the chapter. Definitions can be found in the Glossary. Additionally, the student is encouraged to use the flash card exercise in the chapter's Companion Website as practice for possible quiz or exam questions.

Key words

Bid Bonds

Bonds

Cardinal Changes

Cash on Delivery (COD)

First or second tier subcontractors or suppliers

Little Miller Acts

Mechanics Lien or construction lien

Miller Act

Responsible

Responsive

Sub-subcontractors

Three Cs of bonding

CHAPTER 10

Warranties

A warranty is related to insurance in the sense that a warranty guarantees continued performance of the structure in the manner intended for a period of time. There are, however, significant differences. The manner in which a Contract Administrator approaches project closeout and warranty guarantees can be a significant factor in future work with an Owner and the Owner's representatives.

Student learning outcomes

Upon completion the student should be able to. . .

- **recognize critical steps in the project closeout process**
- **discern Contractor responsibility related to warranty obligations of a project**
- **identify warranty periods in relation to Suppliers and contractual responsibility to the Owner**
- **understand Apparent and Implied Authority from the Owner in relation to inspections**
- **know the concept of commissioning a building in relation to project closeout and warranties**
- **compare rights and duties of all parties in the project closeout and warranty phase of construction.**

10.1 Inspections, project closeout, and warranties

What I began by reading I must finish by acting.

Henry David Thoreau

Remember, a **warranty** is meant to ensure ("make certain") for the Owner that the project will continue to meet the contract requirements through a specified period.

Warranties should not be confused with completed operations insurance. Insurance protects against damage or injury due to improper installation or unexpected occurrence. The warranty is not covered by insurance. It is based on the contract obligations for the continued performance of the structure as approved and certified by the Design Professional and as it was received as completed and paid for in full by the Owner. A claim against a warranty means that something that should have performed for the required contractual period has not. The Contractor (or perhaps a Subcontractor or Supplier) is then expected to "make certain" that the structure or part of the structure returns to its intended performance.

The Contract Administrator must begin the process of project closeout the moment he or she has finished reading the Contract. This will include the specifications that will contain several requirements related to warranty items, operation manuals, and perhaps training sessions for the Owner's personnel who will eventually maintain the structure and its equipment.

* * * * * * * * * *

One of the most important habits a Contract Administrator can develop is to begin project closeout at the beginning of the project and not at the end.

* * * * * * * * *

The Contract Administrator should make a detailed spreadsheet of all the items and requirements to successfully closeout the project. The tediousness of creating the spreadsheet will soon disappear as you find yourself on each project with an incredible "cheat sheet" to guide you from the beginning to the end of the project.

It may seem "jumping the gun" to start thinking about closeout so early, but the moment a subcontract or a purchase order is written, the warranties and operation manuals should be an important part of the overall service and delivery. Have a specific place to store and assemble the material as it comes in. Make any training sessions a part of the subcontract that is sent out.

Also, recognize from the beginning that rapport with inspectors is very important. Some might seem to be too tough, but in the end, they can prove the most valuable in obtaining a **Certificate of Occupancy**.

There will be other inspectors and inspections along the way. These should be anticipated and documented. Often, manufacturers will require certain conditions for the installation of their products.

> **Example:** Paint manufacturers will expect moisture content level in masonry block walls to be below certain levels or they will not guarantee or warranty their product through the warranty period. If the roof has not been installed or leaks throughout the project, it may be difficult to have the walls pass inspection by the painter for moisture content. The painter might then balk at doing the work at all, thus delaying the final completion, or the painter might condition the completion of the work on not providing a warranty.

The Contract Administrator who has anticipated the possibilities of additional inspections per manufacturer requirements and perhaps even scheduled "potential" inspections will be moving efficiently to final completion and closeout.

10.2 Inspections

We cannot rely upon mass inspection to improve quality, though there are times when 100 percent inspection is necessary.

W. Edwards Deming

In construction, inspections are many and often, and generally they are a source of stress for the Contract Administrator.

* * * * * * * * * *

One attitude a Contract Administrator might wish to adopt is:

"The more often informal inspections take place, the easier formal inspections become."

* * * * * * * * *

The Contract Administrator needs to involve the entire project team in inspections. It is easy to recognize we are all inspectors when it comes to safety. However, we need to develop the attitude that *we are all inspectors when it comes to quality*. Every time anyone walks the project, they should have the thought that they are looking at it in regard to the next inspection, whether that is an inspection by the Owner, the Design Professional, one's own boss, or some government entity.

Some might say, well does the first-day laborer really have an interest in inspection beyond making sure things are safe? The answer is "yes." Each first-day laborer should recognize a clean work site makes an important impression on others. No one, laborer or not, should ignore helping to maintain a clean project site while walking from one location to another.

Everyone on the Contractor's team should understand their input might be as mundane as picking up a piece of trash that missed the dumpster, to as important as saving a life, but each person on the team can be an inspector, moving the project a little each day to final acceptance.

10.2.1 Owner as inspector

The Contractor will be required to schedule the formal inspections needed throughout the progress of the work and also pay for any required testing involving material, such as concrete, to be placed into the project.

The Owner has the right to inspect the work in regard to the contract requirements above and beyond the necessary inspections that might take place in regard to safety and building code issues that government officials might mandate.

It is again best to anticipate all potential inspections that might be done informally and not scheduled by the Contract Administrator. Contract language differs on the rights and responsibilities, but Owners normally reserve the right to have a Contractor remove material already installed, in order to inspect something that might be hidden. The cost of removing and replacing may be significant, and so who pays for it is important. It is far better not to have to remove anything in the first place, but often, once it is removed, if the inspection shows the Contractor did something incorrectly, all costs are usually placed upon the Contractor. If the installation, however, is correct, then some contracts read that the costs will be paid for by the Owner.

That might be what the contract states, but it is not the way to "win friends or influence people," and so the Contract Administrator who even informally schedules inspection of what will be concealed installations will be improving the relations and progress of the project.

10.2.2 Rejection by Owner

The ultimate disaster is rejection of the work by the Owner. Of course the Owner has the right to reject work that is not in conformance with the contract documents, but this will often raise the issue of what is and what is not conformance, based on whose interpretation. Interpretation of the contract can lead to a dispute at this point. This is not what is desired, but it is far better to have it arise at the point of an early inspection during installation than to have it discovered later, when the impact of correction might be far greater, and the only recourse for relief might be litigation.

The earlier the discrepancy in interpretation of contract requirements is found, the more likely a butterfly will not become a hurricane. Early detection will allow solutions of compromise, accommodation, and collaboration to be explored.

By contract, however, the Owner usually will reserve the right to take some exceedingly drastic actions if inspected work still remains uncorrected. This can include: (1) Withholding payment, (2) having the work repaired and backcharging the Contractor, or (3) even terminating the Contractor.

The Contract Administrator will need to maintain rapport through what could be difficult negotiations, but the best defense will be a positive offense, and that begins with early and often informal walk-throughs of the project that create clear understandings of expectations on both sides.

Of course, it is possible that an Owner or an agent of the Owner will reject acceptable work. The general rule is an inspector has the right to reject nonconforming work, but the inspector does not have the right to change the specifications without proper compensation for the Contractor's additional work. If the Contractor believes additional work is being required due to an inspection, documentation of the cost of bringing the work to the Owner's standard is very important for any future recovery. It is also important to give polite written notice that you do not agree with the rejection and you may wish to submit costs that you incur for review, with the obvious potential for some form of dispute resolution.

10.2.3 The Design Professional or agent of the Owner as inspector

Architects, Engineers, and Construction Managers may be given the authority to inspect the work of the Contractor and Subcontractors by the Owner.

10.2.3.1 Expressed, Implied, and Apparent Authority

Expressed Authority and **Implied Authority** are sometimes referred to as **Actual Authority**. They differ slightly from **Apparent Authority**, but the Design Professionals and/or Construction Manager on a project have different levels of all three.

Expressed Authority comes from the Owner as defined in the contract with the Constructor. Such Expressed Authority might give the Design Professional and/or Construction Manager the right to inspect the work on a monthly basis and certify a payment application is correct and should be paid.

Implied Authority is an extension of the Expressed Authority. For example, if a Design Professional has the right to certify and approve an application for payment, it is implied that that inspector also has the right not to approve it or portions of it for payment.

Apparent Authority mainly arises through the actions or inactions of the agent that are not denied or contradicted by the Owner. Estoppel can become an important aspect of this

Apparent Authority. If a Contractor repeatedly has been doing something and the Design Professional or Construction Manager has not stopped or criticized, then the Owner may be bound to accept what has been approved through Apparent Authority.

However, Apparent Authority can be dangerous territory for the Contract Administrator. It is best to have any Apparent Authority issues confirmed in writing or made a part of the record at a progress meeting where the Owner is present, or in some other clearly documented fashion.

10.2.4 Delays due to inspections

One of the most frustrating aspects of inspections is the delays that waiting for an inspector can cause. It could be something as simple as waiting for someone to approve the bottom of footings before pouring concrete, but to proceed is not an option. Within reason, the Contract Administrator will have to tolerate the schedule of the inspectors, but documenting delays will also be important. The Owner is going to respond to notices that delays are being caused simply because inspections cannot be expedited in accordance with the project schedule. In fact the Owner and the Owner's agents do have an implied obligation to cooperate with the Contractor. If this does not seem to be taking place, the Contract Administrator should politely point out the consequences of delay and document the effects.

Of course, these delays could also be in response to a different type of "inspection" – such as shop drawing submittals or requests for information (RFIs). The Contract Administrator should document all forms of delay, but the manner in which the problem is approached with the parties will be important in building or tearing down the sense of teamwork.

10.3 Project closeout

What is not started today is never finished tomorrow.

Goethe

Project closeout is the goal from even before the trailers arrive on the project. Once a contract is signed, the true moment of success is when the Owner accepts the building. Hopefully, the next time someone from the Contractor visits the site, it is for a sales call related to future projects for the Owner.

Between Substantial Completion and final acceptance, however, can be many days, weeks, even months of activity to settle the details that would more easily have been completed during the regular project schedule.

10.3.1 Substantial Completion

The point at which the Owner can use the structure for its intended purposes has been the traditional point in the project known as Substantial Completion. Contract Administrators might be so happy to reach this point that their guard will be down. The Owner should not move into the structure unless the Owner is prepared to provide the necessary property insurance. The Contractor is not in a position to drop its own insurance on the project, but the Contractor does not also want to assume responsibilities that are actually the Owner's.

If there are any Liquidated Damages agreed to in the contract, Substantial Completion is usually the date on which such damages will stop accruing. However, some contracts are now setting two Liquidated Damages dates. The first is the date of Substantial Completion. Then the Liquidated Damages are reduced to another number and will continue to accrue until final completion.

10.3.2 Final payment

Once all the minutiae have been addressed, including all punch list items have been completed, as well as all inspections, and training attended to, the Contract Administrator is looking to receive that final payment from the Owner. It is important to realize no payment, even the final payment, is confirmation that everything is perfect. Latent defects and issues of warranty might still arise, but the feel-good moment is what has been the goal ever since the exhilaration of being awarded the contract had dissolved into the realization the project, with its many challenges, had to be built.

The Owner is almost certainly the final authority for acceptance, but the Design Professional and the Construction Manager, if one has been involved, will almost certainly be part of the final certification and sign-off process.

10.4 Warranties

Art is never finished, only abandoned.

Leonardo Da Vinci

Upon acceptance, the warranties can be the last vestige of Contractor involvement in the project. Hopefully for the Contract Administrator there will be no issues of warranty that need to be addressed, but keep in mind that how warranty issues are addressed can be a marketing plus or minus. If the Contract Administrator has handled the punch list effectively and the time between Substantial Completion and final acceptance has been brief and uneventful, the Owner has been left with a positive feeling. Efficiently handling a warranty issue can reinforce that feeling, but remember:

> You may never get another chance to change a bad impression

One point related to warranties that should be understood by the Contract Administrator is that warranties vary in time and scope. What a manufacturer warrants may be less than what the Owner has contracted for. Careful cross-referencing of subcontracts and purchase orders with the scope of the specifications is important.

It is also important to recognize that the typical one-year warranty is seldom what happens. Immediately, one has to recognize that some warranties extend much longer, such as a roof warranty. However, with roofs as well as most materials, one must recognize that the manufacturer's warranty for products may be conditional upon specific installation methods and materials being used. Without attentive care, installation in the field can negate a manufacturer's warranty.

10.4.1 Implied warranties

There are two types of warranties. The first is not found specifically in the contract. These are the implied warranties. The Contractor warrants the structure will remain together during the warranty period. Normal "wear and tear" may "scratch" the structure, but it should not fall apart. In addition there is the implied warranty of title that the material and equipment were lawfully transferred to the Owner. For example, the material was not stolen or also sold to someone else.

Relative to an implied warranty are the ultimate expectations of the Owner. If an Owner calls a Contractor back after a year and a day, does the Contractor say, "Sorry, the warranty has expired"? Certainly that is not the response if the Contractor wants continued business with the Owner.

A CASE IN POINT

No matter what your experience, you should always expect the unexpected.

A man had been in charge of construction up and down the east coast of the United States. He had dealt with mega projects on time and met standards of quality for major clients for years.

When it came time to build for himself a new house, he figured there would be no problem. He contracted with a home builder with a good reputation. The work progressed with some fits and stumbles, but it progressed. The time came to move in, and all that seemed to remain was to put the garage door on. With that the project would be complete, and the builder be due his final payment. With some prodding, the work was done. After more difficulty than he had expected, the finally pleased Owner of the completed house paid the Contractor.

The next morning he went to work, proud and happy his new house had finally been completed. When he drove home from work that evening, he was amazed to turn into his drive to find the garage door missing!

With singleness of purpose, the home owner set out as a posse of one and found his garage door installed on another property. That Owner was also about to pay for final completion. Eventually the whole thing got straightened out, but not before a good lesson learned to "never consider yourself so smart you can't be fooled."

That's what my grandfather learned.

Since it is not expressed in writing, the strength of an implied warranty usually is in the case law developed through the common law that was described as the foundation of the United States legal system.

10.4.2 Expressed warranties

Expressed warranties are those that have been contractually defined. The Contract Administrator has spent considerable time tracking and confirming the various expressed warranties of the Subcontractors and Suppliers. If something does not function as intended it might be an installation problem or it could revert all the way back to the manufacturer of the item. A prompt inspection by the Contract Administrator or someone assuming those duties on a warranty call will ease the Owner's mind that the issue will be taken care of promptly.

Some issues are more important than others. For example, in the case of air conditioning in summer or heating in winter this could be very important. But all warranty items should be treated as a priority by the Constructor, because they almost always are so treated in the mind of the Owner.

Figure 10.1 Engraving of London by Claes Van Visscher

One of several "London Bridges" through the centuries that have taught children early in life the importance of warranty work:

London Bridge is falling down,
Falling down, falling down.
London Bridge is falling down,
My fair Lady
Build it up with wood and stone,
Wood and stone, wood and stone
Build it up with wood and stone,
My fair lady

In the case of the London Bridge, depicted in the Claes Visscher engraving, the original "engineering" of the bridge had not anticipated the many additional structures that would be erected on it. As with many designs, "improvements" over time can add stresses to existing structural members as well as strains on electrical, plumbing, heating, and air conditioning designs.

The original Designer of the six-centuries-old London Bridge is long gone, but with buildings currently expected to serve a thirty-year life span (based on return on investment to the Owner), renovations and additions can tax the existing systems and prove a challenge even during the warranty period. This can be particularly true if the contract has called for a performance specification of any of the equipment that has been installed, and an early change in occupancy of the space has changed the ability of the equipment to meet the performance criteria.

Again, however, the Contract Administrator needs to understand whether the warranty is expressed or implied, future relations with the Owner and the Owner's representatives will greatly depend on how well one responds to needs before, during, and after the contractual warranty period.

Handling Subcontractors on a call-back basis may depend on continued good relations with the Subcontractor. Taking the lowest Subcontractor on a one-time basis may impede the efforts of the Contract Administrator for effective prosecution of warranty work.

Some Owners are writing into the contract "extended warranties." These can be two years or even twenty years in length. Working with reputable Subcontractors and Suppliers who will be around when you need them can be very important.

When reviewing a contract with a Subcontractor or Supplier, the Contract Administrator should:

- confirm the warranty period coincides with the warranty required by the Owner's contract
- thoroughly review the prime contract to determine there are no other warranties expressed or that a Precedence Clause does not change the order in which the documents should be interpreted in relation to warranties
- immediately notify the Owner and Owner's representative of any issues between manufacturers' warranties and those set forth in the documents. Although this would ideally have been done before submitting the bid, or at least before signing the contract, the Contract Administrator may be able to rely upon good rapport with the Owner and Design Professionals, or even the doctrine of impossibility related to a design ambiguity.

10.5 Commissioning

It is the courage to continue that counts.

Winston Churchill

More and more Owners are requiring by contract that the Constructor continue past final acceptance and maintain the building for a specific period of time. This is known as **commissioning**. With more and more structures being built with extremely technical equipment and a trend for improved green construction, the technology within the building can require specialized experience. Traditional electrical, plumbing, and HVAC issues are being coupled with life safety, cogeneration, sustainable systems, building security, and many more products to complicate the daily operation of maintaining the building so it functions as designed. "Intelligent buildings" are requiring very smart building managers.

Rather than see this as a burden, the Contractor should welcome the continued presence. This puts the Contractor in a very good position for future work within the building related to renovations, and it also can provide early access to future construction plans of the Owner in other locations.

Previously we mentioned that project closeout must start when the contract is signed. However, "commissioning" must start at the beginning of the bid phase. For the most part, commissioning will actively involve the MEP trades. It is important to verify that those giving the General Contractor pricing for the project have included any costs that will be required to maintain the equipment and structure during the commissioning phase.

As the systems being installed become more sophisticated, the coordination between trades will become more important, and sometimes more difficult. Early in the development of "intelligent" buildings, the various systems did not communicate well together. Owners, however, are expecting more and more interfacing of their systems.

For example, motion-sensitive lighting by the Electrical Contractor needs to "talk" to the Owner for other reasons than saving electricity. Computer chips in the lighting for retailers need to tell the traffic patterns of shoppers in order to more effectively integrate displays. Those same chips in restrooms can be used to alert the need for replacement of paper products or the need for periodic maintenance of flushless urinals. And motion-sensing devices need to be effective in conference rooms to not turn off the lighting simply because the occupants in the meeting have not moved sufficiently to re-activate the sensors.

If the coordination of all the systems and the duties involved has been well defined during the estimating phase, then the Contract Administrator's task is still complex and requires diligence, but the team is in place with the duties known. Coordination and follow-up by the Contract Administrator can become a great benefit to the Owner and a great marketing approach for future work for the Constructor.

Figures 10.2 and 10.3 Two views of the banking floor (second floor) of the PSFS Building during punch list inspection. Built in 1932, the PSFS building had many unique features for its time that made maintaining the building an initial challenge
10.2 Author's Collection; 10.3 Photo credit: Charles W. Cook

The concept of "commissioning" a building is really not new. Well before the concept of "commissioning" a building beyond the completion and final payment milestone, Owners recognized the need for expert assistance in the transition from construction to occupancy and operations. The PSFS building, with several (at that time) unique features, including push-button elevators and central air conditioning, needed expertise that the maintenance staff would not initially have once the building was turned over to the Owner. Frank Cook, the "clerk of the works" whom George Howe had hired to oversee the construction, was asked to stay on, and he did so as the "building Superintendent" for a period of time until others could familiarize themselves with the operations and maintenance of the building.

Also, on a much smaller scale, homeowners and shoppers are familiar with the concept. Often when people buy technology, equipment, or machines at retail stores they are offered extended warranties for after the manufacturer's warranty expires. In some cases stores offer installation assistance. It is not unusual for a homeowner to have a yearly maintenance contract on their heating and air conditioning systems. If something goes wrong, the Contractor is supposed to immediately respond to put the equipment back into operation.

With "intelligent" buildings the systems are far more complex than one finds in the usual residence, but the concept of paying for service to maintain the working order of the system bought by the Owner is the same.

"Nullum tempus occurrit regi"

Before leaving the subject of warranty, it should be noted that warranty claims and latent defect claims can become confused. Although a Contractor might assume after a one-year warranty has expired that the burden of repairs is no longer a contractual obligation, the possibility still exists that an Owner can come back against the Contractor, the Design Professionals, and the Construction Manager for defective work.

A 2012 ruling in the State of Connecticut used an old ruling based on *"nullum tempus occurrit regi"* ("no time runs against the king"), which states that there is no defense for a Constructor or parties associated with the design and construction based on time. If latent defects are discovered, the expiration of any time limits contained in the contract is null and void.

This is based on Connecticut's common law, which dates back to colonial times when the "King" had no limit on the power to expect justice.

In this case, the State of Connecticut sought repairs and damages for the University of Connecticut's School of Law library. The project had been completed in 1996, but the defects that included design errors, improper construction of walls, windows and roofing, incorrect waterproofing in the basement, along with inadequate HVAC systems had required over $15 million dollars in repairs. The repairs had been completed in 2008, obviously several years after any warranty had expired.

The State had brought suit against twenty-eight defendants for the various and multiple issues of defective work. All had attempted to assert a defense that time had expired on their warranty or any other contractual obligations.

The initial court to hear the case agreed with the twenty-eight defendants.

However, on appeal, the Supreme Court of Connecticut overturned the previous ruling and asserted that the *"nullum tempus"* dates back to and is carried over as common law from the colonial period and English law.

This is, of course, another example of how one can never be sure how the courts will interpret contractual obligations. Latent defects will most likely supersede an expiration of a warranty defense. *In the end*, nothing can replace having a superior and diligent team re-checking each team mate's or partner's contribution *from the beginning*.

10.6 Chapter closeout

From the beginning of *insure, assure, ensure*, the Contract Administrator has traveled a long way, whether through reading these chapters or in the delivery of the project to final closeout for the Owner and the Design Professionals. When an artist begins a creation there is a great deal of enthusiasm for what might be. Along the way there comes a time when that enthusiasm diminishes, sometimes even dries up and blows away with the constant pressure of realizing that what initially seemed so simple actually is not. At that point the artist can abandon the work or continue on past the stress and doubts. For those who continue on, the project becomes the prize. Every work does not lead to a *Mona Lisa*, but then again we would not all be smiling if it did. Whatever our contribution to the vertical or horizontal construction of our world might be, from the beginning our goal should be to walk away when done with the pride we earned when first we knew we had won the chance to do something new, something very special.

That is the Contract Administrator's challenge, and it is also what the Contract Administrator must help the whole team to continue to believe along the way. The key is not that they all have to believe in you. They must believe in each other and themselves as a whole.

Exercise 10.1

Take a moment
Another thoughtful word play

A few chapters ago we examined three words: *insure, assure*, and *ensure*. Here is another phrase you might want to keep in mind throughout your career, and even life:

Aspire to inspire before you expire.

Take a moment to write below at least one asset, trait, ability, or virtue you currently have that can help you help others to achieve their best, even in the difficult circumstances of a construction project. This does not have to be a world-renowned or gold medal talent. It just has to be something you are good at doing well.

How can you help others by using this talent?

Discussion of Exercise 10.1

Hopefully you took some time to determine what you might use to inspire others. Whatever your talent or virtue, it can become a most important and effective asset for you and your team. We will see in future chapters how essential our individual skills are to the team, so if you did not spend enough time on the exercise above, take a little more time now to find some strength(s) you can bring to the team.

INSTANT RECALL

- Warranties have limitations, and the Contract Administrator should read carefully any disclaimers related to warranties.
- The Contract Administrator must begin the process of project closeout the moment he or she has finished reading the contract.
- We need to develop the attitude that *we are all inspectors when it comes to quality*.
- The Owner usually will reserve the right to take some exceedingly drastic actions if inspected work still remains uncorrected. This can include: (1) withholding payment, (2) having the work repaired and backcharge the Contractor, or (3) even terminating the Contractor.
- Expressed Authority comes from the Owner as defined in the contract with the Constructor.
- Implied Authority is an extension of the Expressed Authority.
- Apparent Authority mainly arises through the actions or inactions of the agent that are not denied or contradicted by the Owner.
- When reviewing a contract with a Subcontractor or Supplier, the Contract Administrator should:
  - confirm the warranty period coincides with the warranty required by the Owner's contract
  - thoroughly review the prime contract to determine there are no other warranties expressed or that a Precedence Clause does not change the order in which the documents should be interpreted
  - immediately notify the Owner and Owner's representatives of any issues between manufacturers' warranties and those set forth in the documents. Although this would ideally have been done before submitting the bid, the Contract Administrator may be able to rely upon good rapport or even the doctrine of impossibility related to a design ambiguity.
- It is important to realize that no payment, even the final payment, is confirmation that everything is perfect. Latent defects and issues of warranty might still arise.
- More and more Owners are requiring by contract that the Constructor continue past final acceptance and maintain the building for a specific period of time.

YOU BE THE JUDGE

How long should a warranty on the design last?

Skidmore Owings and Merrill (SOM) had designed a landmark tower for Business Men's Assurance Co. of America (BMA). The building was completed in 1963. In 1985, 4x4 slabs of marble that covered the outside of the tower began to fall off. Investigation of the problem indicated the mechanism for attaching the slabs to the concrete behind them was defective. The metal strips connecting the marble to the concrete were weakened by the wind until they finally snapped, similar to the way a paperclip continually bent back and forth will eventually snap.

The SOM contract with BMA stated SOM would assist BMA in "checking of contractors' and manufacturers' shop drawings, approval of material samples, issuance of certificates of payment and full-time supervision of work by an architectural superintendent on the site who shall be responsible for the coordination, performance and completion of all architectural, structural, civil, mechanical, and electrical engineering work in accordance with the approved drawings and specifications. It is understood, that, although in supervision of construction we will use our best efforts to protect you against defects and deficiencies in the work of contractors, we will not guarantee performance by them of their contracts."

SOM, in response to the claim, asserted many defenses, the chief of which was the statute of limitations. BMA sought over $5 million in damages and repair. Should the case be allowed to proceed, and, if so, is this a matter involving the Contractor or the Marble Subcontractor?

Find in favor of: BMA or SOM (circle one)

Find against: BMA, SOM, Contractor, Subcontractor (circle one or more)

Amount(s):

IT'S A MATTER OF ETHICS

Think to yourself or discuss with others whether a Contractor should remain responsible for equipment that was supplied, installed, and operating properly for the period covered by the manufacturer's warranty, but had failure after the building was accepted but not past the Owner's contractually stated warranty period. Ask yourself: (1) Does it matter if the Owner delayed the completion of the project beyond the control of the Contractor? (2) Does it matter if a *force majeure*, such as a hurricane, beyond anyone's control, delayed the project? (3) Does it matter if there was only one day left on the warranty period? (4) Are there any circumstances under which the Contractor would be justified in not recognizing the warranty?

Going the extra mile

1. Study the "You be the judge" exercise above, write your opinion, and then go to the Companion Website to determine how your judgment agrees (or not) with the actual judgment by the court.
2. **Discussion exercise I** (with your classmates): Where do you think technology will one day take the sustainable construction of buildings? What future do you see in making buildings even more "intelligent"? It is from your "pie in the sky" visions that the successes of tomorrow will come. It is not difficult to recognize that some billionaires today started in college with an idea. And in some cases, those ideas were not initially embraced by the faculty or other students.
3. **Discussion exercise II:** We have already reviewed the warranty clauses in AIA, ConsensusDOCS and EJCDC in Chapter 6. Go back and review, while discussing with classmates, which one is fairer, considering the issues related to warranty discussed in this chapter.
4. In the Appendix I Sample Subcontract there are at least two paragraphs that place warranty responsibility on the Subcontractor. Do they sufficiently cover the possibilities related to Commissioning? If not, what would you do to improve the Subcontract? Go to the Companion Website to compare your answer with some thoughts on the subject by the author.

5. Effective use of the correct vocabulary improves communication and promotes respect among colleagues. The reader should become familiar with the key words listed below that have been used in the chapter. Definitions can be found in the Glossary. Additionally, the student is encouraged to use the flash card exercise in the chapter's Companion Website as practice for possible quiz or exam questions.

Key words

Actual Authority

Apparent Authority

Certificate of Occupancy (C/O)

Commissioning

Expressed Authority

Implied Authority

Warranty

CHAPTER 11

Change orders

Certainly the most prevalent issue facing a Contract Administrator or Constructor, change orders can be a source of profit or a source of great concern. How to handle the often conflicting issues of time and cost can be crucial to a project's success.

Student learning outcomes

Upon completion the student should be able to. . .

- **know the importance of the change order clause**
- **recognize the steps in the approval process for change orders**
- **describe the effects of time in relation to the change order process**
- **understand the full impact of cost related to changes**
- **determine the responsibility for initiating and approving changes**
- **recognize the possibility for deteriorated relationships on the project can result from how the change order process is handled among all parties.**

11.1 The change order clause

If you do not change direction, you may wind up where you are heading.

Lao Tzu

Technically, once a contract is signed the Constructor is required to build only exactly what is in the contract – the original promise. If there were no means of changing the "promise," the Owner, as well as the Design Professionals, would not be able to change directions, and everyone would wind up exactly where they were originally heading. That may not be best for anyone.

For that reason change order clauses have been put into contracts which acknowledge the Owner has the right to make changes to the original contract – the original promise. Although there is no penalty for making such changes, changes usually cost something, and they often affect the schedule. So making changes is often a necessity, but they also can be a source of tension. The Owner has a limited budget, and the Owner needs to occupy the space within a specific time frame.

Then, depending on what has already been discussed as Expressed, Implied, and Apparent Authority, the Owner may transfer the right to make changes to others.

Figure 11.1 Structural steel installation
Photo credit: Charles W. Cook

The Contract Administrator, whether on the Contractor or the Design Professional side of a project, will view changes differently depending on who might be responsible for the cost of the change and the effect it has on the schedule, particularly the critical path of the project.

During the design phase, changes in the contract are usually a matter of a stroke of a pencil, or more commonly a pass-through on a computer rendering. BIM software is making the detection of ambiguities, flaws, and points of "crashes" between installations of different elements more apparent during the design phase. However, **fast tracking**, which is when parts of the structure are already being put in place while later items of construction are still being designed, removes the luxury of planning and replaces it with what will eventually become fewer options. As the foundations are poured and the steel is erected, the opportunity to make changes decreases, and each new element that is put in place raises the potential that if something needs to be changed the options are less, and/or more expensive.

Still, Owners may request changes based on differences in need that arise after the contract has been signed. Sometimes Owner personnel will want changes once they are able to physically see the structure, have incorporated some of their materials or equipment into part of the space, or see the flow of items and personnel in a way they had not imagined. In such cases, the Contractor and the Design Professionals will generally not be held responsible for the cause of the change, and therefore the main consideration and potential tension may be only over how much the change will cost and if it will affect the schedule.

Design Professionals may institute a change because something does not look, fit, or work correctly. There may need to be clarifications to the specifications. Inspections may reveal the need to add something to the design because something else will not look, fit, or work correctly. Any of these reasons may be very valid, but they can also be a reflection on the work of the Design Professional(s). In such a case, particularly on a Design-Bid-Build project where the Designers and Constructor are separate entities, the potential for harmony among the project team can be in jeopardy.

Constructors can also initiate changes for many reasons, including pointing out ambiguities in the contract documents, differing site conditions, or the need to accelerate the construction process due to delays. Again, such a change might create more tensions between parties of the project team, depending on who is ultimately the reason or cause for the change. Remember, what a Constructor sees as **Latent Ambiguity** might seem to the Design Professional to be a **Patent Ambiguity** that the Constructor should have brought to the attention during the bid phase, or they should have understood the correct interpretation without requiring a change.

Contractors have a reputation for wanting changes, but not all changes are good:

- small changes seldom have a significant markup to make up for lost time or the overhead of processing the change
- large changes sometimes consume more time in arriving at an acceptable price and in the process they impact the efficiency of putting other work in place at the estimated units
- all changes can affect the schedule in ways that are difficult to document and therefore the true costs are not compensated for in the final pricing.

Because so much tension and "who struck John" finger pointing have continued through the decades, Owners and Design Professionals have continually reworked the change order clauses within contracts to protect their positions in the process. The most difficult of these clauses are the exculpatory clauses discussed previously, which are intended to shift responsibility for such issues as delays, differing site conditions, hazardous materials, and ambiguities and defects within the plans and specifications. The Contract Administrator for the Contractor must be aware of such clauses and do the best they can to maintain a just and reasonable dialogue in such situations.

Also the Contract Administrator needs to pay attention to how changes are to be submitted. Notice of the requirement for a change is usually required to be submitted in writing within a specific period of time (e.g. ConsensusDOCS 200 between Owner and Contractor allows 14 days from the time the need for the change was known, and the AIA A201, ed. 2007, requires 21 days' written notice to the Architect).

In most cases a flaw or ambiguity in the contract will be **"construed against the drafter"** of the document. This will usually be the Owner, who supplied the documents and with whom the Contractor is bound by contract, but the Owner will probably have recourse against the Design Professional for errors and omissions. Finding fault and the need for a change, therefore, can start everyone down a proverbial "slippery slope."

There are two important caveats for the Contract Administrator to be sure of before relying solely on the presumption that every ambiguity will be construed against the drafter:

1. The ambiguity is not an obvious one – a Patent Ambiguity – that any reasonable Contractor would have recognized and made allowances for already.
2. The Constructor has an interpretation of the contract that is both different and reasonable from what the Owner and/or Design Professional intended.

A CASE IN POINT

More and more Owners are looking for the "team" to resolve changes before construction actually begins.

The current design for what the public has known as Freedom Tower, but which, by change directive, has been named One World Trade Center, underwent several changes in design well before any construction began. These changes included the shape of the structure, the base cladding changing from prismatic to glass and steel design, the size and configuration of certain floors.

Figure 11.2 One World Trade Center, also known as "Freedom Tower"
Photo credit: Charles W. Cook

During construction, a major change was announced. It was determined that due to other construction in the area the building's main loading dock would not be completed on schedule. This would mean occupancy would be severely hampered if not impossible, so five temporary loading bays were created.

The most crucial change, however, may not be as significant structurally as it may be symbolically. The top of the structure was originally to be an enclosed sheath or radome of interlocking fiberglass panels. That design, however, was changed, and the top of the structure is an exposed antenna. The significance of the antenna is that it symbolically raises the overall height of the building to 1,776 feet – an important number for Americans related to the founding of the nation.

The insignificance of the antenna, however, is usually that an antenna is considered an appendage rather than structurally part of the building. The change in the design could have affected the "official height." No one was certain whether it would be accepted by the Counsel on Tall Buildings and Urban Habitat (CTBUH), which is the lead body in determining what buildings are the tallest in the world. Although Freedom Tower/One World Trade Center will not be the tallest in the world, the developers did aspire to have it be the tallest in the western hemisphere, but it would need the 408-foot antenna to do that.

For a happy beginning to the story of the completed structure, the CTBUH did accept the additional height of the antenna after completion of the building.

11.1.1 Resolving changes before construction begins

As mentioned in "A case in point" above, Owners are looking for resolution of "changes" before construction begins. This can even be accomplished during the design phase, depending on the team approach.

11.1.1.1 Building Information Modeling

As discussed in a previous chapter, the technology to provide 3-D plans of the final structure has given Designers an opportunity to discover "crashes" where different physical features impact on each other. In order to accomplish this, Architects and Engineers are often working closely with critical trades, including structural, electrical, mechanical and plumbing trades. Early detection of problems means fewer issues to resolve during the construction process.

11.1.1.2 Integrated Project Delivery

Owners have also promoted greater teamwork among the different parties by proposing the Design Professionals work closely with the Constructor, key Subcontractors, and Suppliers during the design and construction process. Some Owners have proposed a set budget with a sharing in savings among the parties if the cost of construction comes under the budget. Bringing the parties together during the design phase may negate some cost savings generated during the bid phase of construction, but greater savings are often realized when delays are minimized and fewer ambiguities mean fewer cost overruns.

However, for most Contract Administrators a more traditional approach to project management and change order administration will be the case. The design phase will be completed by the Design Professionals, and the Constructor will be brought in to make the vision a reality. In such a case, there may be significant changes to process.

11.1.2 Determining cost

Most contract change order clauses have been written with the understanding it is not always easy or even desirable to arrive at a lump sum price before work on the change needs to take place. For that reason, there are usually several methods suggested for arriving at an agreed price:

- an agreed-upon lump sum addition or deduction from the current contract price
- a cost based on units submitted with the bid proposal or subsequent to the contract signing and agreed to by both parties
- costs based on the actual hours of personnel and equipment involved, along with extra material put in place and subcontract work directly involved in the change, along with an additional agreed-upon markup for overhead and profit. Usually the markup for overhead and profit will have been made a part of the original contract, and it often varies based on whether the markup is for the Contractor's own forces, or for material, equipment, or Subcontractor work
- in some cases, time may be more precious than cost to the Owner. Owners or their agents will often issue change order directives before an agreed price is finalized. In such a case the Contract Administrator will need to be very careful to document the actual costs in case the negotiations for a lump sum price break down. However, the more difficult

negotiation may be just over the horizon. If the Owner is intent on continuing work without a price, the schedule must be important, and changes, even ones that are approved to go ahead without pricing, can affect the schedule.

Compare contracts

The student is encouraged to compare the change order clauses. In the AIA 201A, paragraphs under 7.3, the Architect assumes considerable authority in approving and processing the change order. In 7.4 the Architect has authority to make minor changes that do not involve the contract sum or schedule. In 7.3.7, the Architect assumes broad authority to determine costs of a change if not agreed to or put forward in a timely fashion by the Constructor. In ConsensusDOCS, the Architect is eliminated from directing or approving changes in cost and schedule, and the authority is strictly with the Owner dealing with the Constructor. Since final authority almost always rests with the Owner, this may seem appropriate, although the Owner is certainly capable of obtaining additional counsel. Although the AIA contract document details acceptable costs related to extra work, ConsensusDOCS includes and details more items for the Constructor's submission. In the EJCDC Article 10, the Engineer replaces the Architect in approval of changes. Article 10 is briefer than either the AIA or ConsensusDOCS paragraphs on changes, and the EJCDC is less detailed, but clearly requires notice and approval by the Engineer of changes.

Some may consider more better, while others consider less better. See "Going the extra mile" for a discussion exercise regarding this.

11.1.3 Determining changes in time

The Contract Administrator must recognize that any change has the potential to affect the schedule. Not all changes actually do, but each change should be analyzed for its potential effect on the **critical path** of the project.

Even if initially there does not seem to be an impact on the critical path, the fact that some part of the schedule may be affected should be put in writing. Notification *now* could become very important later, especially if some other action might be affected due to the delay, or, more importantly, something should cause the critical path to shift.

The contract will most likely have a notice clause related to delays or changes in schedule. The diligent Contract Administrator is well ahead of the situation if he or she has already developed a spreadsheet of notification clauses while reading the contract. As soon as a change is recognized as affecting the schedule, the Owner and the Design Professionals should be notified. If an agent of the Owner, such as the Design Professional or Construction Manager's Contract Administrator, recognizes the delay, then it is important all parties be brought into the conversation to discuss options, and hopefully solutions to the change in schedule.

In seeking an extension of time, the Contract Administrator will first have to demonstrate how the change will affect the critical path. If the change is directly on the critical path, then the effect is obvious. Calculating the increased time, and putting that into the schedule, will demonstrate how much longer the project will take if no other adjustments are made.

Examples:

1. In the case of delays due to changes in deliveries of material or equipment on the critical path, such as structural steel or air handlers, it is very easy to demonstrate how they impact on the schedule. The length of delay in the delivery is the same as the extension.
2. The cost of a delay due to a change involving personnel will have to be calculated based on the most effective use of people and equipment on the project. Such a change might involve 100 additional "man days," but if the work can be done by ten craft workers over ten days, then the actual extension of time would be only ten days. Of course, the Contract Administrator might have to demonstrate that twenty craft workers would not be able to reduce the delay to five days. Also, it should become obvious at some point to all reasonable Owners that only a certain amount of increase in workers is possible. For example, 100 craft workers could not be found to do the work in just one day.

Presenting such a situation to the project team may allow others to offer suggestions that could improve the schedule and negate the delay or at least some portion of the impact of the delay.

11.1.3.1 Costing changes in time

Documenting the delay can be difficult, but it should be undertaken with due diligence. A formal request for the change in time should be made. Hopefully, it will be accepted. If it is not, the Owner may direct acceleration (**directed acceleration**), in which case that will lead to another change based on how many additional personnel will be required to get back on schedule. Usually such changes do not involve more material, since the material would already be in the contract or in the change for what has caused the acceleration. There might, however, be additional costs related to more equipment, such as mechanical lifts or generators that are needed to be brought on site for additional crews, and, of course, there might be additional personnel and/or equipment required for Subcontractor work.

If the Owner does not direct the acceleration, the Contract Administrator's work becomes a little more difficult. At this point, the Constructor has to document all the costs that are incurred due to the **Constructive Acceleration**. If the costs can ultimately not be agreed to by the end of the project, and if they are significant enough, they may need to be part of a mediation process, or ultimately a court or arbitration case.

What the Contract Administrator will want to demonstrate first to the Owner and the Design Professionals, but ultimately, if need be, to a mediator, judge or arbitrator, is at least the following:

1. There was an excusable delay that was not caused by the Contractor.
2. All contract provisions were adhered to, especially notification of the delay and the notice of the need to accelerate if a change of time was not granted.
3. The Owner and/or an agent of the Owner with the Expressed Authority to do so refused any extension of time and ordered through some means such as threats of Liquidated Damages, withholding payments, or written notice that there would not be an extension of time.
4. Proof of actual costs above the costs that would have been made, had the acceleration not been necessary.

In addition to tracking extra costs that the Contractor incurs associated with constructive acceleration, the Contract Administrator will have to carefully monitor the extra costs of Subcontractors affected by acceleration. Design Professionals have long recognized that some Contractors and Subcontractors often "pad" the actual contract work into the extra accelerated work. In doing so, they increase their profit, and some feel justified in doing so, since the whole issue might wind up in litigation, and that means some costs will likely never be paid.

Once again, the Contract Administrator through the hard work of accurately documenting costs and transparent communication can establish trust. Failure to do so will only increase the likelihood of disagreement and difficult negotiations.

11.1.4 Deductive changes

Not all changes are additions to the Contract. There are changes that should obviously reduce the cost of the contract. Such changes can, of course, also reduce the schedule, particularly if the change falls along the critical path of the project.

The concept of Termination for Convenience was discussed previously, but it should also be recognized that such a clause in the contract could also be construed as a deductive change order situation. In fact, if both clauses exist in the contract, and they usually do, it may be beneficial for the Owner to treat a deductive change as a partial Termination for Convenience. Of course, it may depend on the size of the project, what exactly is being eliminated, and whether it would be easier to track the costs of the deduction from a partial termination rather than a simple deductive change.

The Contract Administrator must be aware that both possibilities could exist, and negotiations one way or the other could affect profit. In a small deduct change order, Owners sometimes agree not to deduct overhead and profit, since the actual time of the project may not really be affected and the overhead and profit should continue.

On a large deduct, the Owner may demand that overhead and profit be reduced. Such a reduction has often been agreed to in accordance with the contract. However, on a very large deduct this could be a serious hardship for the Contractor. If the original estimate did not include as high a markup for overhead and profit as the contract includes for change orders (both add and deduct), then the Contractor will actually be giving back more than is in the original project. In such a case it would probably be better for the Contract Administrator to operate under the Termination for Convenience clause, in which the deduction would most likely be based on the schedule of values submitted for the project and documented on pay applications.

Still, relying solely on the Termination for Convenience clause might produce other difficulties. For example, if the Contractor front-end loaded the costs, then the Owner might suddenly recognize that the amount of the credit remaining in the pay applications for what has been terminated would not be sufficient to complete the work. The Owner would want more back. The next step might be "hard ball" related to retainage. The Owner might want to negotiate for a portion of the retainage to be withheld, along with the remaining amounts in the pay breakdown. One thing might lead to another, and the deteriorated relations could affect future work as well as lead to mediation or worse in relation to the final payment. Final payment in such a case might also affect relationships with innocent Subcontractors who had nothing to do with the "cooking the books" pay applications of the Contractor, but the Contractor is now caught in a "cash flow" difficulty.

The conscientious Contract Administrator will be alert to all possibilities and negotiate the most effective process for deductive changes.

Exercise 11.1

Take a moment
How do you react to Murphy's laws?

Some have compared the challenges of daily project management along with both patent and latent flaws and ambiguities to Murphy's laws in operation:

1. If anything can go wrong, it will.
2. If there is a possibility of several things going wrong, the one that will cause the most damage will be the one to go wrong.
3. If anything just cannot go wrong, it will anyway.
4. If you perceive that there are four possible ways in which something can go wrong, and circumvent these, then a fifth way, unprepared for, will promptly develop.
5. Left to themselves, things tend to go from bad to worse.
6. If everything seems to be going well, you have obviously overlooked something.
7. Nature always sides with the hidden flaw.

On the following scale, how do you react to Murphy's Laws?

0 ———————————— 50 ———————————— 100

Murphy was an optimist (it's much worse than that) | Poor Murphy (he missed all the fun)

Discussion of Exercise 11.1

Changes are inevitable. How we react to them and process them can be the difference between successful contract administration and an unpleasant experience all around. Ideally, it might be argued that the best score is 50 on the scale in Exercise 11.1. Approaching everything as though it will get worse in accordance with Murphy's laws is no better than not recognizing the challenges of keeping on schedule and how important changes are to the progress of the work and an on-time completion.

We will shortly see how this opposite-scale approach will become an important tool in helping us all with improving our own skills and attributes to help our team and advance our careers. For now, if you scored closer to one extreme or the other on the scale, think if that in any way would affect your team play.

Of course it is important to realize that in any given circumstance our reaction might be affected, and be more effective, at one extreme or the other, but, overall, balancing the challenge of change(s) with the confidence of a can-do attitude will be a successful path for most circumstances.

11.1.5 Authority to make change

As discussed previously, the Owner may grant to an agent the authority to make changes. The Contract Administrator, however, must be careful. The wording of the contract will express the exact authority any agent might have. Sometimes a Design Professional is given outright authority to make any changes that do not affect cost or schedule. Rarely, but sometimes, an agent may have authority to make changes that do involve cost and/or the schedule, but usually the Owner reserves the final right of approval after review by the Design Professional and/or other agent, such as a Construction Manager.

If, for example, the contract states that the Architect will "prepare" the change order for the Owner, then the Architect is not actually approving the change order. The same would be true if the Construction Manger prepared the change order. If the contract states a change order needs to be signed by the Owner, then proceeding without the Owner's signature can be worrisome. Most contracts caution the Contractor from proceeding with any extra work without written authorization. Sometimes, however, it is necessary to keep on schedule and avoid unduly interrupting the work of the rest of the project. In such a case, the Contract Administrator needs to secure confirmation in some fashion that the written approval will follow.

Relying solely on what is perceived as the **Implied Authority** or the **Apparent Authority** of an agent of the Owner can lead to difficult negotiations later in the project when payment comes due for an item that has not been "signed" for. Of course, as discussed previously, an **estoppel** presentation might be a means of prosecuting a claim for payment, but it is far better not to have to go that far in the first place.

11.1.5.1 Constructive changes

Particularly in Federal work, but also in private contracts, the Owner will make some changes that do not affect the price or schedule that are termed **constructive changes**. These are changes that may take place on a walk-through of the project, during a project meeting, or even over the phone. Once again, the Contract Administrator must recognize who has the actual authority to make constructive changes. An undocumented constructive change made in good faith might later prove to be costly.

> **Example:** Someone directs the Contractor to have the boss's room painted a different color than is on the drawings. The boss moves in and does not like the color. The boss wants what was on the drawings. Now the room has to be repainted the correct color, and the boss's furniture is already in the room. If future work is of importance and current payment is desired, the Contract Administrator has very little opportunity in such a case to say anything more than, "I'll see to it right away."

When someone not authorized to make a change does so, it is important for the Contract Administrator to have that change properly documented. Even if it does not appear to be a change that will affect cost or the schedule, documentation and authorization by someone contractually able to do so is important.

Often these changes are casually initiated and they seem innocent at first. They often arise during a walk-through, a casual discussion about something else, or a phone conversation. Often when an answer is arrived at, the Contract Administrator, or other personnel on the project, is so happy to have the clarification that they proceed without authorization from the proper authority. Such initially innocent constructive changes can include:

- clarification of the plans or specifications that seem to have an ambiguity
- resolution of an existing disagreement
- incorrect technical data transmitted by an end user of the space
- a delay caused by the Owner's personnel preventing access to critical areas
- request to accelerate a section or portion of the project.

The above possibilities might be seemingly easy to accommodate under certain circumstances, but without proper documentation, the Contract Administrator is exposing the company to possible uncompensated costs, as well as no change in the project schedule.

On the other hand, there are instances when an Owner or agent of the Owner with the authority to make a change might informally direct a change and it becomes costly to the Contractor. In such a case, what initially seemed an informal constructive change should be turned into a formal change request. A Design Professional might clarify something at a job meeting, and with the Owner present, the Contract Administrator will understand the item is a change. While doing the work, the Contract Administrator recognizes an impact on cost and should then document and submit for same.

There could also be an instance of constructive changes where the agent of the Owner made unreasonable demands with the threat perhaps of not approving the final pay application if certain items well beyond a punch list were not done. Again, the Contract Administrator should document such costs and submit them if such an action seems worthwhile.

Always, however, the Contract Administrator must balance accommodation versus competition. We accommodate for good relationships now and, hopefully, future favors. We compete when the current situation is worse than our prospects for what might happen tomorrow, whether we win or lose.

11.1.6 Cardinal Changes

Although it might seem strange that a Constructor would turn down work, it can sometimes happen. As mentioned in a previous chapter, a Cardinal Change can be so large or there can be so many changes to a contract, that the scope of the original contract is significantly altered beyond the Constructor's ability to deliver. For reasons of volume, schedule, or even quality, the Contractor is stretched beyond the capacity to fulfill the contract. In such a case, perhaps directed by the bonding company, the Constructor is required not to accept the change(s).

Even worse, of course, is that the changes might require safety issues well beyond the capacity of the Constructor. Unexpected exposure of hazardous materials, or requirements for processes the Constructor has never managed before, are causes for concern about safety.

If a Contract Administrator is aware of Cardinal Change situations early enough, there might be time to seek solutions through formation of a joint venture or involving other expertise. The key will be early involvement of competent experts and good legal assistance, along with open dialogue with a bonding company if one is involved. Failure to communicate timely and openly will almost certainly lead to unwanted results.

11.1.7 Unilateral versus bilateral change order

In government contracts, the Contract Administrator must be aware that the government's Contracting Officer usually has the contractual authority to make unilateral change orders that the Constructor must execute. There is usually a cost and schedule change included in the unilateral change order, but the consent or signature by the Constructor is not required. If the Constructor fails to abide by the unilateral change, the Constructor can be terminated for default. However, since no signature is required, the Constructor is not prevented from seeking further compensation or additional time related to the change through the claim process.

On the other hand, if a signature of the Constructor is obtained, then the change order is bilateral, and there is usually language within the change order document prohibiting further claims for time or money.

Although this unilateral change order process is generally in government contracts, the Contract Administrator must recognize that if relations on a project deteriorate sufficiently many contracts prepared by the Owner have the same consequences. The student might again want to refer to the AIA 201A paragraph 7.3 to see if this might be possible under such a contract.

11.2 The change order approval process

If there is no struggle, there is no progress.

Frederick Douglass

Recognizing and notifying of a change is only the beginning. The entire approval process for a change order can be so burdensome that small changes are often more work than they are worth. Nonetheless, the Contract Administrator has to recognize the importance of the steps along the way to getting paid for any change order.

By contract some of these steps may take longer than others. In some instances, some of the steps may be unnecessary. The Contract Administrator needs to know early in the project, and hopefully before the first submission of a change order, just exactly how the changes will be processed and the time required to do so.

1. A need for a change is discovered or uncovered. This can originate from the Owner, Design Professionals, Construction Manager, Contractor, Subcontractor, Supplier, code inspectors, or craft workers.
2. All parties required by the contract are advised of the situation. At this point, the Contract Administrator should begin to document for compliance with notice provisions of the contract.
3. Potential resolutions are explored by all parties that can appropriately contribute to a solution. At this point the Contract Administrator may act as a funnel related to ideas or as a pass-through, but depending on the complexity of the situation or the number of parties involved in a solution, a meeting in an office or on the site might be more effective than continual remote communication.
4. One or more solutions are determined.
5. Plans and/or specifications are created for a solution and delivered to all parties affected.
6. If there is no increase in cost or effect on time, the change is implemented. If an increase in cost is required, the Contractor will need to advise and submit within the contractual or agreed-upon time frame.
   a The Owner or agent may require lump sum price before work on the change proceeds.
   b The Owner or agent might authorize work to begin and submittal of pricing to follow. The Contract Administrator should be very careful to document such authorization and determine there will not be any difficulty related to payment going forward.
7. The work is priced and the effect on the schedule (if any) is submitted for review by the agents of the Owner.
8. Discussion of the cost and schedule might involve some refinement of either the pricing or the schedule before the agents of the Owner recommend approval of the change order.
9. Formal submission of the change order, usually through the agent of the Owner, is made.

10. Owner signs and approves the change order.
11. Work on the change order is completed in part or whole.
12. Invoicing within the next pay application takes place related to the amount of the change order completed to date.

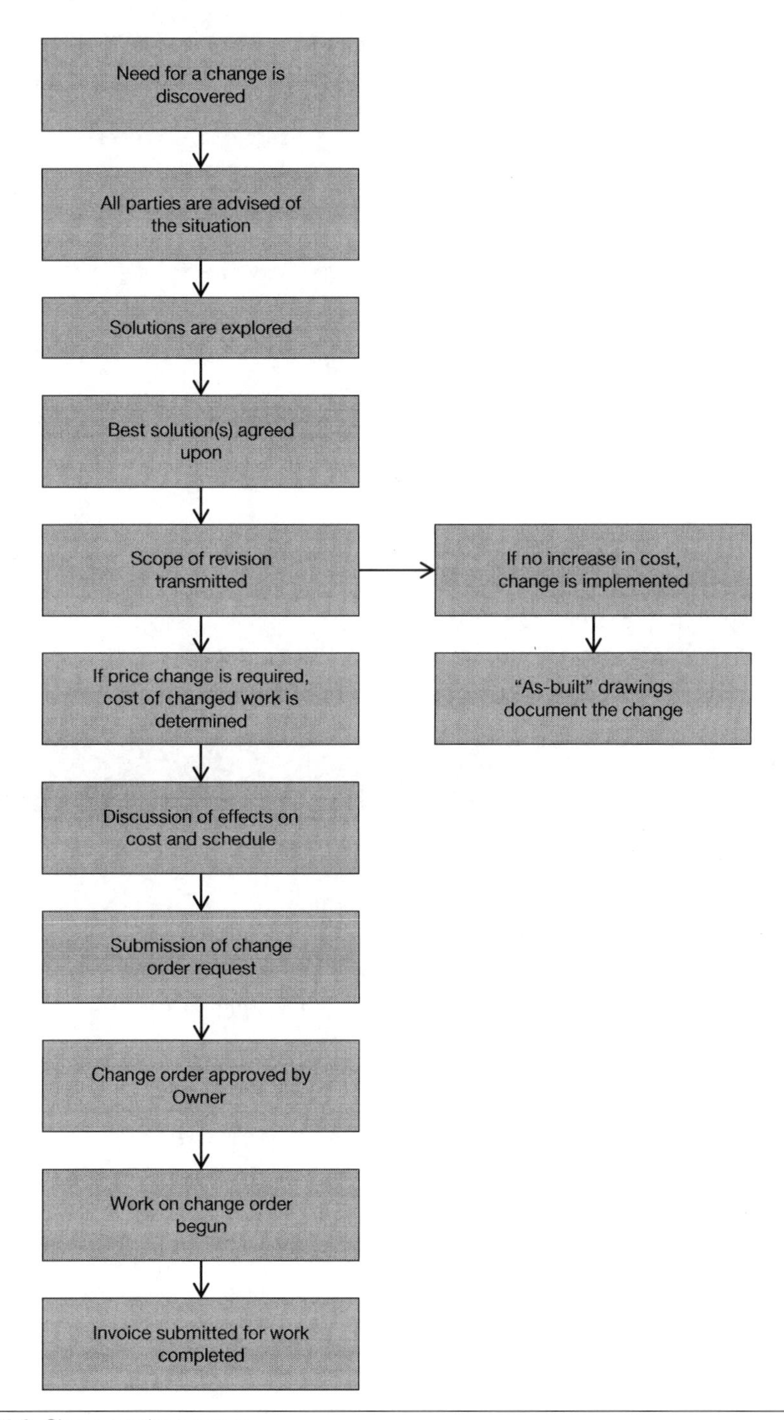

Figure 11.3 Change order process

11.2.1 Change order process delays

The above process can happen very quickly, but it seldom does. In almost all instances it becomes a major task for the Contract Administrator to move the process through to completion, and there are, as mentioned previously, some challenges along the way. Design Professionals may be very sensitive to the costs, since the change might reflect negatively on their work. Owners might have a contingency in the budget, but they might not have room for an extension of time. Subcontractors and Suppliers do not want to be detained in the progress of the work, so they will look for responses faster than they may be forthcoming.

Add to that possible mix of needs the occasional time when the change order might not be considered by the Owner or the Design Professionals to be work that should be compensated, and the Contract Administrator sometimes has some difficult days ahead. Once again, the documentation of notices and the work involved is vital for future submission of the costs and any effects on schedule.

Costing a change order is important, but recognize that any change, seemingly innocent in costs, can have an impact on the schedule, and if it does, it can be more costly than had no change been required. The Contract Administrator must from the start of the process make all parties aware of the impact that the process for the change order has on the overall schedule. Often this is hypothetical in nature, but failure to move the change order process forward can have real consequences on the project schedule.

The Contract Administrator, therefore, should propose, if necessary, alternative change order costing provisions and directives if the change would severely impact on the schedule. Such alternatives may be found in most contracts, such as proceeding on a Time and Material basis. The Owner may not wish to proceed in this way, but honest dialogue on the subject should be put forward and then documented in relation to the impact on the schedule.

11.3 Determining cost

Money often costs too much.

Ralph Waldo Emerson

When computing the actual cost of a change, the Contract Administrator must first look to the contract to determine what is considered an allowable cost. Many people and costs contribute to the success of a project. Not all of them are incorporated directly into the costs allowed in a change order.

11.3.1 Inefficiency of acceleration changes

Although it is often difficult to prove, and therefore often not obtained, the Contract Administrator should put forward, and hopefully the Owner and the Owner's agents will recognize, that there are often inefficiencies involved in accelerating a project. Overtime may mean more money for the worker putting units in place, but those units are often put in place at a less efficient pace than the estimator had figured in the original proposal.

There are many legitimate reasons for this, including poorer working conditions, such as colder temperatures or less natural light. Sometimes setting up artificial light in itself can add to the cost of overtime work. Also, the normal body rhythms of workers usually respond better to eight hours of day work than to longer hours or shift work from 4:00 PM to midnight or midnight to 8:00 AM. Additionally, during shift work, the Owner might recognize the supervision is not the "A-team" that attends to the project during the normal working hours. The Contract Administrator might have no satisfactory solution for this, since the best supervisors tend to demand the better shifts.

11.3.1.1 The Measured Mile

One method for demonstrating or supporting productivity cost changes is what is known as the **Measured Mile**. In this method, the Contract Administrator looks for a period of time in which units were being put in place that was not adversely affected by the change or delay. This period should be removed from either the start or the finish of the unit production process, since both those times are usually slower than during the middle or height of production. Once the productivity of the unaffected units is accurately known, then the adversely affected productivity is compared to these units.

How many units were put in place under the normal circumstance, and how many were put in place under the adverse circumstances gives a clear understanding of the loss in productivity.

Of course, the Contract Administrator will have to demonstrate there were no other factors affecting the productivity loss at that time, such as weather, changes in personnel or number of personnel, lack of tools or material.

11.3.2 The Eichleay formula

Generally, Contractors are not allowed to invoice for the cost of people working off site in a corporate office. For this reason, many people associated with a project will work in elaborate trailer complexes on site. Off-site personnel and the equipment and material used and consumed in an off-site office are considered administrative. They are invoiced in the general category of "overhead."

Nonetheless, in some instances the Constructor's home-office personnel are very much a factor and a financial burden if their costs are not compensated.

One such notable exception to invoicing off-site administration costs could be as a result of a change that created a delay which was totally out of the control of the Contractor and that stopped productive progress of the work on the project. In such a case, at least on Federal work, courts have allowed what is known as the **Eichleay formula** (usually pronounced *eye-KLEE-ah*, although the final syllable should be a long A) to be used to justify off-site general and administrative costs. In 1960 the Eichleay Corporation was able to demonstrate that it had been stopped from working for a period of time by a government-caused delay. It had been contracted to build three Nike missile sites for the Federal government. Based on the nature of the delay, Eichleay showed it had to remain at the ready and could not pursue other work with the project team.

The success of the Eichleay case resulted in others taking up the formula, which has traditionally been named for the founder of the Eichleay engineering company – John Eichleay. Although it is basically a simple mathematical formula, a company seeking such damages should consult an expert, and the attorney representing the Constructor should be thoroughly familiar with the concept and the method of documenting costs. Some courts may not recognize the formula at all, but even if they do, there will be those that challenge the validity of the actual dollars being sought or justified. Additionally, accurate, substantiated accounting records will have to be submitted. It is even possible the accounting will be reviewed in the office of the Contractor. This is not a situation in which the Constructor can give numbers.

The opposition will be closely examining not just the bookkeeping records of the project but the entire company accounting. It is possible that some costs will not be allowed, including advertising, food and entertainment, and education expenses, particularly after the project was awarded, or for projects that were not part of the work in question.

Initially, opposition to the results of the Eichleay formula calculations might be challenged on the basis they are an approximation, and no one can be certain exactly what time in the home office was being spent on any project, particularly the one in question. Generally, however, the fairness of the overall interpretation is reasonable simply because accuracy would be impossible.

The Constructor should also recognize that the time of the delay may be a factor. If the project is basically done, then the argument will be that hardly any overhead at the home office was being devoted to a project that only required a few final (probably minor) items to complete. This then becomes a decision for the Constructor of whether seeking compensation is going to be worthwhile at all. As with any litigation, the cost of seeking the damages has to be weighed against the chances of success and the reward such success will bring both financially and in regard to future business relations.

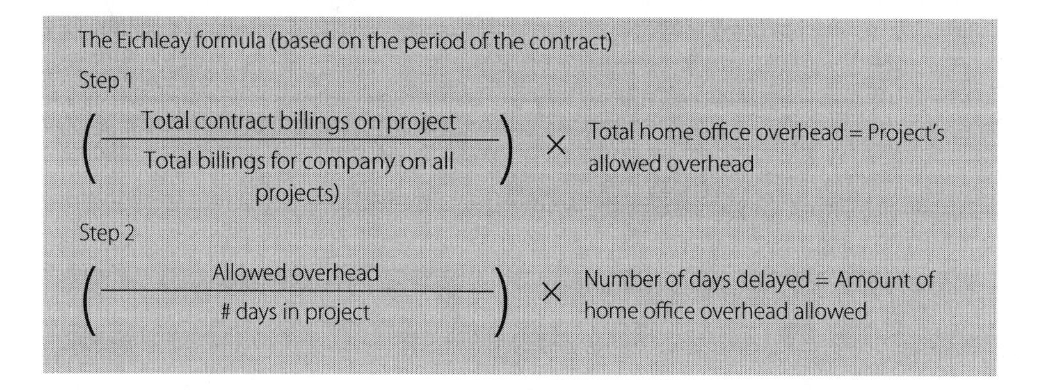

The Eichleay formula (based on the period of the contract)

Step 1

$$\left(\frac{\text{Total contract billings on project}}{\text{Total billings for company on all projects)}}\right) \times \text{Total home office overhead} = \text{Project's allowed overhead}$$

Step 2

$$\left(\frac{\text{Allowed overhead}}{\text{\# days in project}}\right) \times \text{Number of days delayed} = \text{Amount of home office overhead allowed}$$

Exercise 11.2

Take a moment
Calculate the Eichleay formula

Let us take some round numbers to calculate the Eichleay formula. Using the formula in the box above, assume you have a $1 million project. Your project will take exactly 400 days, including the delay of 100 days. During that time the company will also invoice $6 million worth of work on other projects. Records show the total overhead expense taken from the balance sheets of the corporate accounting during that period is $350,000. Calculate what you will be allowed to invoice for the home office overhead.

Discussion of Exercise 11.2

Your answer should have been $12,499, or basically $12,500 would be allowed in additional overhead costs for home office activities.

Step 1. Take $1,000,000 and divide by $7,000,000 [the additional 6,000,000 plus the 1,000,000 of the project]. Then multiply that result [.143] times the total overhead of $350,000. That gives $49,999.

Step 2. Divide $49,999 by 400 days in the total period equals $124.99 per day. That is then multiplied by 100 days in the delay to get $12,499.

The Eichleay formula has been recognized in some public work, but it does require a complete or almost complete stoppage of work by the government, along with a demonstration that other work could not be undertaken by the Contractor through the delay period.

The Contractor will have to decide if it is worth the potential cost of pursuing compensation from a private Owner where the formula might make sense but the public relations would not, and the only probable recourse would be through the court system or arbitration and the likelihood of success rather slim.

11.3.3 Adds and deducts

One of the first methods that will be used to determine add or deduct changes will be Unit Costs. Some contracts have units agreed upon for certain types of work. Many Owners who do put Unit Costs into the contract recognize and allow different unit prices for adds and deducts.

In addition to units defined in the contract, Design Professionals will often look to pay applications for determining unit prices. If there are a known quantity of units already in a contract, then computing the pricing in the pay application should produce a cost that would be reasonably close to what additional units should be worth.

Unfortunately, although unit pricing is a possible method, the Contract Administrator must recognize where in the flow of the project the change occurs. If the situation arises in the middle of production of those particular units, then unit pricing should be fairly accurate in terms of the contract, provided the original estimate was accurate. (*Note*: this is a good reason not to front-end load a project in case units are derived from pay application items later in the project.) However, often the need for the change might be discovered when the rest of the units have already been installed.

Examples:

1. To add an additional cubic yard of concrete one day after the pour is complete, so that a new pad can be added for an additional column base, would cost much more than having done that same cubic yard while the rest of the concrete was being put in place.
2. If the painter has already left the project, to return to paint an added knee wall at a lobby stair is going to cost more than if the painter could have done it while on site.

Sometimes additional work, such as the knee wall painting in the example above, can be put off until the punch list, but it should be recognized and documented that in such a case the work is still an extra. How mobilization for both the punch list items and the extra work is handled will be determined by how well the parties can cooperate and appreciate each other's position.

After looking at Unit Costs, the Contract Administrator will probably be asked by the agents of the Owner to put together a lump sum proposal for the costs. Estimating skills will be important. For the typical Contract Administrator it might be important to talk with the field personnel who would be called on to do the work. They might have insights into potential costs that would not appear immediately just based on pricing the work from a drawing.

Examples:

1. A change might have been uncovered that required other trades to remove work already installed or to do some additional work not shown on the change drawing in order to accommodate the installation.
2. Double handling, remobilization, or replacement and handling of supplies consumed in the extra might not be obvious from simply looking at the change drawings.

If one or more Subcontractors are involved, the Contract Administrator might need to be the "first line of skepticism" related to pricing. Just as the Design Professional might challenge a Contractor's numbers, the Contract Administrator must be prepared to challenge Subcontractor and Supplier numbers. The more confidence the Contract Administrator has in the dollars being given by Subcontractors and Suppliers, the easier it will be to justify costs to the agents of the Owner. This is true for both add and deduct change orders. Although every Contractor must protect the Subcontractors and Suppliers that make the successful completion of any project possible, there is more than a contractual obligation to the Owner and the Design Professionals, also. Just as it is an implied contractual obligation for all parties to cooperate, it is also assumed all parties will treat each other fairly.

If the lump sum is not approved or, more likely, if "time is of the essence" to start work on the change, the Owner might authorize the change costs to be determined by the Time and Material cost method. This sounds like a good approach, and it is, provided there will not be arguments over the split between actual change work and what was to be done as contract work.

The Contract Administrator will have to keep careful documentation of all costs associated with the change, but even then, there can be issues that one needs to balance between good will and getting paid what is due.

Examples:

1. You over-dug for the footings of a warehouse due to your own survey error. The Owner decides to use the additional space that will be allowed and directs the Engineer to detail a new column location and changes in the beams. These changes will require some extra costs related to the steel fabrication. Do you charge for the additional excavation, or even the additional concrete, since you would either have poured the excess regardless or have more formwork to do in order to correct the error effectively?
2. The Owner's agent knows you overstocked the project with drywall. A new corridor wall is added. If you charge for the material should there also be a labor credit for not having to remove the excess material?

Once again, just as in accumulating the actual costs for the lump sum, the Contract Administrator should not be documenting costs in a vacuum. Several trips to the field and/or close communication with individuals supervising and doing the work will help ensure that the costs are being recorded accurately.

Most accounting software for payroll will allow the Contract Administrator to assign a specific number for changes. The assignment of a different number is usually not the true challenge. What becomes a major issue is having the field personnel accurately assign the right time to the right number. Of course this is usually true of contract work just as it is true of changes; but with a change, the Contract Administrator has an additional level of approval that goes beyond the typical pay application. The agents of the Owner, and ultimately the Owner, must be convinced the time and the material listed are accurate.

* * * * * * * * * * *

Another habit the Contract Administrator might want to develop is that, at the beginning of any change order that involves, or might eventually involve, Time and Material justification, the Contract Administrator goes into the field with whoever is documenting the Time and Material and discusses the process of recording the costs for the extra. This will truly be "a stitch in time saves nine" approach to change order documentation.

* * * * * * * * * *

11.4 Responsibility

You must be the change you wish to see in the world.

Mahatma Gandhi

Of course there can be difficulties at all stages of contract administration, but responsibility related to change orders can be the start of deteriorating relationships among the parties involved in the project. No one wants to be wrong. No one wants to have fault placed on them. No one wants to learn they made a mistake. No one wants to know something is going to cost more than expected. And no one wants the project to take longer than necessary.

Whether you are an Owner learning of serious problems, a Design Professional recognizing some flaws, or a Contractor discovering you have to "eat" something you never expected and did not have in the original estimate, changes can hurt someone.

At some point in the steps of the change order scenario detailed above, the Contract Administrator has to recognize that finding fault is not finding a solution. The sooner the solution attitude can permeate the entire team, the more likely relations can remain stable and, ultimately, productive.

The Contract Administrator's approach to changes, therefore, becomes a major factor in the successful completion of the project and the continued teamwork of all parties to do so. In a Design-Build approach where both the design and the construction are under the same corporate entity, the internal blame game should not leave the company's boundaries, but in a Design-Bid-Build project delivery method the impulse will usually be to blame others. External excuses become the easy way to explain deficiencies, ambiguities, and non-compliance.

In some instances the Design Professionals under separate contract with the Owner might be liable for errors and omissions. For the Constructor to brush that off as "that's not my problem" is short sighted. Although insurance may cover the issue, no one likes to have claims made against their insurance. Moreover, the Constructor should be very careful about errors and omissions, since more and more the entire shop drawing process is drawing Contractors, Subcontractors, and Suppliers into the design process. What might initially appear as a design error on the part of one of the Owner's agents could in fact come back to something submitted by the Contractor.

The Contract Administrator's greatest asset in the change order process is to begin by following all the terms and conditions of the contract and then creating an atmosphere of open dialogue to find solutions. In this process the more individuals involved, the more likely more solutions can be brought forward. With more choices, the potential is greater to find an acceptable solution that is also less costly for whichever party must bear the costs. The attitude to see "how well we can get past this" should please all parties, and it will certainly be recognized by all those who are looking to work together toward a successful outcome. And those, in fact, are the people you want to work with again.

A CASE IN POINT

We must be aware how we all set the tone of collaboration and teamwork.

One of the greatest partnerships in architecture and life is Bob Venturi and Denise Scott Brown. The house Bob designed for his mother was featured on a United States postal stamp series of great twentieth-century Architects.

Along with her husband, Denise has been a prolific writer and great influence on current and future generations of Architects. Their planning and designs have been hailed both nationally and internationally, but when they were awarded a contract in a particular city in the western United States, they were amazed to be suddenly inundated with phone solicitations from law firms.

Figure 11.4 Vanna Venturi House
Photo credit: Robert Venturi

Finally asking why so many attorneys were soliciting their business just because they had been awarded a new contract, they were told, that is how Constructors do business in that part of the country. They intend to make their profit on the Architect's errors and omissions. The Constructors view the change order process as their best way to make money, particularly on an out-of-state Architect.

As Denise pointed out to me, "forewarned was fore-armed." In "Going the extra mile" you will have an ethics discussion topic on this practice by Constructors.

INSTANT RECALL

- Change order clauses are put into contracts, acknowledging that the Owner has the right to make changes to the original contract – the original promise.
- In most cases a flaw or ambiguity in the contract will be "construed against the drafter" of the document.
- There are usually several methods suggested for arriving at an agreed price, including lump sum, Time and Material, and Unit Cost.
- The Contract Administrator must recognize that any change has the potential to affect the schedule.
- As soon as a change is recognized as affecting the schedule, the Owner and the Design Professionals should be notified.
- In seeking an extension of time, the Contract Administrator will first have to demonstrate how the change will affect the critical path.
- If the Owner does not direct the acceleration, the Contract Administrator's work becomes a little more difficult. At this point, the Constructor has to document all the costs that are incurred due to the Constructive Acceleration.
- What the Contract Administrator will want to demonstrate first to the Owner and the Design Professionals, but ultimately to a mediator, judge, or arbitrator, is at least the following:
  - There was an excusable delay that was not caused by the Contractor.
  - All contract provisions were adhered to, especially notification of the delay and the notice of the need to accelerate if a change of time was not granted.
  - The Owner and/or an agent of the Owner with the Expressed Authority to do so refused any extension of time and ordered through some means such as threats of Liquidated Damages, withholding payments, or written notice that there would not be an extension of time.
  - Proof of actual costs above the costs that would have been made had the acceleration not been necessary.
- The conscientious Contract Administrator will be alert to all possibilities and negotiate the most effective process for deductive changes.
- The Owner may grant to an agent the authority to make changes.
- Relying solely on what is perceived as the Implied Authority or the Apparent Authority of an agent of the Owner can lead to difficult negotiations later in the project when payment comes due for an item that has not been "signed" for.
- Recognizing and notifying of a change is only the beginning. The entire approval process for a change order can be so burdensome that small changes are often more work than they are worth. Nonetheless, the Contract Administrator has to recognize the importance of the steps along the way to getting paid for any change order.
- The more confidence the Contract Administrator has in the dollars being given by Subcontractors and Suppliers, the easier it will be to justify costs to the agents of the Owner.
- The Contract Administrator will have to keep careful documentation of all costs associated with the change, but even then, there can be issues that one needs to balance between good will and getting paid what is due.
- The Contract Administrator has to recognize that finding fault is not finding a solution.
- The Contract Administrator's greatest asset in the change order process is to begin by following all the terms and conditions of the contract and then create an atmosphere of open dialogue to find solutions.

YOU BE THE JUDGE

Contract Management, Inc. vs. *Babcock & Wilson Technical Services*

A project to rehabilitate water lines for the U.S. Department of Energy in Oak Ridge, Tennessee, involved a Subcontractor – Contract Management, Inc. (CMI) – hired by Babcock & Wilson Technical Services (B&W). After the contract between B&W and CMI had been executed, Congress reduced the funds available for the project, thus necessitating a reduction in scope. Four of the five pipes on which CMI was to work were eliminated, and this led to a dispute over how the subcontract overall price would be determined.

B&W computed the reduction to the contract based on the line items submitted by CMI. This eliminated $769,105 from the contract, leaving a total contract of $1,506,582. To justify its method, B&W cited the case of Eugene Iovine, in which the same method had been used to calculate the reduction of fire alarms and smoke detectors.

CMI argued that the Eugene Iovine case had been different, since each line of the four to be installed could have been awarded to different Contractors, whereas the CMI/B&W contract was a total lump sum for the entire project to one Subcontractor.

CMI further stated that it objected to the deletion of the line items as a method of recalculating the costs of the one line, because its general and administrative costs had been spread throughout the other items that had been deleted in their entirety, and the solicitation breakdown had not contained a separate line item for general and administrative or overhead and profit costs.

B&W countered that by contract, add and deduct changes to the contract should include overhead and profit.

CMI countered that the changes actually increased the overhead by minimizing the work, changing the sequencing, and pushing the work out longer, due to the delay.

CMI sought a reduction change in price of $445,957, rather than the $769,105 taken by B&W.

How do you rule? In favor of CMI or of B&W (circle one).

Why? (Include whether you accept the argument that CMI was further delayed.)

IT'S A MATTER OF ETHICS

In the story about Bob Venturi and Denise Scott Brown, the specter of making a profit on architectural errors and omissions was pointed out to be a common practice among some Constructors. Think to yourself, or among friends or classmates discuss both the contractual ramifications of such a practice (e.g. does it fall under patent or latent ambiguities if the Constructor knows about the errors?) and the ethical implications of treating others fairly. Do bad contract documents make bidding the project tighter more effective when one anticipates multiple add change orders? Is there a down side to this practice not related to contract law or ethics? With profit margins so tight and competitive bidding so difficult, does this practice become necessary for some Constructors to stay in business?

Going the extra mile

1. Try another Eichleay request for home office overhead. The project is delayed by 237 days. The project itself will be billed out at $5,351,767 and require 499 days to complete, including the delay. The total company billings during that period other than the delayed project in question will be $62,404,231. How much home office overhead can the company submit for, due to the delay? Go to the Companion Website to check your answer. Use $2.5 million for total office overhead during the period.

2. Do the "You be the judge" exercise above, and then go to the Companion Website to determine how your ruling agrees or does not agree with the court's ruling.

3. **Discussion exercise I** (with your classmates): Discuss the challenges of the change order process shown in Figure 11.3. Particularly consider what challenges the Contract Administrator will face if the Owner directs the work to proceed before there is an agreed-upon price for the change or a recognized effect on the schedule.
4. **Discussion exercise II:** Based on the different approaches to changes in the AIA 201A, ConsensusDOCS, and the EJCDC, which do you think is a better approach? Does having more detail regarding the change order process make it easier, better, fairer? Does having less give more opportunity to submit accurate and reasonable requests, or is it an opportunity to inflate costs or the schedule of extra work?
5. Turn to Appendix I and locate one or more paragraphs dealing with changes in the contract. If your company had signed this contract, what would concern you most? How would you prefer the contract be rewritten? Check your answer on the Companion Website, and compare the subcontract with ConsensusDOCS subcontract.

6. Effective use of the correct vocabulary improves communication and promotes respect among colleagues. The reader should become familiar with the key words listed below that have been used in the chapter. Definitions can be found in the Glossary. Additionally, the student is encouraged to use the flash card exercise in the chapter's Companion Website as practice for possible quiz or exam questions.

Key words

Apparent Authority

Constructive Acceleration

Constructive changes

"Construed against the drafter"

Critical path

Directed acceleration

Eichleay formula

Estoppel

Fast tracking

Implied Authority

Latent Ambiguity

Measured Mile

Patent Ambiguity

CHAPTER 12

Differing site conditions

Often the first and longest-lasting situation facing a Contract Administrator is handling differing site conditions. Costs for differing site conditions can become extremely high and often are not fully known when initially uncovered. Also project delays are usually another major concern once a Differing Site Condition is uncovered. Issues of responsibility can create an atmosphere of opposition and contention at the start of the project or upon discovery of a Differing Site Condition that carry through until the end if the situation is not effectively handled.

Student learning outcomes

Upon completion the student should be able to. . .

- **analyze the conditions for both Type I and Type II Differing Site Conditions**
- **assess responsibility for differing site conditions**
- **organize procedures for notifying and administering Type I and II Differing Site Conditions**
- **understand risk and responsibility related to contract clauses regarding differing site conditions**
- **be able to explain why timely notice is important to the validity of a Differing Site Condition clause and the Contractor's claim.**

12.1 Differing site conditions

Expectation is the root of all heartache.

William Shakespeare

Nothing can be more confounding at any time during a project, but particularly at the beginning of the work, than to find or uncover something that was never expected. Almost always this can lead to additional costs for someone and a delay or increased effort to stay on schedule. Differing site conditions usually refer to uncovering something underground or related to the soil, but it could be something else that has been hidden, such as asbestos above a ceiling.

Occasionally a Contractor might also find that nature itself changes the conditions, and a dry site becomes inundated with water or a raised water table creates pumping and dewatering that was never included in the estimate. Just how cooperative the Owner will be in regard to acknowledging extra work will depend on contract clauses as well as the Contract Administrator's initiatives. Differing site conditions are not a ***force majeure*** in which nature in the form of storms, earthquakes, and other acts beyond the control of the

Constructor (or some acts of humans, such as war or strikes) makes project work or portions thereof impossible to complete. Differing site conditions can sometimes be considered a gray area in which some people, including the Owner or Design Professionals, might argue the Constructor should have known such a condition existed. Sometimes worse than that, even if the Constructor could not have known, the Constructor still bears responsibility through **exculpatory clauses** in the contract that relieve the Owner and Designers from all responsibility and place the burden upon the Contractor. In such a case, the Contract Administrator might very well believe the world has suddenly turned against him or her.

Once again, the more people notified immediately and brought into the discussion to find a solution, the better chance there will be that a successful conclusion will come about. In doing so the Contract Administrator should first provide a clear understanding of what type of Differing Site Condition exists.

12.1.1 Private versus public differing site conditions

A Differing Site Condition can arise on any project. However, the vast majority of claims and issues related to differing site conditions arise out of public contracts. In fact, in many private contracts, the language precludes claims for certain types of differing site conditions. Also in private work the term differing site conditions is often referred to as "unforeseen site conditions" or just "unforeseen conditions."

Nonetheless, a Contract Administrator should recognize that approaches to differing site conditions on a private contract can, and should be, very different than what alternatives exist on a public contract.

As early as the bid proposal, a private contract submission can usually contain a qualifying clause stating one or more issues are not included in the price. For instance, the bid proposal might state all excavation is based on certain soil conditions, or demolition does not include removal of hazardous substances. In private work, this actually allows the Owner to accept a lower price if he/she feels the gamble is worth it. In some instances, private bid proposals might contain a "contingency." The Constructor is telling the Owner that a certain amount is in the contract to be used for a specific issue that might be encountered. At the end of the work on that issue, the contingency might be adjusted upward or downward. Such stipulations within a private bid proposal usually lead to discussions between the Owner and Constructor prior to the final contract signing.

However, to include a contingency or an exclusion on a public proposal would almost certainly disqualify the Contractor as nonresponsive.

Once underway with construction, if a Differing Site Condition does arise, the private contract may contain clauses that place the burden on the Contractor, but with a good relationship existing between the Owner and the Constructor, it is possible the Owner might recognize that the Differing Site Condition was not something a Constructor should be burdened with without additional compensation. Approaching the situation from an open and reasonable perspective, the Contract Administrator might find solutions that will satisfy all parties.

A public contract does not have such latitude in interpretation. The clauses within the contract must govern how the Owner pays taxpayer dollars to the Contractor, and all changes have to be justified within the terms of the contract. Certainly the Constructor can put forward a claim for additional compensation based on differing site conditions, but that requires both timely notice and careful documentation. Generally, stating a claim will be made on a public works contract is recognized as part of the process when a dispute arises that cannot be settled within the existing contractual terms. Such a notice does not affect other work on the project or the relationship of those working together, provided blame is not openly ascribed to individual(s) on the project, thus creating tension between those attempting to complete the rest of the project.

To state in a project under a private contract that a claim will be made, however, often does affect the relationships throughout the rest of the project, and can usually affect opportunities to work on future projects.

A CASE IN POINT

There was a time when my attention was diverted by other projects not running as well as they should. One of our estimators had thought he had learned the secret of being the lowest bidder. He put three proposals into the same private Owner using a newly discovered Specialty Contractor who had amazingly low prices.

We were awarded all three projects, and it was not long before the Owner learned how we had been low bidder, and we learned how the Specialty Contractor operated. Working on the assumption that Type I Differing Site Conditions would become a change order, the Specialty Contractor presented an almost endless stream of requests for increases in price. The Designer and Owner, however, did not view any of the claims as more than patent ambiguities that a prudent and experienced Specialty Contractor should have recognized.

The Specialty Contractor threatened, and ultimately did present, a claim for most of the work he considered Type I claims. Since the Specialty Contractor did not have a contractual relationship with the Owner, his only recourse was to sue us. We protected the Owner against the claims, but were faced with considerable costs in the process of defending them.

Prevailing against the Specialty Contractor was in fact a pyrrhic victory in which winning was ultimately losing, since we never worked for that private Owner again.

12.1.2 Site investigation

Going back even before the bid proposal for public or private work, the Constructor should be well aware of the need for a thorough site investigation. Recognizing the potential of any differing site conditions during the pre-bid walk-through is important. It is at that time that such issues should be brought forward so the conditions can be noted and all parties will be bidding on the same situation. Although there can always be a condition that was not anticipated, if the Constructor is found to have been negligent or had not investigated the site prior to bidding the project, the chance of recovering costs for what seems to be a Differing Site Condition becomes very slim.

Site investigation will be discussed again later in this chapter in regard to certain clauses that are meant to protect the Owner from claims by Constructors in regard to differing site conditions.

12.2 Type I Differing Site Conditions

The whole difference between construction and creation is exactly this: that a thing constructed can only be loved after it is constructed; but a thing created is loved before it exists.

Charles Dickens

In many instances the Owner and the Design Professionals have waited a long time to see their vision fulfilled. Certainly, the Dickensian "love" has worn off a bit by the time the contract is finally signed by the Constructor, and the real thrill becomes the actual construction of the structure. Seeing the vision become a reality is the next real excitement for everyone. Anything that stops or detains progress is not going to be enjoyed at this point in the construction process, especially if it might cost more money.

We will discuss contract language in a later section of this chapter related to responsibility, but first it is necessary to understand exactly what the Differing Site Condition entails. Basically, the industry considers two different types of differing site conditions.

Type I Differing Site Conditions are conditions that differ significantly from what was indicated in the contract. The important distinction for Type I is that the condition is different from what was indicated. Obviously this can be a very difficult issue to handle diplomatically, because what the Contract Administrator will usually be stating is that this could be the fault of the Owner's plans and specifications, in which case, of course, the Owner will then want to understand if and why the Design Professionals made an error.

Examples:

1. The drawings indicate that according to test borings, there is no rock throughout the site. Of course this can have consequences both ways, if footings or piles need to be on bedrock or driven to a certain "refusal."
2. What was indicated to be a drywall wall to be removed turns out to be drywall over masonry block on a load-bearing wall.
3. Dimensions on the drawings for new rooms will not fit in the existing confines of the building.

Whoever is estimating the project, along with the Contract Administrator, should recognize that what a Constructor views as a Type I condition may not be considered such by the Owner or Designer. Usually information provided in the bid documents is for information only. There is almost always a phrase clarifying that the Constructor is not to rely upon certain information and should investigate the site before submitting a proposal.

12.2.1 Reasons for Type I Differing Site Conditions

There are many possibilities for Type I differing site conditions, including:

- The subsurface conditions are different than the conditions stated in the contract based on test boring data. There might be clay instead of earth, fill instead of clay, rock instead of earth, water instead of dry conditions. Such conditions can have a major effect not just on excavation but perhaps on the driving of piles, dewatering of caissons, depth of foundations, reaching required compaction levels, and replacement of excavated material that was supposed to be used from on-site material.

- The water table might be higher than indicated on the drawings. Again, dewatering is only part of the challenge. The Contractor must coordinate a revised method of putting the foundations in place, and the foundations will also probably need to be redesigned.
- Interior demolition reveals hazardous materials, such as lead paint or asbestos. Generally such issues are not shown on the drawings but addressed during the bid walk-through of a site. As long as it was documented that no hazardous substances existed or were anticipated, this would be a Type I Differing Site Condition. The Constructor will be required to stop immediately and arrangements must be made for removal by a qualified Contractor.
- Demolition uncovers that different structural elements are involved than shown on the drawings. This could include a bearing wall that was not noted as such, differences in thickness of floors, walls, or ceilings, passages too narrow for the installation of new material or equipment, hidden walls, ceilings, roofs that subsequent renovations covered.

In all these instances, and many more that might arise, there is an implication that something was missed during the design. Over many years, Owners and Design Professionals have recognized the difficulty of getting everything correct. How that has been handled contractually will be discussed later in the chapter, but for the Contract Administrator, the moment a Type I Differing Site Condition is discovered, certain steps should take place, even though this might be an instance where "shoot the messenger" is the first reaction.

12.2.1.1 Documenting the Type I Differing Site Condition

It is important that all personnel, particularly the field personnel, recognize that the Contract Administrator should be notified of any Differing Site Condition immediately. Sometimes enterprising craft workers will want to test their own abilities at resolving problems, and this can lead to even greater problems. Everyone on the construction team should know to alert the Contract Administrator the moment any Differing Site Condition is recognized. As with all potential change orders, notice to the Owner and the Owner's agents is important, and timeliness matters.

Once the Contract Administrator has been advised of a Differing Site Condition:

- Notify all parties to the contract and the Owner's agents of the Differing Site Condition, as well as anyone else, such as a Subcontractor or Supplier, that might be impacted.
- Request everyone meet about the issue. If at all possible, it is important to have this meeting in person at the site of the differing condition. Generally, tensions can more easily mount over a phone conversation than in person, and also "seeing is believing."
- Fully document all aspects of the difference. Make sure the contract is thoroughly examined for all details related to the condition, and then take photographs of the condition in the field. In addition to the photographs create a narrative of the situation, how it was discovered, and how it differs from what was shown to be the condition on the drawings or contract.
- Brainstorm with as many people as possible for as many solutions as they can imagine in order to find the best resolution.

Again, the Contract Administrator has to recognize that a Type I Differing Site Condition situation can put the Design Professionals in a difficult position. Seek solutions that can help everyone, but at the same time carefully document one's own position. The documentation of the Differing Site Condition is extremely important. In order to prove the Differing Site Condition beyond any exculpatory clauses in the contract that would relieve the Owner and/or the Design Professional(s) from responsibility, the Contract Administrator will have to demonstrate that:

1. notice of the differing condition was given in accordance with the contract;
2. the conditions shown on the plans and drawings are significantly different than what exists on the site;
3. the difference requires a significant change in approach that will impact cost and/or schedule.

Underpinning all the discussion, of course, is the assumption that the Contractor relied upon what the Owner provided. If the Differing Site Condition can be shown to be something any prudent Contractor would have recognized before submitting the original bid, then the Contract Administrator will be faced with a **Patent Ambiguity** or **patent flaw** situation and will have a much more difficult time justifying additional costs or time.

12.3 Type II Differing Site Conditions

Type II Differing Site Conditions are not directly traced to the contract documents. They arise out of conditions that are unexpected in the normal events of construction. The Constructor encounters events or conditions that are not normal and would not have been expected by a prudent or experienced Contractor.

Such a situation might seem to eliminate the finger pointing to who is at fault, but based on contract language, the Constructor must be careful. It is possible the Owner and Design Professionals have placed responsibility for such hidden or unexpected conditions on the Constructor.

Differing site conditions often arise out of site work and excavation.

Examples:

1. A Contractor expects to reuse excavated soil elsewhere on the site, but upon start of excavation underground contaminants are found to have permeated the soil, making it unacceptable for reuse.
2. Subsurface conditions are uncovered that prevent the use of heavy equipment.
3. Underground utilities are encountered that were not properly identified prior to digging.
4. Unanticipated historical finds requiring archaeological work are uncovered.
5. For those who may work overseas, uncovering an unexploded bomb can still be a hazard. Even in the twenty-first century, some bombs and unexploded grenades and other munitions are discovered in Europe, and mines and IEDs are unfortunately found in many other parts of the world.

12.3.1 Reasons for Type II Differing Site Conditions

Just as with Type I, there are many possible reasons for a Type II Differing Site Condition, and they can be similar to Type I in what is encountered (e.g. water, hazardous substances). The important distinction is that with a Type II Differing Site Condition, the drawings do not mislead the Contractor, but the condition is radically different from what a prudent Contractor would expect:

- The subsurface conditions were unusual for the typical excavation in the area. This could be that underground conditions were extremely different and unusual for where they are encountered. It could also be the result of previously unknown human activity – backfill of areas not known to have been a dumping area. This becomes particularly important related to impacting the schedule when archeological remains are uncovered, and the entire excavation needs to be shut down and experts brought in to analyze and perhaps hand-excavate the site.

- In some instances, burial remains not known to have existed on such a location or believed to have been removed previously have caused a stoppage in construction.
- A utility line that was not marked when the notice of excavation was announced.
- Subsurface water is found, such as an unknown stream or a higher than expected water table. Note how this differs from Type I. In Type I the contract documents gave a specific level of the water table. In Type II the contract documents did not mislead the Contractor, but the prudent Contractor had reasonable expectations the water table should have been lower. (See Exercise 12.1 below.)
- Once exposed, existing load-bearing material is found to be inadequate for what is to be added to the structure.
- Previous work is exposed or discovered that is not in conformance with existing codes.
- Exposure of existing materials and/or conditions will negate manufacturer's warranty for application of new materials.
- Hazardous materials are discovered in an area or structure not previously known for such substances. Of course these could include asbestos or lead paint, but excavators could also uncover hazardous materials under the earth and what was expected to be clean fill turns out to be a contaminated brown field.

Example: Remember the story about the solid rock in the middle of the site to restore what was believed to have been a previously existing saw mill. There were no borings and the Constructor was advised that the subsurface conditions were its responsibility. However, if the site had previously been where a saw mill had existed, the Contractor was entitled to rely on the fact that no rock should have been there. For the Contractor it is clearly a Type II, but to the Owner, the situation might appear to be a Type I where the Design Professionals should have been more certain that the location was in fact where the previous saw mill had been. Had we pursued a Type I, we would have been trapped by the exculpatory clause of "No compensation for rock excavation."

Of course, at times there is a delicate division between Type I and Type II. Some might think the Design Professionals should have exposed some conditions and noted on the contract documents what would be encountered, but the key division between Type I and Type II is that in Type II the situation is not expected. Even this, however, can be an area of discussion or even debate.

Another issue a Contract Administrator must take into consideration is whether a prudent Constructor will have recognized the possibility of a Type II Differing Site Condition. For instance, excavating in a known historic area, uncovering archaeological remains is a distinct possibility. Also, an experienced Contractor would generally know what underground soil conditions exist in an area. It is important to recognize what is reasonable to expect and what is unexpected by a reasonable Contractor.

Exercise 12.1

Take a moment

A Constructor contracted with the Corps of Engineers to construct a levee along the Mississippi River. They needed to work off a barge that required a certain level of water to float the barge. The Corps controls the water levels through locks and dams and provided the Contractor with the expected water levels by date. Shortly after work started, the water level dropped as shown on the documents supplied by the Corps. As a consequence the Contractor failed to meet the contractual obligations until the barge could be properly floated into place and the work completed. The Contractor then sought additional time and damages for the longer duration of the project. The Corps denied the claim on the basis it had provided the Contractor with the anticipated water levels. The Contractor stated it had reasonably expected the Corps would manage the heights of the water levels so the work could proceed on time.

Is this a Type I or Type II claim? (Circle one)

How do you rule? For Corps / For Contractor (Circle one)

Discussion of Exercise 12.1

First, it is a Type I Differing Site Condition. The reason it is Type I is that the contract documents clearly showed the water level would fluctuate. The reason it cannot be considered a Type II is that a reasonable and prudent Contractor should know the water level would vary because the Corps of Engineers not only stated it in the contract documents, but the water level consistently varies and this should not have been a surprise to the Contractor.

That fact is also the reason to award in favor of the Corps. There was nothing contrary to the contract documents that the Contractor could rely upon. The water level had varied in accordance with the contract documents.

In approaching any Differing Site Condition, therefore, the Contract Administrator needs to recognize that each party on the team might have a different perspective on who is responsible for what. Documentation is again very important.

12.3.1.1 Documenting the Type II Differing Site Condition

The steps outlined for documenting Type I Differing Site Conditions are just as valid for Type II. However, an important additional step at the beginning of the process should be added:

- Consult with all individuals familiar with the situation who had input prior to submission of the proposal and signing of the contract about the expected conditions for the site.

Although this might be useful even on a Type I Differing Site Condition, it is very important for the Contract Administrator in a Type II condition to recognize that what has been encountered is unusual and/or unexpected.

12.4 Responsibility for differing site conditions

It is understanding that gives us an ability to have peace. When we understand the other fellow's viewpoint, and he understands ours, then we can sit down and work out our differences.

Harry S. Truman

Some contracts contain a benign or favorable differing site conditions clause relative to the Contractor's responsibility, but uncovering either a Type I or Type II Differing Site Condition can be a challenging moment in the administration of a contract. Such a situation can be very costly for an Owner both financially and in terms of the schedule for completion and permanent use of the structure. For that reason some contracts have evolved that favor the Owner's position and place some additional burdens and responsibilities on the Constructor.

A CASE IN POINT

There are some who believe the Chinese expression for crisis (*weiji*) is composed of two words: "danger" and "opportunity."

Although this may be more motivationally than linguistically accurate, it does provide some perspective to analyze some stories.

Bowen Engineering had contracted to build a wastewater treatment plant in Owensboro, Kentucky. Although the soil borings had suggested there might be some soft earth, the actual conditions after award of contract proved to be quite severe. Excavating a small portion as a test showed the soil to be silty clay that pumped considerably when under pressure.

Notifying the Owner immediately to the possibility of a Differing Site Condition, the Bowen team did not rest there. Of course, by contract it was possible the Owner might consider the soil conditions Bowen Engineering's responsibility. The Bowen team, along with the Owner, analyzed all possible alternatives. In the process they lost a week to planning as they went over all possible alternatives.

Arriving at a different method than originally planned, the team actually came up with a way to save a month over the contract schedule, proving what the Chinese *might* believe – that a crisis can be both *dangerous* but also an *opportunity*.

One immediate issue, however, related to Type I Differing Site Conditions will be whether the responsibility should fall on the Designer(s). The Spearin Doctrine will clearly place errors in the design documents upon the "drafter" (the Owner in the case where the Owner has supplied the documents to the Constructor, even though the Architect actually created the drawings). If there is a discrepancy between the contract documents and the actual conditions, the Owner may be faced with additional costs, but the Owner may look to the Design Professional for restitution, based on errors and omissions on the part of the Designer.

12.4.1 Notice requirements

As with any change, the Constructor will be expected to provide timely notice to the Owner and the Owner's agents whenever a Type I or Type II Differing Site Condition is encountered. Since there might be some repercussions related to the Design Professional's responsibility, the Contract Administrator might be hesitant to state that costs and schedule could be affected by what has been uncovered, but unless the Contractor notifies all parties in the way the contract requires, the recovery of costs or extension of time might be jeopardized.

Within a specific period of time, the Contract Administrator must also be prepared to submit a detailed explanation of what the costs and schedule impact will be. Of course, it may be impossible to give exact amounts or time if the full scope has not been determined, but giving a reasonable estimate will be important.

Going forward, the Contract Administrator will need to provide accurate pricing for the additional work required. This can be quite challenging if the same personnel and equipment will be used for the changed condition as were going to be used for the previously "presumed" condition. In some instances sharing details of the original estimate might be necessary, but a good pay application breakdown might be another appropriate starting point.

Example: If the total cost of the original work can be "lifted out" of the original breakdown, then an agreed-upon lump sum change order, or a change order based on Unit Costs, or Time and Material change order might be written to completely replace the deduction of the original work.

One additional burden for the Contract Administrator related to differing site conditions is the effect an increase in schedule might have on the overall contract costs. The Contract Administrator might be faced with significant additional field costs if the project is delayed. It also could be a time to apply the **Eichleay formula** for home office overhead.

Whatever decisions the Contract Administrator makes in regard to a Type I or Type II Differing Site Condition, strained relations between some members of the project team are quite possible, even probable in some instances. Trust will be the true foundation of keeping everyone moving toward the final goal together. That will start with honesty and openness on the part of the Contract Administrator.

A CASE IN POINT

Archeological finds are cause for an immediate stoppage of work. A determination will be made as to what should be done in relation to the artifacts. Certainly for crews interested in collecting a paycheck so they can pay bills and put food on their families' tables, the possibility of stopping work and being laid off is not pleasant. The Contract Administrator, however, must be careful crews do not treat uncovering an artifact as an inconvenience or, worse, try to ignore it or cover it up.

The story of the Liberty Bell Pavilion is an example of how thoughtful use of such a delay can provide a better finished structure when everyone works together. Below, the uncovered foundation from the excavation of the slave quarters has been preserved.

Figure 12.1 Foundation from the excavation of the slave quarters, Liberty Bell Pavilion, Philadelphia

Photo credit: Charles W. Cook

Figure 12.2 Reconstruction of President Washington's Chief Executive House in Philadelphia
Photo credit: Charles W. Cook

Constructing a new "pavilion" for visitors to view the Liberty Bell was delayed by what should have been a known Type I Differing Site Condition. It was common knowledge that George Washington, as the first President of the United States, had brought with him slaves from his Virginia plantation.

To be situated just south of the restoration site of George Washington's residence when Philadelphia was the capital of the new United States, the new Liberty Bell Pavilion's excavation exposed the foundations of Washington's slave quarters. The tragic history of early America in relation to the treatment of enslaved African Americans was in direct and unpleasantly ironic contrast to the inscription on the Liberty Bell:

> Proclaim Liberty throughout the land unto all the inhabitants thereof.

Construction was stopped so as to excavate with brush and trowel what would usually be done by a backhoe and loader. In place of a dump truck, a box with a wire screen sifted the dirt away from even the most minute artifact.

The delay actually allowed the City and the Federal government to come to an understanding of how to proceed. Many wanted the new pavilion relocated. In the end, the Pavilion was completed where it had originally been intended to be built so it could more closely associate the Bell with Independence Hall, where it had originally hung and tolled "Liberty," but not to all the inhabitants throughout the land. Although perhaps an inconvenience no one wanted, the delay allowed both the Design Professionals, the Owner (the National Park representing all Americans), and the Contractor to more appropriately construct the area. The President's House was only partially finished, a symbol that we now recognize our new nation had a long way to go to truly achieve the ideals we had declared in July of 1776. And the slave quarters were respectfully made a part of the whole story – a story that even in its imperfection might inspire each new generation of Americans that the *building* of America remains the task of each one of us.

12.4.2 Exculpatory clauses

The most grievous of all Differing Site Condition clauses that shift responsibility to the Contractor are the exculpatory clauses. Unfortunately they appear in more and more contracts presented by the Owner for the Contractor to sign. And they are usually signed, because the Contractor wants to do the work, and in a moment of excitement that a new project will come one's way, the fear of assuming Differing Site Condition issues is far from the Contractor's mind.

However, exculpatory clauses can be and often are written to exonerate the Owner from any and all errors or misleading representations and statements on or in the contract documents. The Owner will warrant the documents for information purposes only. This is an obvious response to the **Spearin Doctrine**, which places responsibility for errors in the contract onto the Owner or the entity that provided the documents to the Contractor. By signing a broad-form exculpatory clause the Contractor may be assuming all costs for anything encountered unexpectedly that requires additional work, as well as potential Liquidated Damages if the contract time is extended.

How much a court will enforce such clauses is up for debate, but that debate will be by lawyers in front of a judge. That will be much farther in the future compared to the first discovery of the Differing Site Condition, and it will be expensive. The immediate impact of such a clause is to cause a great deal of pressure on the Contract Administrator either to find an amicable and financially viable solution or to start extensive documentation to be used in mediation or litigation.

12.4.2.1 Preparing to negotiate beyond exculpatory clauses

It is sometimes said the best offense is a great defense. This can be true of exculpatory clauses related to differing site conditions, but the defense needs to play first. In other words, during the bidding phase, any potential differing site conditions should be explored as much as possible. If there is a potential Differing Site Condition that could not be verified and was not addressed in the bid phase, then one possibility in private work is to qualify the bid that the Contractor does not have that Differing Site Condition covered in the bid and will want it to be excluded as a condition of the contract. Again, in private work, differing site conditions are usually considered "unforeseen conditions," but to qualify the bid, the Constructor is alerting the Owner to the cost of such an "unforeseen condition."

In such a case the contract needs to either reference the bid document as part of the contract or preferably call out the exclusion within the signed contract. Otherwise, if not referenced or included, discussions and even written statements prior to the contract will not be considered when interpreting the contract as the true agreement between the parties at the time of signing. As discussed in a previous chapter, conversations and understandings reached prior to the contract signing but not incorporated into the contract would be considered parol evidence.

It is important to again note, however, that such a qualifying statement in a public bid proposal would render the bid "unresponsive."

The Contract Administrator must recognize in relation to private work that a great deal will depend on the rapport and relationships one has established with the other members on the project team, especially the Owner and Owner's agents. It is also important to have strong and competent Subcontractors and Suppliers who fully understand the consequences related to exculpatory clauses and how to put forward cases in relation to what was shown, expected, or reasonable.

In public work, even the best relationships may be strained, since often the Owner's representative will not have the authority to look past the exculpatory clause. If this is the case, careful, detailed documentation and expert counsel will be required if a claim is to be pursued in the future related to the Differing Site Condition.

12.4.2.2 Site investigation exculpatory clauses

For the Constructor, the most grievous of the exculpatory clauses is usually related to site investigations. It generally seems unfair that an Owner and often the Design Professionals have had access to the site for a considerable length of time, but the burden of what may be underground or hidden within the structure becomes the responsibility of the Contractor, who probably has had a much shorter time frame during the bidding process to determine what might be a hidden condition. Nonetheless, a clause absolving the Owner and Design Professional from responsibility for hidden conditions is often inserted into the contract.

At the time of pre-bid meetings, the prudent Constructor will try to make a part of the record any possible issues related to underground obstructions or other hidden conditions. Courts may recognize that a Differing Site Condition could not reasonably have been known and therefore overturn the exculpatory clause based on **conscionability**, but once again, that takes time and money, and there is no guarantee.

In other words, hoping to base one's defense on having been conscionable will be relying upon a judge believing you were at least as honest, ethical, prudent, scrupulous, and moral as the Owner, if not more so. It sometimes works, but there is no guarantee. In fact, an exculpatory site condition clause might be upheld on the basis the Contractor did not make a reasonable and prudent investigation of the site.

A CASE IN POINT

James John "Jack" Cook (no relation of mine), doing business as C&C Development Company, won a case of differing site conditions when renovating a fish hatchery. The court ruled in favor of the Contractor and against the State for failure to disclose the existence of underground water and for misrepresenting the moisture content of the soil.

The State appealed the ruling to the Oklahoma Supreme Court in *Cook* vs. *Oklahoma Board of Public Affairs*.

The State Supreme Court reversed the decision of the lower court on the basis "the Oklahoma Board of Public Affairs and the Oklahoma Department of Wildlife Conservation did not owe a duty to disclose a subsurface condition to bidders on a construction project when its contract contains a clause that expressly disallows additional payment or extensions of time for completion of work if the Contractor fails to acquaint himself with all conditions relating to the work."

The Oklahoma Board of Public Affairs was able to demonstrate Cook had not attended the two pre-bid meetings and that the other Contractors who bid the work had been able to determine the moisture conditions of the soil. The State also noted the only site visit conducted by Cook was by his son after the pre-bid meeting, and that consisted of a drive through the site accompanied by a manager of the hatchery.

Cook also contended he had a right to rely upon the Engineer's soil borings provided by the State.

The Supreme Court disagreed, providing a lesson for all Constructors and Contract Administrators related to due diligence in preparing a bid that might contain a Differing Site Condition claim.

As previously mentioned, therefore, the Contract Administrator will need to meet to discuss with the estimating team and anyone who had input into the potential site conditions prior to submission of the bid. The more the Contract Administrator can show that the company took a careful, reasonable, and experienced approach to the submission of the bid, the better the chance is that the Owner and the Owner's agents will recognize that compensation for the situation might be in order. On a government project that had a public opening, the closeness of the bids might also show the probability that no one else had figured on the Differing Site Condition. On private work, this might be more difficult to demonstrate, but it might very well be something the Owner would recognize.

12.4.2.2.1 The Contract Administrator and the estimating staff

In many companies, particularly the larger Constructors, the estimating staff is different and separate from the actual project team. This can cause a communication gap for the Contract Administrator responsible for the project and the change orders that may arise from it. It is also a point for potential conflict within the company.

A staff that is dedicated to the estimating process will have long since moved on to other estimates when a Differing Site Condition becomes an issue on a project they had estimated months before. The timeliness of approaching estimators becomes quite important. Asking questions about "history" while an estimator is trying to submit a new bid proposal can cause considerable stress.

In addition, approaching an estimator on something that might be construed as an error on his or her part might never be a good thing to do. Human nature is such that no one likes to be told they did something wrong, and even the suggestion will get the conversation off to a poor start.

One effective way to partially eliminate this difficulty is to have a very comprehensive bid turn-over meeting prior to the project team assuming responsibility for a new contract. In order for this to be effective the Contract Administrator should *thoroughly* review (with emphasis on "thoroughly") the contract documents and determine all possible issues, particularly any exculpatory clauses. At the turn-over meeting, the Contract Administrator can then ask the questions related to how the site condition clauses may have been approached in the bid process. The more thorough the Contract Administrator has been in the reading of the contract documents, the deeper and better the recognition and understanding of the challenges will be for all involved in the discussion.

This approach is an additional burden on the project team, and particularly the Contract Administrator, but it will improve understanding of the commitments going forward and probably allow better communication in the future, should the need arise to revisit a particular issue.

Exercise 12.2

Take a moment
Prepare to interview the estimator

Go back to Chapter 7 to re-read the story of the grist mill. If you had been the Contract Administrator who was informed as soon as the rock was uncovered, list below what questions you would have asked the estimator in order to prepare for the submission for a Differing Site Condition claim.

Discussion of Exercise 12.2

In one sense, the above exercise is a "head fake" – perhaps sending you in a different direction than necessary. The Differing Site Condition claim would ultimately be rejected by the State, but thoroughly researching the estimator's intentions would lead to the conclusion that the contract called for something to be "rebuilt" that had not existed on that site in the manner shown on the contract drawings. The best way to determine that would be to truly understand why the estimator had ignored the need to take any site borings. That might be revealed through a lot of questions, or just one simple one: "Why weren't you worried about the possibility of rock on the site?"

12.4.3 Relief from differing site conditions

The Contract Administrator must rely upon two forms of "reasonableness" for relief from differing site conditions:

1. The reasonableness of the Owner is an important element in securing relief from differing site conditions. Often the input of the Design Professionals can be helpful in this regard, but that is not always certain. And, in fact, if errors and omissions might be involved, the input of the Design Professionals might be counter to what the Constructor needs. Reasonableness of the Owner may have been demonstrated early in a "benign" Differing Site Condition clause. On the other hand, reasonableness might be demonstrated after uncovering a genuine Differing Site Condition even though there is an exculpatory clause absolving the Owner. Some private Owners (usually not public authorities bound under requirements of the contract) will recognize that placing the costs and schedule burdens onto the Contractor would be unreasonable.
2. The reasonableness of the Constructor's own approach is also of importance. If the Contract Administrator is unable to convince the Owner, then the courts or arbitration panel will look to the "reasonableness" of the Constructor when bidding and executing the project. Litigation or arbitration is certainly not the desired approach, and it should always be examined in relation to the prospects of success and what that gain might be, balanced against the potential loss in more than just money, including in relation to future work. In short, the reasonableness of the Constructor's effort in relation to the Differing Site Condition should be weighed with the *reasonableness* of how one obtains, if at all, relief from the costs of the Differing Site Condition.

Example: In *Hollerbach* vs. *United States*, Hollerbach was contracted to repair a dam in Kentucky. The backing of the dam was represented on the drawings to be of a certain material, but when work began it was found to be much different, and more difficult to proceed with the repairs. The Owner relied upon the requirement that the Contractors prior to bid were to make their own site investigations and any representations were not to be relied upon. The Supreme Court ruled that a positive statement regarding the conditions of the contract must be taken as true even though there are other clauses within the contract requiring the Contractor to make independent investigation of the facts.

12.4.3.1 Disclosure

One specific relief for a Constructor facing costs of a Differing Site Condition is that the Owner does have an implied responsibility to disclose conditions that would be different than what is expected under normal construction, or different from what is shown on the contract drawings or in the specifications.

In the case of knowing something is different than expected and not disclosing it to a prudent and reasonable Contractor, a court would likely recognize the right of the Constructor to recover. Although perhaps not intentional misrepresentation, this is a matter of not being conscionable.

In the event the Owner clearly knew that what existed was different than what was shown to be the case in the contract, then the courts would usually recognize that as a deliberate or intentional misrepresentation and would place the burden of costs and change in the schedule upon the Owner. Of course, working with Owners whom a Constructor can trust is the ideal way to avoid such disputes, but there have been cases in which Constructors have been deliberately deceived.

12.4.3.1.1 Unintentional misrepresentation

We have also previously seen in the case of the saw mill that an Owner's drawings were in error because, based on the contract documents, the saw mill could not possibly have existed where it was supposed to be "reconstructed." Relief in that case was based on an acknowledged mistake with no intention to deceive, but a clear misrepresentation of the contract intent and conditions. Both public and private Owners, therefore, can recognize that the contract as written would require far more work than a reasonable Contractor (and perhaps a reasonable Owner) would willingly undertake. Termination of a contract for convenience might be the remedy in such a case. Or perhaps a **Cardinal Change** to the contract would be the result. In such a case, the Contract Administrator might need to contact the bonding company if it is a bonded job.

12.4.3.2 Spearin Doctrine

If a Constructor does need to pursue litigation as a means of recovering costs related to a Differing Site Condition, the legal justification will probably refer back to the Spearin Doctrine, discussed earlier. The Constructor has the reasonable position of relying upon the Owner's drawings. In accordance with the contract, the first interpretation of the contract documents will probably be made by the Owner's Design Professional(s). This, however, might easily be a conflict of interest, since the Design Professionals are probably the ones responsible for the drawings and specifications. Naturally, the Design Professionals will not want to find themselves at fault, but just as reasonably, they probably recognize their intent from their own perspective. How acceptable the Contractor's perspective is depends on a fundamental principle of common law:

> No matter how different in whole or part an interpretation might be from the intent of the author of the contract documents, the Contractor needs to show a reasonable interpretation based on a reading of the contract as a whole.

In other words, the Contract Administrator needs to recognize that in addition to a reasonable interpretation, the contract will also be read in accordance with the **precedence clauses**. The "reasonable" interpretation must hold up under the scrutiny of "did another clause in the contract that was intended to take precedence clarify the situation?"

Reliance on the Spearin Doctrine alone will not be a panacea for relief from differing site conditions. The prudent Contract Administrator intent on maneuvering through the obstacles of exculpatory clauses and differing site conditions may want to keep that strategy in the background to be used when and if necessary. In the meantime, the Contract Administrator will want to be looking for a "reasonable" solution that all parties can live with together.

In the end, perhaps that is the most important lesson of all for the Contract Administrator. Forged sometimes from the heat of differing site conditions, the Contract Administrator's approach to all project disputes is to find a way through or around the dispute to a successful resolution in which the parties may have differed, but the team is stronger in the end for the task remaining. Everyone continues toward the goal of achieving a successful completion in which the Owner will be pleased, and everyone else can walk away with pride in the accomplishment.

INSTANT RECALL

- Differing site conditions usually refer to uncovering something underground or related to the soil, but they could be something else that has been hidden, such as asbestos above a ceiling.
- Type I Differing Site Conditions are conditions that differ significantly from what was indicated in the contract.
- As with all potential change orders, notice to the Owner and the Owner's agents is important and timeliness matters.
- Once the Contract Administrator has been advised of a Differing Site Condition:
  - Notify all parties to the contract and the Owner's agents of the Differing Site Condition, as well as anyone else, such as a Subcontractor or Supplier, who might be impacted.
  - Request everyone meet about the issue. If at all possible, it is important to have this meeting at the site of the differing condition in person. Generally, tensions can more easily mount over a phone conversation than in person, and also "seeing is believing."
  - Fully document all aspects of the difference. Make sure the contract is thoroughly examined for all details related to the condition, and then take photographs of the condition in the field. In addition to the photographs, create a narrative of the situation, how it was discovered, and how it differs from what was shown to be the condition on the drawings or contract.
  - Brainstorm with as many people as possible for as many solutions as they can imagine in order to find the best resolution.
- In order to prove the Differing Site Condition beyond any exculpatory clauses in the contract that would relieve the Owner and/or the Design Professional(s) from responsibility, the Contract Administrator will have to demonstrate that:
  - notice of the differing condition was given in accordance with the contract;
  - the conditions shown on the plans and drawings are significantly different than what exists on the site;
  - the difference requires a significant change in approach that will impact cost and/or schedule.
- Although this might be useful even on a Type I Differing Site Condition, it is very important for the Contract Administrator in a Type II Differing Site Condition to recognize that what has been encountered is unusual and/or unexpected.
  - Consult with all individuals familiar with the situation who had input prior to submission of the proposal and signing of the contract about the expected conditions for the site.
- Within a specific period of time, the Contract Administrator must also be prepared to submit a detailed explanation of what the costs and schedule impact will be.

- The most grievous of all Differing Site Condition clauses that shift responsibility to the Contractor are the exculpatory clauses.
- The Contract Administrator will need to meet to discuss with the estimating team and anyone who had input into the potential site conditions prior to submission of the bid.
- One effective way to partially eliminate the difficulty of seeming to cast blame on the estimator or estimating staff is to have a very comprehensive bid turn-over meeting prior to the project team assuming responsibility for a new contract.
- The Contract Administrator must rely upon two forms of "reasonableness" for relief from differing site conditions:
  - The reasonableness of the Owner is an important element in securing relief from differing site conditions.
  - The reasonableness of the Constructor's own approach is also important. If the Contract Administrator is unable to convince the Owner, then the courts or arbitration panel will look to the "reasonableness" of the Constructor when bidding and executing the project.
- In the case of the Owner knowing something is different than expected and not disclosing it to a prudent and reasonable Contractor, a court would likely recognize the right of the Constructor to recover.
- Reliance on the Spearin Doctrine alone will not be a panacea for relief from differing site conditions.

YOU BE THE JUDGE

Granite Construction Co. vs. *Texas Department of Transportation*

ATS Drilling (ATS), a Subcontractor to the joint venture Granite Construction Company and J. D. Abrams, L.P. (collectively, Granite) sought compensation for a Differing Site Condition on a Texas Department of Transportation (DOT) toll road project. Granite Construction Co. put forward the claim on behalf of the Subcontractor, because there was no contractual agreement between the DOT and ATS.

ATS sought $800,000 in additional compensation for shoring up the drill shafts that collapsed, based on differing site conditions than ATS had anticipated. ATS insisted the varying geological conditions were not what was represented on the contract drawings. It further asserted that although DOT had included in the bid documents the boring samples analysis, it had not included the introductory comments clarifying the analysis. That information would have made the varying conditions more understandable and ATS would have submitted a higher price for the work. ATS pointed out that DOT had no legitimate reason to withhold some information while making part of the report available, and since the information provided was misleading, DOT should be held responsible for the additional costs.

DOT pointed out that the bid documents stated: "The bidder shall examine the site work and satisfy himself as to the conditions which will be encountered relating to the character, quality and quantity of work to be performed and materials to be furnished." DOT also pointed out that the bid documents cautioned each bidder not to rely upon the accuracy of the documents.

Find for:

In the amount:

IT'S A MATTER OF ETHICS

In the case of the saw mill, we were fortunate to have an Owner that recognized the situation was outside the bounds of any exculpatory clauses. In essence, it was a Cardinal Change, but it took some serious discussions to move the issue outside of differing site conditions to a total change of the contract terms, since we were not recreating a previously existing saw mill.

Not all differing site conditions are resolved as well. In fact, can some things be legal by contract but unethical? Think to yourself, or among friends or classmates discuss, how much an Owner and Design Professional should rely upon exculpatory clauses that protect them from Type I Differing Site Conditions that are different from the contract documents, or Type II Differing Site Conditions that are unusual or totally unexpected for such a project. Is it easier to rely upon exculpatory clauses for Type I or Type II? Why? If the Design Professional knew of the condition but did not tell the Owner, is the Owner responsible? Can the Owner then rely upon the exculpatory clause? If the Design Professional knew and told no one, can the Design Professional use the excuse that it was a patent flaw the Contractor should have been aware of, or is the exculpatory clause still the best defense for the Designer?

Going the extra mile

1. Do the "You be the judge" exercise above and see how your decision agrees with the actual court decision that is discussed on the Companion Website.

2. **Discussion exercise** (with your classmates): Discuss how best to prepare for and handle Type I Differing Site Conditions and Type II Differing Site Conditions. Do exculpatory clauses make Type I or Type II more difficult to negotiate? What other team members do you need to help with input and strategy related to each Differing Site Condition?
3. Effective use of the correct vocabulary improves communication and promotes respect among colleagues. The reader should become familiar with the key words listed below that have been used in the chapter. Definitions can be found in the Glossary. Additionally, the student is encouraged to use the flash card exercise in the chapter's Companion Website as practice for possible quiz or exam questions.

Key words

Cardinal Changes

Conscionability

Eichteay formula

Exculpatory clauses

Force majeure

Patent Ambiguity

Patent flaw

Precedence clauses

Spearin Doctrine

Type I Differing Site Conditions

Type II Differing Site Conditions

CHAPTER 13 Schedule

Once a contract is signed and the final cost known, including the contingency budget, the Owner's chief concern is schedule. Often a Contract Administrator's chief concern is cost. This can cause conflict of opposing interests if not carefully recognized and kept in full view at all times.

Student learning outcomes

Upon completion the student should be able to. . .

- **recognize the importance of schedule in delivering the project for the Owner**
- **distinguish different types of delay, including compensable and non-compensable**
- **understand the intent and issues of Liquidated Damages**
- **analyze the different types of acceleration**
- **evaluate the key elements of a delay claim**
- **outline the different costs associated with a delay claim.**

13.1 The schedule

The best thing about the future is it comes one day at a time.

Abraham Lincoln

It has been mentioned before that once a contract is signed, the most important issue for an Owner is usually the schedule. With the contract defining the costs, and recognizing there is a certain contingency for change orders, the one element that can affect the Owner's plans the most is usually time. Failure to move into or use the structure as intended on a timely basis can have consequences on the reputation as well as the finances of the Owner.

A highway that is shut down longer than expected or a bridge that fails to join two shores on time, a hospital that needs patients in beds that are not available, or a supermarket that has scheduled to ship produce for a store that is not ready for the public can be a financial drain, but a late completion can also be a career bust for whomever the Owner assigned to get the project done on time.

The schedule is so important to the Owner that in most contracts it carries more provisions than the cost of the contract. In fact, it can detail other costs that might accrue should the scheduled date for completion not be met.

Recognizing the pressures of other parties on the project team can be a first step for the Contract Administrator to fulfill the Owner's schedule. It can also boost the Contract Administrator's self-motivation to do what is necessary to keep the project on schedule as much as possible.

Even before the Constructor actually puts trailers and people into the field, there can be important steps to take to help improve the ultimate delivery of the project on time:

- confirming the starting date and what will be necessary to begin the project in relation to equipment, materials and personnel
- validating the estimate with the parties that are going to actually construct the project
- validating the schedule itself. Remember, whether done using the Critical Path Method (CPM) or a simple bar chart, the schedule is the assumption of someone who has significant knowledge of the project but has never completed the project to prove it can be done in the time allotted
- studying all long lead items along the critical path as well as any that might affect float
- highlighting any intermediate milestone events as well as Substantial Completion dates that will be critical for project delivery
- creating a spreadsheet of critical contract provisions for the Contract Administrator to recognize items and events that can impact the successful compliance with contract conditions
- making sure the pay application process is understood and fully organized, from Subcontractor through Owner's agents
- listing all deliverables for final completion.

In the world of construction, the schedule is just as important for the Contractor as it is for any other party involved. The sooner the Contractor is done, the less chance there is for a drain on profits. Still, Owners and Design Professionals are not always convinced Contractors know how important the schedule is.

It is not the purpose of this text to dwell on the many different types of schedules or scheduling softwares that are available. Depending on the size and complexity of a project, the schedule may be figured on bar charts or in software packages that update and interface with payroll and invoicing. In between there may be three-day to two-week look-ahead schedules developed by the Superintendent or other field personnel. Ultimately, the Contract Administrator on very complex projects will want a scheduling software that interfaces with BIM packages that include 4-D (schedule) and 5-D (cost). At this point, however, the Contract Administrator must recognize that he or she cannot do it all. The more detailed the schedule is to be, the more need there will be for significant input from others, and the more likelihood the project will need an experienced professional to develop the project schedule.

Once that is developed, however, the Contract Administrator should seek as much input as possible from others involved in the construction for validation of the schedule. Accepting a schedule without comment could "validate" by silence a schedule that cannot actually be met for one or more reasons. The other advantage of seeking input from others is that once others have validated the schedule they have bought into the schedule for themselves and what they are to contribute to the whole.

* * * * * * * * * *

If the Contract Administrator develops the habit of having others validate the schedule it will (1) reassure everyone the schedule can be met, and (2) get their buy-in to accomplish their part of the contractual obligations.

* * * * * * * * *

No matter how simple or complex, whatever schedule is developed will become the first place everyone must turn to understand the impact of any issue or action in relation to project completion. The Contract Administrator will have to use the base schedule to develop a schedule impact analysis for discussion and determination of what change might need to be incorporated into the schedule.

13.1.1 Contract Administrator in the middle

Each Contract Administrator must walk a fine line or maintain a delicate balance, whichever metaphor is more appropriate. By their boss, the Contract Administrator will almost certainly be judged by the project's profit or lack thereof. By the Owner, if the schedule is paramount, the Contract Administrator will be either a hero or a scapegoat, depending on how effectively the schedule is met or even improved upon. We often hear that no one can serve two masters, but the Contract Administrator must always be cognizant of cost in relation to schedule. A negative impact on the schedule will usually require increased costs to "get back on track."

How a Contract Administrator handles the challenges of cost and schedule will often depend on personalities. If the project team, including the Owner and the Owner's agents, work well, with everyone's interests in mind, then each crisis might become an opportunity. However, this is not always the case on every project, and the Contract Administrator will have to determine how to handle each situation – always with the understanding that his or her own company is in business to make a profit.

13.1.2 Time is of the essence

Because Owners want to stress the importance of the schedule, **"time is of the essence" clauses** are usually part of all contracts between the Owner and the Contractor. They have become so standard, however, that they have lost some strength and impact. How important time is to the Owner is actually something that should be determined even before the bid proposal is submitted.

The pre-bid meeting is a good time for the Constructor to understood fully the needs of the Owner related to the schedule. What end-date pressures the Owner's representatives might see on the horizon are important indicators of just how much "time is of the essence" clauses are going to impact the relationships of the parties. They might also reveal the likelihood of directed or **Constructive Acceleration**, as discussed in a previous chapter. Since getting in and getting out as quickly as possible, while meeting all other contract requirements, should always be the intent of the Constructor, knowing the Owner's needs related to schedule is essential to recognizing how important keeping the Owner informed about potential delays will be.

13.1.3 Liquidated Damages

One very serious clue to how important the schedule is and how lost time will be enforced is the insertion into the contract of **Liquidated Damages**. At first, the Contract Administrator might view Liquidated Damages as a penalty imposed on the Contractor for each day the project is not completed. However, Liquidated Damages cannot be a penalty or they are not enforceable.

Liquidated Damages are set by the contract, and the Constructor, by signing, has agreed to them as being the assessment on a daily basis of what the cost will be to the Owner for every day the Owner cannot occupy the space for its intended purpose. This practice may come the closest to "duress" that a Contractor faces when signing a contract, since the likelihood of actually proving the costs as accurate is very remote. Owners set a Liquidated Damages price and the Contractor agrees to it at a time when a delay in the project is viewed as remote and the thrill of more work for the crews and staff is greatly anticipated.

If Liquidated Damages exist in the contract, any delays become a severe challenge for the Contract Administrator. All notice provisions related to delays are imperative. The Contract Administrator must also document that the delay was an excusable delay and the Contractor had no responsibility related to the cause of the delay. If the Owner fails to grant an extension, this documentation will be critical if Constructive Acceleration becomes necessary. Of course, this may eventually lead to a claim, but when an Owner does not grant an extension of time and does not approve increased costs to meet the contract schedule, the Contractor is forced to add crews, work overtime, revise phasing of the project, and sometimes put in increased equipment. There may be other areas that increase the cost, and the Contract Administrator under Constructive Acceleration will have to determine at the end of Constructive Acceleration how important submitting the extra costs will be. The Owner has already indicated there will be no compensation for the extra costs, but that does not mean a claim should not be put forth. Usually the amount of extra costs set off against future work with the Owner determines the validity of submitting costs for the Constructive Acceleration the Owner refused originally.

13.1.3.1 Avoiding Liquidated Damages

We talked about a good defense when confronted with differing site conditions. The first defense against Liquidated Damages, however, is a good offense. Staying on top of the schedule to everyone's satisfaction is very important. Whether using a simple bar chart or more complex software with Critical Path Method software, paying attention to the details of events and milestones is essential. Making sure long lead items are not just ordered, but continuously tracked, is a start. Updating schedules is another important asset. Continually inviting field input into **look-ahead schedules** will also be a major help. Follow-through on RFIs so they do not sit on someone's desk consuming valuable float is also very important. Updating project schedules before or during project meetings can also alert everyone to the challenges ahead. Design Professionals can be a major contribution to the team's success by prioritizing answers and approvals.

In addition, the Contract Administrator must also not be intimidated by the prospect of Liquidated Damages. When the project is properly documented, there is the possibility the imposition of such damages is inappropriate. Such circumstances might include:

- Liquidated Damages generally should cease at the time of Substantial Completion. However, some Owners are now actually imposing two tiers of Liquidated Damages. The first tier of Liquidated Damages is for days prior to Substantial Completion. The second tier is for a reduced Liquidated Damages amount to be imposed between the date of

Substantial Completion and the final acceptance of the project. This number is definitely subject to review, since there is no way an Owner or a Contractor could have known at contract signing exactly what the Owner would not be able to do in or with the structure. Therefore any Liquidated Damages amount past Substantial Completion is merely a guess, and as such is most likely inaccurate. It also might border on being a penalty, and therefore, be disallowed by law.

- Upon closer examination, the damages originally agreed to by contract might not actually represent a true loss to the Owner. This could also include the reality that the Owner is using a part or even all of the building while imposing Liquidated Damages.
- The date for assessing Liquidated Damages needs to be adjusted based on delays out of the control of the Contractor. If the date cannot be adjusted, the Contract Administrator should seek authorization for directed acceleration. If the Owner refuses to authorize costs for the acceleration, then the Contract Administrator may have to undertake Constructive Acceleration. In such a case compensation for the additional costs might require taking the Owner to court or arbitration, but it is a standing court principle that Liquidated Damages against a Contractor will not be allowed if the sole cause of the delay was due to the Owner or the Owner's agents.
- The amount of the Liquidated Damages must be something that would be impossible to demonstrate. In some instances, it could be shown that the actual loss would have been far less than the damages being imposed.

Examples:

1. A movie theatre might require additional compensation for presumed loss of food and beverage sales, but it cannot justify ticket sales greater than 100% of its capacity.
2. Tolls collected on a bridge can be confirmed once the bridge is open for business.
3. Costs paid for extended rentals at another location because a tenant was not able to move into a completed space can be accurately documented.

- Depending on the cause of the delay, the Contractor usually has flow-through or flow-down wording that would place responsibility for a delay upon any Subcontractor that caused the delay. Certainly the Contract Administrator does not want to fight a two-front war, with the Owner on one side blaming the management of the project, and the Subcontractor on the other side using all the excuses possible and then some to deny culpability. Nonetheless, sometimes an Owner's wrath is justified if a Subcontractor or Supplier has been the cause of a delay. Sometimes Subcontractors and Suppliers will want to cast blame on the Design Professionals for ambiguities in the documents, or slow responses to shop drawings, RFIs, or change orders. Sometimes these objections by the Subcontractor or Supplier are justified, and sometimes they are not. In placing responsibility on a Subcontractor or Supplier the Contract Administrator may have the financial solution to the Liquidated Damages, but the personal and corporate damage to the Constructor's reputation might take more effort to resolve.
- It is possible to use the project meeting minutes effectively in seeking extensions of time. In addition to complying with contractual notice provisions, the determined Contract Administrator should also use the project meeting minutes to document and confirm any delays that might negatively impact the completion date. Sometimes this will produce controversy, but if it is maintained in an atmosphere of open communication where both parties can express their side, the opportunity for effective dialogue may head off

entrenched positions based solely on the contract. Generally, it is much easier to deny a request in writing than it is in person, particularly when the request is politely put forward with facts and figures.

13.1.3.2 Bonus and penalties

Although Liquidated Damages cannot be a penalty, but a genuine attempt to put a dollar figure on what the Owner will suffer if the project is not completed on time, there is another method of imposing penalties for not reaching completion as scheduled. This also requires, however, the inclusion of a bonus for each day or period of time the project is finished ahead of schedule. When such a bonus/penalty situation exists, the Contract Administrator cannot relax in an attempt to finish on time or to give notice and document any delays. Extension of time will move the bonus schedule also. As the project progresses, the Contract Administrator with other key personnel and Subcontractors should ascertain the chances of improving on the schedule. It might even be worthwhile to incur some extra costs to finish early and collect a bonus greater than the costs expended.

13.1.4 Start-of-project delays

Just as a Constructor can be delayed during a project, there can also be delays during the very start of the project. Often Owners will have to restrict access to all or part of a site for a period of time for reasons totally out of the control of the Constructor. If this does occur, the first issue needs to be an understanding of how this delay has affected the completion date.

Sometimes a Constructor will be so happy to finally get started, the delay is not recognized, and failing to give proper notice could lead the Owner to believe the project will still be completed as originally scheduled. A delay of a day might not be so drastic, but put a week's delay into Liquidated Damages at the back end, and it can hurt. When weeks turn into months, the issue can become even more difficult.

A Constructor may have bid a project based on certain weather conditions prevailing at least through the in-ground portion of the work. A delay might push the work into the winter or the rainy season, causing considerably more effort to get out of the ground. Concrete might need different additives or take longer to finish. Or if the building cannot be enclosed in time, winter weather could require far more temporary heating costs.

For the Contract Administrator it might be difficult to approach the subject of delay right at the beginning, but there is actually no better time, from a negotiation standpoint, than before major construction costs have been incurred on the project. After the project has begun, the billing process puts the Owner's payments sixty days (or more) behind the work being put in place.

13.1.5 Float

We have mentioned a few times related to a schedule the subject of **float**. Events that are off the critical path will have durations in which the start time can "float" and not affect the completion of the project. Delays along any part of the schedule containing float are considered **non-critical delays**.

One important aspect for the Contract Administrator is who controls or "owns" the float. The Contractor wants to be the one who can use the float for whatever purpose or advantage. This could be for cash flow reasons. Committing to purchasing some items might not be best, depending on the time of delivery and payment to a Supplier relative to the payment application process. Or, scheduling personnel on a project could be more effective if there is not an extreme fluctuation in number of craft workers needed at different times. A smooth flow of production with the same crews is usually better than needing more people one day and fewer the next. Staggering the use of certain pieces of equipment might be more effective than increasing the number of equipment and having some sit idle for extended periods.

There are many reasons a Constructor wants to use the float, but float can disappear if Owners or Design Professionals use it first. This is often related to the approval process or answering RFIs slowly. The Contract Administrator wants to keep information flowing smoothly and efficiently. If it does not, a delay in what might have been off the critical path could be the reason the critical path actually shifts to a new series of events and milestones. In such a case it would be termed a **critical delay**, because it now affects the critical path.

Even if the critical path is not affected, however, the Constructor may be put in a difficult position. The general rule, when assessing who caused a delay, is to ignore who used the float. Float is often seen as belonging to the party that uses it first, and, as mentioned above, if there are samples, shop drawings, or RFIs in need of processing, the float on those items can disappear. Owners and Design Professionals will justify using float on the basis "float belongs to the project."

Therefore, simply because something is submitted, the Contract Administrator cannot consider the item resolved. Friendly but persistent follow-up is required to keep the items flowing quickly so the float will ultimately be available for the Contractor's use.

Remember, Owners, Design Professionals, and Construction Managers are also busy. They have several items to deal with and often multiple projects and distractions. Politely keeping your project's issues on a "front burner" is important, but not guaranteed without effort on the Contract Administrator's part.

* * * * * * * * * *

A technique each Contract Administrator should develop, and have others on the team also include in as many RFIs as possible is, in addition to stating the question or issue, to provide a solution. Often those RFIs that have a solution (one the Constructor can live with) get answered sooner than those that do not, since RFIs without a suggested solution will require time to research, and are often set aside for when the Design Professional has more time (and when will that be?).

* * * * * * * * *

In addition to composing RFIs carefully and keeping them on the "front burner," the Contract Administrator should have a frank discussion with all parties involved about the use of float on the project. It is important to remind everyone that the schedule is for construction and rightly belongs to the Constructor to utilize as effectively as possible. Good project schedules have built-in approval process times. If everyone can agree to specific approvals and the time or sequencing needed for issues involving the Owner and Design Professionals, then it will be easier for the Contract Administrator to handle the float to the benefit of the Contractor. It is important that the Owner and the Owner's agents recognize that the schedule imposes time constraints on them so that the final completion date can be accomplished by the Constructor.

13.2 Types of delays

You may delay, but time will not.

Benjamin Franklin

The Contract Administrator needs to understand that there are **excusable delays**, and then there are **non-excusable delays**. Whether a Constructor will be compensated for a delay will depend on the delay being excusable, but even then, compensation may not be forthcoming.

13.2.1 Non-excusable delays

Although a Constructor might not agree, a non-excusable delay was caused by the Constructor. These can also be the result of Subcontractors under the control of the Constructor. Failure to properly equip the project, or lack of personnel to prosecute the work, are two of the more common Constructor caused delays. Another would be the failure to submit materials or information for approval on a timely basis. The Constructor cannot expect to be compensated for the costs of such delays.

Difficulty might arise, however, when the Constructor believes the Owner or Owner's agents were partially responsible for the delay. Failure to approve submittals by the Design Professional might be viewed by the Constructor as more of a problem than the late submittal.

Documenting the process will be important for the Contract Administrator if the delay ultimately becomes a factor in the project schedule, but seeking compensation for what the Owner sees as a non-excusable delay is almost always more work than would have been required to properly and efficiently monitor the flow of submissions for approval.

A CASE IN POINT

Of course, a Contract Administrator's most pressing non-excusable delay is their own workforce. There is no compensation for lack of productivity. How productivity is monitored has been a challenge since before the building of the Pyramids. An interesting twist on production came in with the construction of skyscrapers.

The Empire State Building provided a perfect example and an effective solution. At its peak, over 3,000 workers were on the site each day. Getting workers, such as the ironworker shown Figure 13.1, to their workplace in the morning and back down at night was one challenge, but attempting to have them all exit for a half-hour lunch break would have been impossible and totally nonproductive.

The solution was to have a nearby restaurant owner set up, as the job progress required, a restaurant on the 3rd, 9th, 24th, 47th, and eventually the 64th floors. The condition imposed by the Starrett Brothers construction firm was that the food needed to be of the same quality and price as would be found in the restaurant down at street level.

Figure 13.1 Ironworker during the construction of the Empire State Building
Photo credit: United States Government

13.2.2 Excusable delays

Although a delay may be what is termed "excusable," it still might not be subject to additional costs for the Contractor. Excusable delays can be broken into two categories:

- compensable
- non-compensable

13.2.2.1 Non-compensable delays

The good news about a non-compensable delay is that the Constructor is entitled to a time extension. The bad news is that the Constructor will not receive additional costs for a non-compensable delay.

Non-compensable delays are considered **Acts of God** or a ***force majeure*** – from the French, meaning "a superior force." These are events outside the control of anyone. Acts of God or a *force majeure* would include extraordinary weather events, such as a hurricane. Additionally, an Act of God might be an earthquake, or perhaps a major fire. Some man-made events can also be considered a *force majeure*, such as war, a serious epidemic, or a strike.

These are all reasons for an extension of time. However, the traditional approach is that the extension of time is granted without additional compensation. The theory behind such an approach is that the disruption caused by the "superior force" stopped work, and when the work resumes, it can begin where it left off. If there is damage done to the project, such as a hurricane might do, then insurance is meant to compensate for that, not the Owner.

However, the Contract Administrator might want to document and express some concern for legitimate costs not covered in a project that has experienced a *force majeure*. These could include expensive issues if the duration of the interruption is significant:

- equipment rentals that could not be used during the shutdown
- additional costs for or scarcity of materials that have escalated in the wake of the *force majeure*

- potential for increased labor costs due to shortages of craft workers or contract expiration dates
- acceleration expenses required to get the project back on schedule

13.2.2.2 Compensable delays

Assuming a delay does not take place within a "float" area of the schedule, there are delays that are no fault of the Contractor and thus could require both an extension of time and additional compensation. "No fault of the Contractor" is critical for the Contract Administrator to be able to demonstrate a **compensable delay**.

If the Owner or the Owner's agents are responsible for errors in the drawings or have failed to process approvals or answer RFIs, then the Contractor might have cause for a compensable delay. In some contracts, the Owner is responsible for specific items, such as equipment in a production line. If that equipment needs to be installed before the building can be enclosed, then there is cause for an increase in time, and perhaps additional money to bring back crews to finish the close-in of the structure. Related to not supplying the correct items or space, the Owner or Owner's agent might have failed to properly coordinate other Prime Contractors on the project. The failure of one Contractor to finish a space on time will usually negatively affect the production of the other Contractors.

It should also be recognized that one change order may not affect the time and require additional personnel to execute the work. This could be true for each and every change order an Owner authorizes. By itself, a change order might not require additional personnel and no more time. However, put a hundred such change orders together, and that would no longer be the case. In such a case, the Owner or the Design Professionals have caused so many change orders to be issued that the available personnel could not finish the project without an extension of time and some costs would also probably be affected due to the extension.

Owners can also be responsible for failure to authorize work or changes in the work, lack of or improper inspections, and failure to make payments on a timely basis. Unfortunately, if relations have deteriorated to this point, it is not likely the Contract Administrator will be able to repair the damage. Litigation or arbitration will probably be the recourse if the loss is significant enough to warrant that approach.

In the chart opposite, certain typical delays are mentioned, and what the Contract Administrator will initially be required to demonstrate, provided the Contract Administrator has given proper notice regarding the delay. Of course, it is quite possible that there will be exculpatory clauses excusing the Owner from responsibility for such delays, but notice having been given, the prudent Contract Administrator will still prepare the Constructor's side of the issue.

13.2.3 Demonstrating delay costs

The first challenge for the Contract Administrator will be to show the actual length of the delay. This will be especially important for projects with Liquidated Damages. The Contract Administrator should start with the original schedule. This is one reason why constant updating of the schedule during construction is so important. From the beginning of the delay to the resumption of activity can then be demonstrated. Of course some other work might also have been undertaken, and these activities should also be analyzed in relation to the overall project schedule. Once the events are re-examined and aligned a new critical path can be produced, showing the new completion date and the critical path to get there.

Table 13.1 Potential compensable delays

Cause	Impact/delay cost
Error in contract documents	Constructor must prove it was a latent flaw and had an effect on the schedule's critical path.
Owner denied site access or contractually required space, material, equipment, etc.	Constructor depended on access or items in order to fully meet the schedule requirements.
Owner's agents did not approve submittals or RFIs in a timely manner	Constructor's submittal log clearly shows proper timing of all submittals and dates of resolution or continued lack thereof.
Owner did not properly manage other key Contractors on the project	Particularly in Multi-prime Contracts, but whenever the Owner has independent control over a portion of the work, the Contract Administrator must demonstrate any impact others within the Owner's control have had on the project. Ideally such work by others has been included in the project schedule with a start and completion time noted.
Owner or Owner's agents do not inspect in a timely fashion	Constructor must document completion of items required for inspection, request for inspection, and date when inspection actually occurred.
Owner temporarily suspends work	Constructor must demonstrate contract work was current at the time of the suspension and that the suspension affected the completion in a way independent of any other delays that the Constructor might have caused.
Owner issues a change to the contract	Constructor shows that in addition to price, the schedule will be affected due to phasing, delivery, or some other reason outside the control of the Constructor and not contained within the pricing of the change. For example, additional shifts, overtime, or premium costs for delivery were not included in the cost of the change as submitted.

Being armed with the new schedule and the duration of the delay is a beginning, but the Contract Administrator must next recognize delay costs are not easily demonstrated. If the Owner agrees a delay was excusable and compensable, seldom is the work required to do the actual contract work billable. That would be under the original contract breakdown. What are billable are all the costs that were incurred during or because of the delay. Those may be more difficult to demonstrate, particularly if the documentation did not start immediately upon the start of the delay.

Another complication is that the crews might have been delayed on one part of a project and were shifted to contract work on another part of the project. There may be inefficiency costs that are very difficult to demonstrate, but the overall costs of that crew would not be billable in total.

Generally, material will again fall under the original contract rather than a cost billable under the delay, but idle rental equipment might be easier to demonstrate as an additional cost.

One area that is demonstrable but often a point of contention with Owners is the cost of the supervisory personnel. From Field Engineers to Superintendents and other managers in the field, the extension of the contract means they will be costing the Constructor more money than was figured in the original estimate. The same is true of office trailers and any other rentals, or monthly items such as phone service or office equipment.

Again, this can be a contentious subject, but depending on the contract related to billable items, such items might be included in the delay costs.

As the Contract Administrator is gathering the costs for a delay, it is important to keep in mind that such costs will probably be scrutinized. Failure to be open with the Owner will generate suspicion and distrust, and that can lead to difficult times both on the current delay submission and in the future.

13.2.4 Concurrent delays

There is always the possibility that two issues might delay a project at the same time. These are called **concurrent delays**. The first rule of thumb is that whichever delay occurred first will have to be resolved before any delay occurring simultaneously, but beginning after the cause of the original delay is considered a delay affecting the schedule.

One difficult issue that might arise between the Owner and the Contractor in relation to concurrent delay is that one is excusable, but the other is not excusable. If the non-excusable delay lasts longer than the excusable delay, then the Owner will likely argue that the Contractor is not entitled to damages for delay, particularly if the Contractor was the party responsible for the non-excusable delay.

> **Example:** The Contractor requests a substitution for a mechanical item that is readily available and works as well as the specified item that is no longer manufactured. If the Contractor is delayed by the Owner or Owner's agent for approval of that piece of mechanical equipment at the same time that the Contractor has failed to complete the concrete pad and masonry block walls for the equipment, then a concurrent delay in the work is taking place. The Contractor may have thought the concrete pad and walls were not important if the specified item was not available. However, once it is approved, the pad and walls should have been in place according to the schedule. If they are not, then there is a non-excusable delay that will last longer than the excusable delay, and the Owner is justified in not granting an increase in time.

Another issue for the Contract Administrator will arise if the concurrent delays were reversed and the excusable delay was longer; then the Contractor may be entitled to the length of the excusable delay minus the length of the non-excusable delay. Documenting actual costs will again become a challenge.

Example: Using the above situation, let's reverse the order. The Contractor asks for a substitution, and the Owner's agent needs to do some research on the substituted equipment. At the next project meeting, the Owner's agent is still "looking into the matter," but the Owner points out that if there is no change in size, which the specifications state there is not, then the pad and walls should still be completed. According to the project schedule they are currently behind schedule. The Contractor agrees, and immediately begins work on the pad and wall. They are completed before the next meeting, at which time the Owner's agent does agree to the substitution. The Owner should then grant the Contractor an extension of time, which should be the net difference between the total delay for approval of the substitution minus the days it took to complete the pad and walls after the original date scheduled for their completion.

13.3 Acceleration

There are plenty of recommendations on how to get out of trouble cheaply and fast. Most of them come down to this: Deny your responsibility.

Lyndon B. Johnson

It is always possible the Owner cannot extend the contract completion date. In such a case, the Constructor will be required to implement some form of **acceleration**. Who is going to pay for the acceleration is based on who is responsible for the delay.

An excusable and compensable delay should be the responsibility of the Owner. A non-excusable delay will be the responsibility of the Contractor.

The gray area will be the excusable but non-compensable delay. The contract might at first seem to support "no damages for delay" in the case of weather or other Acts of God, for instance, but the Contract Administrator will have to point out that if the delay is "excusable" then the cost of acceleration is not "damages based on the delay," but extra costs to move what would be the extended date back to the original contract date.

The Contract Administrator will also have to be aware of any exculpatory clauses such as "no damages for delay," but it will be important to shift the focus if at all possible. If not, documentation of the extra costs should be maintained in what might develop into a case of Constructive Acceleration, to be further discussed below.

There are basically five elements the Contract Administrator will need to demonstrate when turning an excusable delay into a Constructive Acceleration claim:

1. There was indeed an excusable delay on the project.
2. The Contractor gave proper notice of the delay and requested an extension of time in compliance with the notice clauses of the contract.
3. The extension of time was denied.
4. The Owner required the Contractor to finish by the originally scheduled completion date.
5. The Contractor incurred extra costs in order to comply with the Owner's demand.

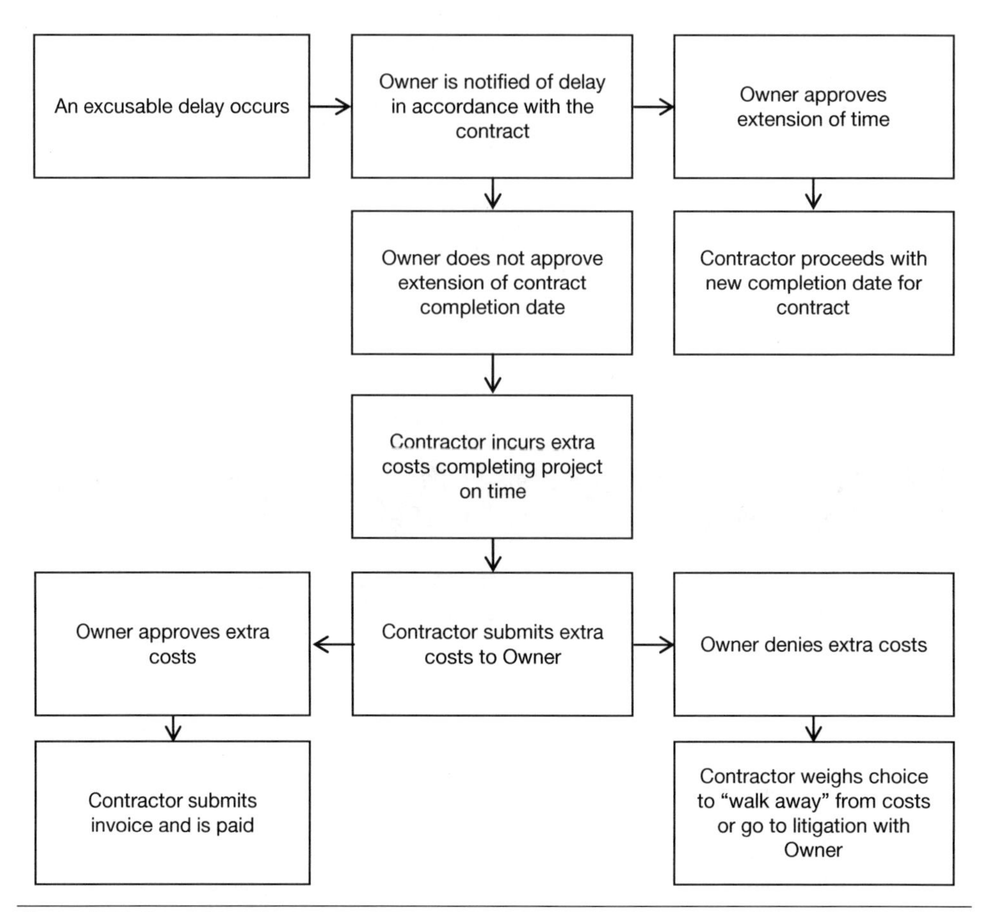

Figure 13.2 Excusable delay process

13.3.1 Documenting acceleration

Openness in documentation is going to be important in supporting any claim for costs associated with a delay. The first place the Contract Administrator must look for the difference in cost will be the original estimate upon which the contract was based. In addition, any Unit Costs contained within the contract may need to be analyzed.

The Contract Administrator will need to review, often in tandem, the original estimate and the original schedule. Beyond these two essential documents, the Contract Administrator will want to review several other sources of information to truly understand the effect and costs of the delay:

- conversations with all field personnel directly involved with the delay and its effects on productivity
- with Subcontractors and Suppliers, the cost of the delay and any acceleration costs associated with expediting project completion
- daily work reports, particularly in relation to weather or any other cause of the delay, and the effect on the personnel on site and productivity
- payroll time sheets and/or manpower loading reports from the time of the delay through completion of all work associated with the delay

- meeting minutes, project correspondence, RFIs and e-mails to or from any parties associated with the delay
- photographic records of the project related to before, during, and after the delay.

The Contract Administrator will have to become a sleuth, investigating the scene of a potential crime and gathering all evidence that might support the case for acceleration and document the extent of the costs. The more thoroughly the Contract Administrator investigates the cause and results, the more confidence others will put in the final accounting of the actual costs of the delay.

13.3.2 Directed acceleration

If the Owner agrees to pay for the acceleration, the Constructor will begin what is termed **directed acceleration**.

A CASE IN POINT

By 1940 over 27,000 military and support personnel were scattered around the Washington, DC, area in almost two dozen buildings. A contract to build the world's largest office building was bid in July of 1941, and construction began in September of 1941 in what was then basically a swamp southwest from the Capital and near Arlington Cemetery.

At one point 327 Designers working on the project were turning out up to 30,000 drawings and clarifications per week. John McShain became the lead Constructor, along with Wise Contracting Company, Inc., and Doyle and Russell as additional Primes. The actual design of the building changed from a square to the iconic Pentagon shape early in the process, but the design completion was not until almost a year after construction had begun.

However, one dramatic event would forever alter the acceleration of completion on the project, and that took place on December 7, 1941, almost three months to the day after construction had actually begun on the site.

That, of course, was the "day that will live in infamy," when the Japanese attacked Pearl Harbor – a true *force majeure*. Instead of it causing a delay, however, the military very much needed the project completed as soon as possible. From that point onward, completion of the Pentagon became a top priority and a case of acceleration. Initially, John McShain is reported to have had to argue for the increased costs. Being successful in doing so, he changed the acceleration from constructive to directed. Round-the-clock shift work, overtime, supervision, some equipment, and additional office personnel to expedite design changes all added to the costs of the directed acceleration.

I have been told by those involved that John McShain, himself, relieved his Superintendent to allow him a two-week vacation because of the strain of the schedule, along with a significant increase in the size of the structure. I was also told that John McShain was proud that during those two weeks they set a record for concrete put in place.

After those two weeks, however, John McShain, at the insistence of Mary, his wife, took a vacation himself.

Certainly, the accelerated schedule, combined with the increased size under the war conditions, put everyone under a great deal of stress, but I was told by John McShain's daughter, Polly, that her father had always considered one of his biggest challenges was feeding the crews (up to 15,000 workers at the height of construction). The site being located in the middle of a swamp, the prospect of taking time to "go out to lunch" was out of the question.

Figure 13.3 The Pentagon under construction: 6,500,000 sq. ft. of floor area (five floors above ground and two below); five ring corridors per floor, totaling 17.5 miles. The finished structure is assigned six zip codes by the U.S. Postal Service. During its construction an event took place that both increased the scope and shortened the schedule.
Photo credit: United States Government

13.3.2.1 Costing directed acceleration

Directed acceleration is not an open check book from the Owner.

Directed acceleration will usually require either an increase in the number of craft workers on the project or overtime for those who are already on the project. Normally additional material will not be required for the acceleration, since only the schedule but not the scope of the contract has been affected. However, additional equipment might be needed to be brought onto the site, or some equipment that is charged on an hourly or daily basis might be included in the overtime costs if used for extended hours.

What the Owner will want to see on the acceleration invoice is the additional costs that occurred due to the acceleration. In the case of overtime, that cost is only the overtime differential, since the standard rate would have been the cost of doing the contract work during normal working hours. In the case of additional crews being brought on site, if they work during normal working hours, technically none of their time is billable, since it is all on contract work and not at an increased rate. However, if the Contract Administrator can demonstrate inefficiency related to the increased personnel, then that factor would be an extra cost.

One factor that all Constructors must consider is how acceleration can be handled efficiently. At the end of any project even without acceleration, there is a tendency to overload parts of the project just to get it done. Different trades will be vying for space to complete their work, often interfering with other trades completing their work. This is what is termed by some as **trade stacking**, and it is not only inefficient, but can also lead to some quarrelsome encounters as nerves and tempers become frayed.

Sometimes more personnel or longer hours is the easy answer, but not the best answer. How all the events and milestones of the project inter-relate may require careful study to determine what really will be the best solution for effective acceleration. In all likelihood the project is in trouble time-wise. That could also mean it is in trouble financially. Making the best decision for the Constructor while maintaining the satisfaction of the Owner is a challenge that can make or break the project.

There are many reasons why productivity may decrease during acceleration. If possible, the Contract Administrator should attempt to demonstrate this (perhaps through use of the Measured Mile) or request Time and Material rather than lump sum for accelerated work. This, of course, would require removing all the contract work as a deduct. If the cost breakdown is inaccurate, this solution will not be acceptable to the Owner or Design Professional.

13.3.3 Constructive Acceleration

We have previously discussed Constructive Acceleration, but it is important to always remember that the Owner may not extend a completion date, and based on contract language the Owner might believe the Constructor is responsible to pick up all additional costs of the acceleration. The Owner may be basing that judgment on the actions or inactions of the Constructor that were the cause of the delay. Or the Owner may be relying upon exculpatory clauses that place "blame" for the actions of others onto the Contractor for resolution.

In such circumstances, the Contract Administrator is not without recourse, but it will take time and documentation. Part of the Owner's strategy might be to negotiate a final cost based on the threat of Liquidated Damages also being assessed, and the pressure of the completion date is merely part of the negotiation process. Certainly the Contract Administrator will have time to recognize just how sincere the Owner and the Owner's agents are, but whether part of a negotiation process or eventually to become embroiled in mediation, arbitration, or litigation, the Contract Administrator will be required to fully document the delay and the costs.

- First, be sure to comply with all written notice requirements in accordance with the contract. Even if it appears there is no right of recovery, submit the notice in accordance with the time frame for other notices in the contract.
- Fully document the extent and cause of the delay. It will be important to demonstrate the Contractor was not responsible for the delay. In addition, the Critical Path Method or the schedule should have been validated before the project began. If it was not, it is a bit late in the middle of the project to discover it was not a good estimate, but it should still be investigated. There might ultimately be a claim of "impossibility" put forward. For instance, even with exculpatory clauses, the Owner may have put forward change orders that made reaching completion even with Constructive Acceleration impossible.
- Be sure to secure the cooperation of all Subcontractors and Suppliers that might be involved in the acceleration. This will require communicating with them the path to successful resolution as you see it, whether you think it will be a negotiated settlement or involve something from mediation through litigation. When discussing the possibilities and seeking cooperation of others, remember the same rules for what is an extra will apply to Subcontractors as apply to the Contractor. As demonstrated previously, much of the labor, usually all of the material, and some of the equipment costs are covered by contract and

not an extra. Acceleration is not an opportunity to make up for mistakes in the estimate by loading the project and billing for everything from the start to the finish of the acceleration.

- If, on a Multi-prime Contract, other Prime Contractors are the cause of the delay, be sure to coordinate as best as possible your own work while fully documenting the effect others are having on your own production and completion.
- Keep accurate records of all labor and equipment costs, and be ready to differentiate between time on contract work and costs extra to what was included in the estimate. These costs might also include inefficiency factors.
- In some instances there will be costs associated with materials if Suppliers impose escalation costs for expediting shipments of materials. This could, however, lead to further discussion and possible rejection by the Owner, since any material coming from off site should have been ordered for delivery based on the original contract date of need. The Contract Administrator will have to show how that was not possible or reasonable, due to the delay.
- Submit all extra costs in the same manner as a change order is submitted in accordance with the contract. From this point negotiations might proceed, or the Contract Administrator will recognize the Owner is holding firm in the no-compensation position, and the next step will be whatever form of dispute resolution is stipulated in the contract.

If a form of dispute resolution does become necessary, the Contract Administrator will need to involve an attorney. In fact, advising an attorney early in the process is a good idea. Some coaching and advice on a major acceleration claim from the beginning could help frame the claim better.

13.3.4 Cost theories of presenting a claim

Up until this point, we have stressed the accurate assembly of the true costs of a delay. The point behind such an approach is that the openness of the communication should assist good players on all sides to come to an effective resolution. This should be done in the hope of avoiding legal costs on top of the burden of the project costs for all parties involved. There are, however, two other methods of demonstrating costs that have mixed reviews when it comes to validity and certainly accuracy. Still, some court cases have justified these approaches.

13.3.4.1 Total Cost Method claim

If it is certain the Owner is responsible for the damages resulting in the delay, but the actual extent of the damages is difficult to provo, then it could be argued that the difference between the actual costs and the original estimate is the amount the Contractor should be compensated for the delay damages. This is called the **Total Cost Method.**

This modified cost method can also be used by Subcontractors seeking compensation against a Contractor, and the method does not need to be restricted to just delay claims.

> **Example:** A General Contractor fails to properly excavate and shore for foundations. Pressed to keep the project on schedule, the Concrete Subcontractor does extra work to make the site safe for the workers. Since excavation and the shoring associated with it was not part of the concrete subcontract scope, the Concrete Subcontractor litigates for the difference between the original contract amount and the final contract costs.

The great fault with such a method that even those unfamiliar with construction will recognize is that the original estimate may not have been accurate. The final costs might be what the original estimate should have been. This has led to another claim method.

13.3.4.2 Modified Total Cost Method

The **Modified Total Cost Method** is used to show good faith on the part of the Contractor. It is acknowledged that there are discrepancies between the original estimate and the actual prosecution of the work that are the responsibility of the Contractor, such as less efficiency and lower productivity on some cost items. In essence, the Contractor is giving back some of the costs based on self-inflicted errors or inefficiencies.

What the court will be looking for in such an equation is not only how big is the difference between the original estimate and the final project costs, but also how generous is what is being subtracted from the final project costs that the Contractor is taking responsibility for. In such a case the court may not be ruling so much on fact as it is on fairness.

The Contract Administrator in both these approaches may be fighting a two-front war. Subcontractors and Suppliers might be seeking their own damages against the Contractor, and it will be a difficult balancing act to maintain allies rather than make enemies. Nonetheless, the Contract Administrator for the Contractor must recognize that some of the Subcontractor costs, or perhaps costs from a Supplier, need to also be modified before presenting on the basis that usually no one is perfect and estimates that become the original contract prices are often wishful thinking of the past that have become clouded in the reality of the present.

13.4 Looking back to look ahead

Time is money

Benjamin Franklin

As mentioned in the beginning, in almost all projects, the schedule is paramount for the Owner. If "time is money," the schedule is gold, and it needs to be viewed as treasure for the whole team. Although this treasure is not buried, there are many challenges buried within the schedule. The Contract Administrator's task is to uncover each of those challenges before they become problems.

It is not the subject of this text, but planning and scheduling for the modern Contract Administrator has been greatly enhanced by both technology and old methods restored under new catch phrases. Some researchers believe construction is the only major industry in America that has suffered a loss in productivity during the past century. Restoring the planning and scheduling that made possible the construction of major works in the past is just beginning, but focusing on the wonders of BIM, Lean, first and last look-ahead planning methods incorporated into 4-D Integrated Project Delivery (IPD) will assist the Contract Administrator in building the team and keeping the focus on success.

In the meantime, the successful Contract Administrator will have to continue to treasure the Owner's schedule as if it is his or her own.

A CASE IN POINT

What goes around comes around. Stories about both the Pentagon and the construction of the Empire State Building suggested that feeding the crews was an important consideration for keeping the craft personnel close to their work. Coming forward to the twenty-first century, as the Freedom Tower was rising in New York City at the site of the former World Trade Center, an enterprising chain sandwich shop was lifted to where the workers were rather than have the workers wait for the construction elevator to move them up and down (a wait that for most would exceed more than the half-hour lunch break).

Sometimes the best strategy for Contract Administrators is to study what worked in the past.

Figure 13.4 Construction elevator on the side of One World Trade Center
Photo credit: Charles W. Cook

Exercise 13.1

Take a moment
Technology – friend or foe?

List below the pluses and minuses of some instrument of modern technology you use (e.g. cell phone, computer, GPS):

Plus	*Minus*

Discussion of Exercise 13.1

In all likelihood the younger the reader, the more positives and less negatives one found with technology. And perhaps that is true overall, but subtle differences are taking place, and they do affect one's productivity.

It has been pointed out that the truly best companies do not adopt any new form of technology until they have a specific use for it.

Certainly the computer has made our life much easier, but it has also changed how we do things. Most of us now do our own typing. Is that better? Time-wise, yes, but that also means we receive a lot more written communications, because everyone else is typing and sending over the internet. Is that better? Perhaps, yes, if we have the time to read it all. On the other hand, that has gradually worn away our own planning time, since someone can get to me at the last minute. Is that good? Perhaps in an emergency it is good, but is everything becoming an emergency when others do not plan ahead for themselves and rely upon me always being available for answers at the last minute?

Those who remember the slide rule recall you had to at least think of the answer before you arrived at one. Today, we accept answers that perhaps we should not. We are losing brain functions by relying on computers to do the work.

There is nothing wrong with technology. It is here to stay and expanding possibilities in the future, but the Contract Administrator will be faced with a greater workload unless the technology is used to benefit the finding of solutions and resolutions rather than to burden the workload or tasks ahead.

INSTANT RECALL

- Once a contract is signed, the most important issue for an Owner is usually the schedule.
- In the world of construction, the schedule is just as important for the Contractor as it is for any other party involved. The sooner the Contractor is done, the less chance there is for a drain on profits.
- Liquidated Damages are set by the contract, and the Constructor by signing has agreed to them as being the assessment on a daily basis of what the cost will be to the Owner for every day the Owner cannot occupy the space for its intended purpose.
- The Contract Administrator must also not be intimidated by the prospect of Liquidated Damages. If delays are properly documented there is the possibility the imposition of such damages is inappropriate.
- Just as a Constructor can be delayed during a project, there can also be delays during the very start of the project.
- Delays along any part of the schedule containing float are considered non-critical delays.
- There are many reasons a Constructor wants to use the float, but float can disappear if the Owner or Design Professionals use it first.
- The general rule, when assessing who caused a delay is to ignore who used the float. Float is often seen as belonging to the party that uses it first.
- The Contract Administrator needs to understand that there are excusable delays, and then there are non-excusable delays. Whether a Constructor will be compensated for a delay will depend on the delay being excusable, but even then, compensation may not be forthcoming.
- Excusable delays can be broken into two categories:
  - compensable
  - non-compensable

- The first challenge for the Contract Administrator will be to show the actual length of the delay. This will be especially important for projects with Liquidated Damages.
- It is always possible the Owner cannot extend the contract completion date. In such a case, the Constructor will be required to implement some form of acceleration. Who is going to pay for the acceleration is based on who is responsible for the delay.
- If the Owner agrees to pay for the acceleration, the Constructor will begin what is termed directed acceleration.

YOU BE THE JUDGE

Asset Recovery Contracting, LLC vs. *Walsh Construction Company of Illinois*

Asset Recovering Contracting (ARC) contracted with Walsh Construction Company (WCC) to do demolition work in the conversion of an office building in downtown Chicago to residential condominiums and retail space. ARC claimed that delays in the project caused a severe cash flow problem. It walked off the project and afterward it was forced to declare itself bankrupt.

The initial schedule required ARC to finish within a four-month period of time. Delays stretched the schedule to fourteen months. These delays were from many sources, including the Chicago Fire Department, several tenants demanding off-hours demolition, and the financial problems of the Owner.

A lower court ruled that ARC was not entitled to the $2.3 million it sought in damages because it had failed to raise any objections during all of the delays.

ARC then appealed to a higher court. ARC wanted the court to recognize a material breach in the contract based on the long delay.

WCC contended the language of the contract made it clear the "Contractor shall have the right to direct reasonable adjustments to the sequence and pace of the Subcontractor's work."

ARC pointed out that parol evidence should be considered because the date the work on the project had actually begun was several months before the contract was signed. ARC put forward that such evidence demonstrated the contract itself could not be interpreted without knowing what the parties had said between the time when construction began and the contract was signed.

ARC testified it had never expected the delays and they were extremely unreasonable, so the "no damages for delay" in the contract should be disregarded.

WCC argued that since the delays were known to ARC and it still signed the contract with the shorter schedule, then it had accepted the delays without qualification. In fact, ARC had known of both the Fire Department delay and the delay caused by the tenants before the contract was executed. It was pointed out that the Illinois Supreme Court has recognized that delays of over a year are still covered by the No Damages for Delay Clauses in construction contracts.

WCC further denied responsibility because ARC had breached the contract by leaving.

How do you rule:

Amount:

IT'S A MATTER OF ETHICS

Owners who expect more and give less can make difficult partners in the construction process. Some Contractors find working with such Owners an excuse to "put it to them" whenever the situation arises. If, for example, an Owner refuses to allow an extension of time, how accurate should you be in separating contract costs from the actual costs of acceleration? If you intend to take this matter before an arbitrator that might very well "split the difference," and only give a portion of what you claim, do you think you should inflate your claim regarding the costs of acceleration? How do you include costs related to less productivity due to overtime or trade stacking? And how do you think you will justify that in relation to a judge or arbitrator that is not entirely familiar with construction the way you are?

Going the extra mile

1. Beginning with this chapter and continuing through to Chapter 19, we will begin a series on the subject of Ethics. Go to the Companion Website and watch the Introduction of the Ethics video.

2. Do the "You be the judge" above and then go to the Companion Website to compare your ruling with the actual court ruling.
3. **Discussion exercise** (among friends or classmates): Discuss how effective you are in managing your time. Up to this point, the text has put a lot of responsibility on the Contract Administrator. Whether that task is filled by one or more individuals, the responsibility is enormous. So too is the pressure of tasks on the time available. How much do you think you can handle? When do you need help? Do you set specific goals for yourself, and after you have a goal, do you analyze it to break down the priorities necessary to accomplish the goal? Then do you put those priority milestones into your daily, weekly, and monthly schedule?
4. If you would like more information on BIM, you might go to the website www.gsa.gov, or try one of the leading software Suppliers, such as Autodesk (www.autodesk.com). For more on **Lean Construction**, try the Lean Construction Institute at www.leanconstruction.org. And for Integrated Project Delivery try searching at either www.aia.org or www.agc.org.
5. If you want more information on scheduling by experts in the industry, you might try going to www.whi-inc.com and searching on schedule, CPM, or the white paper "Should You Pursue a Construction Claim."

6. Effective use of the correct vocabulary improves communication and promotes respect among colleagues. The reader should become familiar with the key words listed below that have been used in the chapter. Definitions can be found in the Glossary. Additionally, the student is encouraged to use the flash card exercise in the chapter's Companion Website as practice for possible quiz or exam questions.

Key words

Acceleration

Acts of God

Compensable delay

Concurrent delay

Constructive Acceleration

Critical delay

Directed acceleration

Excusable delay

Float

Force majeure

Lean Construction

Liquidated Damages

Look-ahead schedules

Modified Total Cost Method

Non-compensable delay

Non-critical delays

Non-excusable delays

"Time is of the essence" clause

Total Cost Method

Trade stacking

CHAPTER 14

Liens

Each state has a different set of lien laws, so it is important to understand the situation within any given territory, but closing out a project or even parts of a project can be delayed for the lack of cash flow and payment. Keeping everyone current from Owner to Suppliers is an unheralded task, but vitally important.

Student learning outcomes

Upon completion the student should be able to. . .

- **understand what a Mechanics Lien is**
- **differentiate between a lien and a payment bond**
- **recognize other terms for a Mechanics Lien**
- **understand the importance of releases in relation to liens**
- **analyze the concern of the Owner related to liens.**

14.1 Liens

The lack of money is the root of all evil.

Mark Twain

In the last chapter we concentrated on the Owner's essential need to remain on schedule. For everyone else, money becomes the reason for doing the project in the first place. Some will surely be inspired by intrinsic motives such as pride in workmanship and the feeling of accomplishment, but in the end, the *sine qua non* is money.

Therefore, failure by the Owner to pay can be a serious issue. Most Constructors put a Pay when (or if) Paid Clause in their subcontracts, so Subcontractors and, further down the line, the Sub-subcontractors may have to wait a considerable amount of time to be paid, but they still want to be paid. If there is a payment bond on the project, and the lack of money is the "root of all evil," according to Mark Twain, then the Subcontractor can file against the bond. However, if the Owner is responsible for the lack of payment, there is a different issue entirely. The Contractor and all the Subcontractors have done considerable work to create or renovate a structure. If the Owner has use of that structure without paying for it, there is a terrible injustice.

The solution has been around for centuries. It may have originated in maritime law that permitted anyone who had furnished labor or material for a ship to **lien** that ship until their efforts were paid for in full. Presumably the ship would not be allowed to leave port without such resolution of existing debts. From such laws, it is believed that builders were given the same right to lien properties they worked on if they were not subsequently paid for the improvements they had made. Although such property could not sail away, it also could not be sold to anyone until the debt was satisfied. The law, itself, is based on the concept that an Owner of property should not be "unjustly enriched."

In the United States, it is believed such common law from England was promoted by both Thomas Jefferson and James Madison. Maryland may have been the first state to adopt lien laws, but today all fifty states have some form of lien law, but that is also a challenge for the Contract Administrator.

Each state has some variation on lien laws, and a competent attorney should be consulted related to each specific state and the requirement for filing a lien, as well as the prospects of success related to the circumstances. Changes are always taking place from state to state. New Jersey passed new statutes in which the Mechanics Lien (also called a Materialman's or Materialmen's Lien in some states) is now strictly referred to as a Construction Lien.

14.1.1 Requirement of improvement

One of the most important aspects of a lien is that in most circumstances it requires actual physical improvement to a property.

Examples:

1. An Electrical Contractor that installs a new panel for increased service to the building would be entitled to file a lien if not paid for the work, since the work represents an improvement to the property itself.
2. An office Supplier that delivered a copy machine would not be allowed to file a lien, since there was no actual improvement to the property.
3. A Kansas Court ruled that under the Mechanics Lien statutes of the state, removal of hazardous materials was considered to be necessary maintenance and not an improvement to the property, so a lien was denied.

Again, an attorney should always be consulted. The state laws and courts within each state may have different opinions.

A CASE IN POINT

For some, lien law is a moving target, and it is imperative that each Contract Administrator be aware of the general concept, but in order to be current, it is best to consult an attorney familiar with the lien law in each state.

In 2012, for instance, a Superior Court in the state of Pennsylvania dumbfounded the construction industry by ruling that a trustee of an employee union was a Subcontractor under the lien laws of Pennsylvania, and thus the benefit funds of unions were capable of placing a lien for payments a Contractor failed to make. The through line of responsibility was based on a collective bargaining agreement that had been made well before the project on which the lien would be placed had ever begun.

The Superior Court ruling stated the lien law's intent was to protect and do the most good for parties that contributed to the improvement of a property, but the end result will make it even more difficult for Owners, lenders, and Contractors to track all the potential entities that could file a lien against a property.

Such is the nature of law. It should be obvious by now, especially for all who have worked diligently on the "You be the judge" exercises, that your logic does not always agree with a judge or jury's decision. You will find that, if your perspective does not agree with what someone across the project trailer table is thinking, there is no guarantee it will match what someone thinks who is sitting on the other side of "the bar" (that physical difference in space between the working part of a courtroom and the public space).

Some people I know try to settle their disputes in a different type of bar. I have found that the project trailer table is much better.

Figure 14.1 United States courtroom
Photo credit: U.S. Government

The gate is open to "cross the bar" between the public space and the working part of the courtroom, but even better is never having to enter the courthouse in the first place.

14.1.2 Lien as opposed to lawsuit

The ultimate value of a lien over directly suing the Owner is that the Owner might not be able to pay even if the suit is successful. The Owner is probably in some financial trouble, and a successful suit might gain 100% of nothing. By placing a lien on the property, the Contractor now has the property itself, as an asset.

Another advantage for a Subcontractor filing a lien, as opposed to suing the Owner, is that the Subcontractor has no direct Contract with the Owner, and therefore the right to sue is very limited unless pursued through the Contractor.

There are, however, some disadvantages to liens. First, if the Owner is in very serious trouble, the lien right might position the lien holder behind others who have priority, such as a bank holding the construction loan and/or the mortgage on the property. Again, the laws of each state might place different orders of priority on each of the loans, mortgages, and liens. It is also possible Federal or state claims will be satisfied before the lien, or even negate the filing of a lien on public property.

If, on the other hand, the lien can take priority over the mortgage, then satisfaction of the lien might wipe out the bank's ability to collect on the mortgage. This is one reason banks lobby heavily against liens taking any precedence over both mortgages and construction loans, and in some states there are methods to reposition mortgages over liens.

It is also possible the Owner can assert a defense, and in some instances they are quite justified.

Examples:

1. The Owner has already paid the General Contractor in full for the work, but the General did not pay the Subcontractor.
2. There is faulty workmanship that requires correction.

Or the Owner might be protected by the General Contractor's having agreed in the contract to protect the Owner against all liens.

14.1.2.1 Timeliness of lien filing

Also, there are notice requirements related to filing a lien. In some states the party filing a lien needs to give notice of intent to file. In all states there is a statute of limitations based on the last day work was done to improve the property. These might vary from state to state, but it generally falls within a thirty- to ninety-day period from the last improvement made to the project. Failure to give timely notice will negate the right to file at all.

14.1.3 Recovery

If the party that has made improvements to a property without being paid decides to file a lien, and that party has advised the Owner a lien will be filed, then the next step is that a lien must be recorded in the township or county office. In all likelihood, the Owner will mount a defense and a hearing will take place. If the claimant who filed the lien prevails, the claimant can pursue having the property go up for sale. This will be handled by the Sheriff, and is therefore known as a **Sheriff's Sale**. The property will then be sold. Once the sale is completed at a public auction, the Sheriff will distribute the funds. Naturally, a sale at an auction may not bring true market value for the property, and so it is important for the claimant to be the first in line before the bank mortgages and loans. Otherwise the effort might produce very little in terms of recovering lost funds.

14.1.4 Release of liens

When bankers see the prospect of a lien placing them behind the lien claimant in a forced mortgage, they get very nervous. Making a loan for construction or holding a mortgage on a property usually involves very large sums of money, and the bankers want that secured as much as possible.

Owners also want protection from liens when they have been paying faithfully what they believe they owe. Owners for themselves and for their bankers therefore usually require a partial release of liens with each progress payment. The wording of these partial releases requires all parties, through the Contractor, to release the Owner from any liens for work paid for or contained in the current application for payment, and sometimes the wording will include future work by all those involved. Most Contractors and their Subcontractors and Suppliers are generally only partially in compliance when they sign these documents, since significant costs such as payrolls and most materials have generally been paid, but depending on the time of the month there can be some payroll burdens such as taxes or union benefits, and some Suppliers that have not been paid.

Still, this solution for both the bankers and the Owners has been acceptable and it adds to an already twofold protection.

1. During construction, payments for work completed lag behind the actual work by at least thirty and usually sixty days or more. If Mechanics Liens are filed during the progress of the project, the Owner and, by extension, the bank would only be exposed to a limited amount, since monies will have been withheld through the slow application process.
2. By contract the Contractor is almost always required to protect the Owner against any claims, including court and attorney fees, that might be filed.

If during the course of payment application approvals the Owner learns of the potential of unpaid invoices from the Contractor to Subcontractors, then it is possible the Owner, upon investigation, might issue joint checks to the Subcontractor.

Second only to a payment bond, the most effective protection the Owner receives is the waiver or release of liens from each Contractor and Subcontractor. These are generally required for each application of payment as a "partial release."

Just as an Owner will require the Contractor to sign a release of lien rights, generally the Prime Contractor will require a waiver of lien rights from each Subcontractor when submitting an application for payment. This waiver might also include waiving the right to file against the bond.

If timing is everything, this is where integrity and the trust between all parties is very important. Subcontractors are giving away rights that they might eventually regret, but usually the wording of such releases works back to acknowledging release from payments already received. For example, in the AIA application for payment, the Contractor certifies that the payments already received have been paid to the parties that were owed on previous applications.

14.1.5 Alternative protection

Contractors and Owners both want protection for different reasons. The Contractor, along with Subcontractors, wants to be sure to be paid, while the Owner wants to make sure he or she does not have to pay twice.

14.1.5.1 Payment bond

Bonds have been discussed separately, but one of the best alternative payment protections for both the Owner and the Subcontractors is a payment bond. Provided proper notice procedures are followed, the Owner can rely upon the Subcontractors being paid any monies due that have been paid by the Owner but the Contractor has not passed through to the Subcontractors. Similarly, each Subcontractor knows, with a payment bond in place, that there is recourse to be paid in full for monies legitimately owed.

In some states a claim against the payment bond can be used in addition to filing a lien against the property. In other states, the existence of a payment bond negates a party's right to file a lien. Again, attorneys familiar with the specific state's laws should be consulted.

Since the cost of the bond is incorporated into the overall cost the Contractor has submitted, private Owners often do not add this additional burden (usually around 1% when it is combined with a performance bond) to the overall cost of the contract.

Public Owners, however, are usually required to have a payment bond in place if the contract is of a certain value.

Both Federal and state governments recognize there could be severe consequences if liens could be placed on public property, forcing a sale of the "people's property." In fact, how do you sell to the citizens what the citizens already own?

Figure 14.2 Most Americans do not realize that during the Truman administration the White House was totally "gutted" for renovations. John McShain, who had built the Pentagon, was the Contractor. The project certainly went well, and of course there could never have been a lien against the White House. It does not belong to the President. It belongs to the people, and must therefore be protected against a lien.

Photo credit: United States Government

The Federal government did not like the idea that public property could have a lien placed upon it. In 1893 the Heard Act was enacted to protect parties supplying labor or material to a project. Recognizing that a Constructor needed some mechanism to enforce payment rights even against the Federal government, the Heard Act nevertheless required the party to wait until six months after the completion of the project to begin the claim. The **Miller Act**, passed in 1935, replaced the Heard Act. As discussed in a previous chapter, the Miller Act requires all Federal contracts for construction work above $100,000, whether it is new construction, alteration, renovation, or repairs, to be covered by a payment bond.

14.1.5.1.1 Stop payments

Another means of securing payment that a Subcontractor or Supplier might use on Federal work is what is known as a **Stop Notice**. In essence a Stop Notice is a lien that is attached to money rather than a structure. The purpose is once again to avoid a lien on a Federal building or a public project, and the unpaid party is seeking to intercept the funds from the Owner before they are paid to the Contractor that is asserted to be delinquent in payments.

If a proper Stop Notice is filed, the government withholds that amount from the Contractor, and if it is sufficient to satisfy the Stop Notice, the money would be paid to the Subcontractor or Supplier. The key word is "sufficient," since by the time the filing is actually made, the amount of money being withheld from the delinquent Contractor might not be sufficient, and there is usually no requirement for the government to make up funds if no funds remain.

The other issue could be that if one Subcontractor or Supplier has not been paid what is justly due, then it is quite probable others have not been paid also. That means there will be a line of Stop Notice applicants, and the remaining funds unpaid will not be sufficient.

In some states stop notices are also permitted on private work. This can be a temporary benefit to a private Owner as another means of avoiding liens against the property, particularly if a construction loan or mortgage has been taken on the project or property.

The key issue, however, for all stop notices is whether or not the claim is valid. For that reason the claimant is sometimes required to post a payment bond that will protect the Owner if the claim is proven to be invalid and any funds have been lost or improperly paid to the Subcontractor in the process of complying with the Stop Notice.

14.2 Choosing a remedy

Precaution is better than cure.

Johann Wolfgang von Goethe

If conditions deteriorate to a point that a lien must be filed, it is important for the Contract Administrator to involve competent legal counsel as early as possible. The process and even the rights of who can file a lien may be different from state to state. If bonds are in place, there may also be a choice to pursue a bond claim rather than a lien filing. The threat of a lien or bond claim can often send shock waves through a project. The actual filing of a lien or claim can be a "last straw," breaking relations and teamwork among the project members.

From the Owner's perspective liens, stop notices, or claims against a bond are very unwelcome. In most cases, the Owner has a lender that has provided funds in the form of a construction loan and/or a mortgage, and the prospect of paying twice for something is not appreciated and can often shut down cash flow until a resolution is achieved.

For the Contractor, not being paid is more than unfortunate; it can lead to misfortune and worse.

For the Subcontractor or Supplier, slow or no payment is the end of all trust.

So, the first remedy for all parties involved in construction is to choose your partners well.

In some instances even that does not work, and trouble arises. In such a case, the Contract Administrator must carefully monitor all payments and the flow of payment from pre-application to receipt.

* * * * * * * * * *

A good habit the Contract Administrator might want to develop is to "walk" each application through the approval process. This should certainly be the case with the first application. Taking it to each individual (with advanced notice and by appointment) and making sure everything is in order, and then scheduling the forwarding to the next approval "station" can actually be worthwhile for all parties, provided a thorough analysis can be made at the appointed time related to each approval.

* * * * * * * * *

Staying on top of the payment process may at first seem "pushy," but it is why everyone is in business. Corporations exist to make money. People are hired to make that possible. Team members should all recognize how important that is for the success of each entity involved in the project.

The final and ultimate key to success is that if the money is flowing correctly, everyone on the team can concentrate on the more enjoyable aspects of finding solutions to the challenges that really matter for the success of the whole.

Remove lack of money or poor cash flow as a "demotivator," and the motivating factors that everyone came into construction for in the first place (to make a difference and enjoy the process) will move each project forward to a successful conclusion.

INSTANT RECALL

- Each state has some variation on lien laws, and a competent attorney should be consulted related to each specific state and the requirement for filing a lien, as well as the prospects of success related to the circumstances.
- By contract, Owners may be protected by the General Contractor, who has agreed in the contract to protect the Owner against all liens.
- There are notice requirements related to filing a lien. In some states the party that is filing needs to give notice of intent to file.
- Usually, in the progress payment process, the application for payment will require a partial release of liens for the work completed to date, upon receipt of that payment.
- If, during the course of payment application approvals, the Owner learns of the potential of unpaid invoices from the Contractor to Subcontractors, then it is possible the Owner, upon investigation, might issue joint checks to the Subcontractor.
- Second only to a payment bond, the most effective protection the Owner receives is the waiver or release of liens from each Contractor and Subcontractor.
- In some states a claim against the payment bond can be used in addition to filing a lien against the property. In other states, the existence of a payment bond negates a party's right to file a lien. Attorneys familiar with the specific state's laws should be consulted.
- The Miller Act requires all Federal contracts for construction work above $100,000, whether it is new construction, alteration, renovation, or repairs to be covered by a payment bond.
- Another means of securing payment that a Subcontractor or Supplier might use on Federal work is what is known as a Stop Notice. In essence, a Stop Notice is a lien that is attached to money rather than a structure.
- If a proper Stop Notice is filed, the government withholds that amount from the Contractor, and if it is sufficient to satisfy the Stop Notice, the money would be paid to the Subcontractor or Supplier.

- The first remedy for all parties involved in Construction is to choose their partners well.
- The final and ultimate key to success is that if the money is flowing correctly, everyone on the team can concentrate on the more enjoyable aspects of finding solutions to the challenges that really matter for the success of the whole.

YOU BE THE JUDGE

Total Industrial Plant Services, Inc. vs. *Turner Industries Group, LLC*

Total Industrial Plant Services (TIPS) entered into a fixed-price subcontract with Turner Industries Group (Turner) to install insulation at a refinery in Montana. The contract amount was $4.2 million. TIPS claimed Turner's demand to increase the manpower resulted in additional labor costs above the subcontract price.

Turner pointed out that winter weather and other delays had set TIPS work back and the contract called for it to perform work "efficiently and promptly."

During the construction, TIPS negotiated with Turner to change the contract from fixed price at $4.2 million to Time and Material work, which Turner eventually paid upon TIPS' completion of the work at $13.2 million. TIPS, however, asserted that it was owed by Turner inefficiency costs of $700,000. TIPS then filed a construction lien against the Owner, asserting that Turner owed $2 million for breach of contract, *quantum meruit*, unjust enrichment, and failure to deal in good faith.

Turner defended its actions, based on changing the contract to a Time and Material payment. However, TIPS asserted it had already completed almost half the contract before the conversion was made from fixed price to Time and Material. TIPS also asserted that Turner had orally agreed to pay for extra costs prior to the change from fixed to Time and Material.

Turner also defended the *quantum meruit* and unjust enrichment charges, based on the fact that specific scope within the original fixed price contract was done in accordance with the contract, and, once converted to Time and Material, all costs were fully paid for in that manner without objection.

Turner requested that the lien against the property be barred because TIPS had filed after the ninety-day limitation. Turner presented evidence that TIPS' work on the Turner contract had ended June 25, 2008. TIPS filed September 24, 2008 – ninety-one days later, or one day too late.

TIPS pointed out that although it was merely a calendar day late (and it was within the month-to-month time frame) it had still been on the site after the June 25, 2008, date when it completed work for Turner. Turner had to concede TIPS was on site, but after June 25, 2008, it was working directly for the Owner and not for Turner.

How do you rule:

In what amount:

IT'S A MATTER OF ETHICS

Payment is critical for all businesses. The Contract Administrator will usually be in a unique position to understand the flow of funds to and from his or her company. How honest should one be with others outside the company related to payments received or payments due? Is it right to blame an Owner for lack of payment? What if you invoiced late through no fault of either the Owner or those you owe the money to after you will be paid? Is it right to use money owed on one project to pay a debt on another? What if that debt is owed to the Federal or a state government for taxes that will take a priority position over all other debts if not paid? If all the answers to the above sound easy, would you pay yourself and/or others on your own payroll with money due another Subcontractor, knowing that if you did not, you would not have anyone working next week?

Going the extra mile

1. Go to the Companion Website and watch Part I of the Ethics video.

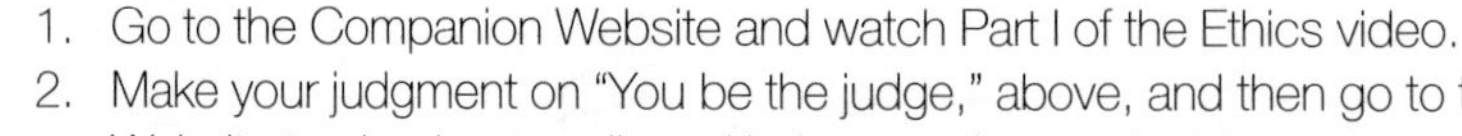

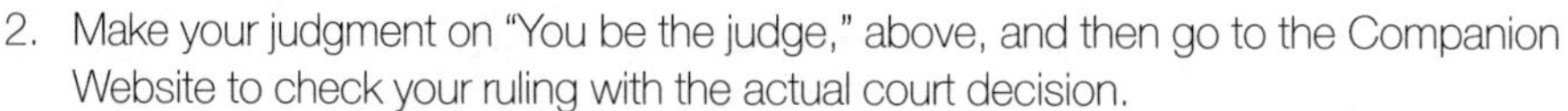

2. Make your judgment on "You be the judge," above, and then go to the Companion Website to check your ruling with the actual court decision.
3. **Discussion exercise I** (with your classmates): Discuss the merits of having a bond on a project over filing a lien. Does the time for filing on a bond make a critical difference related to the longer length of time one has for filing a lien? If you were a private Owner, would you think it was worthwhile to have a bond, even though it might cost more money, or would you rely upon lien waivers and a select bidders' list of Constructors that you believe would not give you reason to wish you had a payment bond with them?
4. **Discussion exercise II:** Go to Appendix I and read the paragraph on liens and claims. How would you improve it? In trying to cover both liens and claims, does it cover too much? Too little of one or the other? Since the majority of contracts R. S. Cook and Associates signed with the Owner were not AIA contracts, but Owner-drafted contracts, each different in lien wording from the other, is there a reason for the strict language of the paragraph on liens and claims?
5. Effective use of the correct vocabulary improves communication and promotes respect among colleagues. The reader should become familiar with the key words listed below that have been used in the chapter. Definitions can be found in the Glossary. Additionally, the student is encouraged to use the flash card exercise in the chapter's Companion Website as practice for possible quiz or exam questions.

Key words

Lien

Miller Act

Sheriff's Sale

Stop Notice

CHAPTER 15

Tort law

Just as important as contract law, tort law can have significant consequences on the future of any project, or even corporation. Contract Administrators must always be careful of negligent acts that can affect others on the project, or even the public.

Student learning outcomes

Upon completion the student should be able to. . .

- **describe and discuss the importance of tort law**
- **analyze the consequences of negligence in relation to actions related to the project**
- **assess what constitutes intentional torts**
- **discern responsibility for strict liability torts**
- **recognize the Attractive Nuisance elements of construction liability**
- **evaluate and judge the actions of contract administration in relation to the legal issues of protecting people and property.**

15.1 Tort law

That old law about an eye for an eye leaves everybody blind. The time is always right to do the right thing.

Martin Luther King

Some Constructors can work for years without a single tort case, but when and if one arises, it can have severe consequences. In studying the contract, a Contract Administrator will not likely find any **tort law** specifically mentioned, and yet ignorance of such laws could be unfortunate, even disastrous. As we have discussed earlier, tort law is part of our lives, and in relation to contract law, our compliance with tort law is implied. In fact, tort law exists whether one or both parties agree to the law or not. Again, with the foundation of law in the United States firmly established in common law, tort law has evolved from the basic concept that everyone deserves the right and freedom not to be injured, or one's property not to be damaged, by another person(s).

Simply stated, a tort "is a civil wrong for which a governing body will impose a remedy." Courts, of course, do not view breaches of contract to be tort cases, but some actions of individuals within a contractual relationship can fall under tort law.

The key difference between tort law and criminal wrongs is that the government will impose a remedy in both, but in tort law, the government expects the injured party to pursue its own case, whereas in criminal law the prosecution will be handled by the government.

> **Examples:**
> 1. An argument develops between two individuals and one decides to get even by backing his truck into the side of the other's vehicle.
> 2. One party to the contract purposely misleads another in relation to the costs for doing some extra work.
> 3. A barrier is removed and not replaced and a worker drops a piece of material that strikes a passerby.

There are countless possibilities of civil wrongs, but generally they fall into three broad categories:

1. intentional – in which the party committing the wrong clearly understands the act to be wrong;
2. negligent – when the actions do not intentionally create harm, but a harm occurs nonetheless;
3. strict liability – when the potential for harm is great and the Contractor has to recognize that the consequences of proceeding could cause some form of damage regardless of who is at fault.

One issue a Contract Administrator must keep in mind is that individuals may react differently to the harm or damages done to them. What might seem trivial to a Constructor might not be to the Owner or the Design Professionals. Or an innocent driver might be more enraged by spray paint blowing in the wind that lands on his or her parked car than the painter who is trying to finish a quarter-million-dollar contract.

Also, as mentioned earlier, the Contract Administrator must recognize sometimes that there is a gray line between the civil wrong of tort law and a potential criminal wrong. Some prosecutors, for reasons ranging from political gain to personal conscience, are turning what used to be tort law civil wrongs into criminal cases. As mentioned in Chapter 2, criminal prosecution in New York City related to crane accidents is setting a new tone for negligent acts in that city. The Contract Administrator must always understand that what might seem unfortunate, even trivial if no one was seriously injured, may not be approached that way by others who are in a position to assert authority.

The best way to handle tort claims, therefore, is to refrain from the intentional, inspect against negligence, and ensure as best as one can (and insure if possible) against strict liability.

15.1.1 Intentional torts

For the most part, **intentional torts** in construction are rarer than negligent torts, which we will discuss next. One form, which fortunately does not occur so much today, is battery. Although assault and battery could turn into a criminal act, physically touching a person can start as an intentional civil wrong. If a criminal case follows, that does not relieve the party that created the battery from the potential of also being liable for civil damages.

A CASE IN POINT

Bert and Victor were stepbrothers, and the two were always arguing, and the arguments often resulted in fights. One day, on a construction site, three floors up on a wood-frame structure, an argument between Victor and Bert grew into more than an exchange of ideas. Quickly the violence escalated, and tragedy would have followed, as Victor pushed Bert a bit too hard, and Bert was falling backward off the upper deck of the site. Before tumbling to the ground below, however, he managed to catch his saw on the nearest column post.

That saved Bert to cool off and live another day, and maybe Victor realized how close he had come to considerable consequences for his unguarded temper.

That's the way my father told the story of two of his uncles, but it was not really about those two uncles. It was about construction sites being dangerous places to work.

So, even though that may seem to be an old story, it repeats itself every day on projects around the country and the world. Maybe it is over something important, or nothing important at all. People can be stressed when deadlines approach and trades and craft workers are stacked on top of each other, blaming the other for being in the way of finishing. Or maybe it is related to union versus non-union strife. Maybe it comes when work is slow and a stressed worker is worried how the mortgage will be paid if the next check is a lay-off check.

According to the Bureau of Labor Statistics, in 2011, 17% of all workplace fatalities were caused by workplace violence.

There might be nothing in the contract that says everyone has to agree, but when we disagree, we need to do so in a way that makes us stronger after we come to a solution.

The tension of cost and schedule in construction is great enough. Fulfilling the contract conditions is enough for any person and team, but the consequences of tort offenses can ruin a great project. This becomes another challenge for the Contract Administrator.

15.1.1.1 Fraud

However, the most common type of intentional torts in construction would be those related to cases of fraud – misrepresentation of facts. Often, fraud cases will arise out of issues related to conditions shown on the document being different than what actually exists on the site. Such cases usually become Differing Site Condition cases, discussed previously, and the defense of the Owner and the Design Professionals often rests on exclusion or exculpatory clauses. There are three critical questions related to proving intention in such a case:

1. Did the Owner and/or Design Professionals know differently from what was indicated on the drawings and contract documents?
2. Did the Owner and/or Design Professionals expect the Contractor to rely upon the information provided?
3. Was the Contractor deceived to the point of relying on the documents, without reasonable recourse to expect something different?

Contract Administrators for the Constructor must also be careful not to misrepresent conditions or issues to Subcontractors. An estimator in the "heat" of bid day might enjoy "shopping" a Subcontractor for a lower price. Misrepresenting conditions can later become an issue that can cause concern and affect job progress for the Contract Administrator. Of course, even after a bid has been submitted, if the Constructor misrepresents to a potential

Subcontractor or Supplier certain conditions that do not eventually prove to be valid, then again, the potential for a tort case exists.

One very important point, however, is that someone might be more engaged in exaggeration than in fraud. Telling someone you believe they can still have a profitable job at a reduced price is not the same as not telling them there is solid rock just below the surface where they have to dig their footings. In short, there are "salesmen," and then there are "snake oil salesmen."

In addition to intentional misrepresentation of fact, there is also the potential for concealment of fact that should have been disclosed

Examples:

1. If an Owner knows there are hazardous substances on site, that should be disclosed.
2. If the Owner is aware of a high water table and schedules start of construction during the rainy season, then prudence dictates that that condition should be discussed well before ground breaking starts.

For some, this may seem a remote possibility, but the alert Contract Administrator will recognize when it happens, and whether pursuing a claim is worthwhile. In the *City of Salinas* vs. *Souza and McCue*, the City Engineer withheld information about unstable soil conditions. He directed an independent testing firm to take borings at specific locations that he knew were not affected by the poor soil conditions. The Contractor was awarded damages for the intentional fraudulent misrepresentation.

15.1.1.1.1 Recovery for damages

Fortunately, many issues of misrepresentation can be handled effectively outside of court, through reasonable discussion and negotiation.

Nonetheless, courts do recognize the seriousness of intentional misrepresentation, and for that reason, in addition to determining actual damages, they also impose **punitive damages**. These damages are left to the discretion of the jury, but they are sometimes figured as a multiple of the actual damages, such as **treble damages**, or three times the amount of proven damages (e.g. hospital bills and lost income from not being able to work plus three times that amount for additional punitive damages).

Exercise 15.1

Take a moment

Calculate how quickly punitive damages can increase damages

A party is found guilty of a tort claim offense, and is held responsible for damages to the other party for $125,000. How much in total will the party pay if punitive damages are assessed at three times the actual damages. Do your quick calculation of what should be a rather large dollar answer:

Discussion of Exercise 15.1

You should have arrived at a total cost a half a million dollars:

Original $125,000

Punitive $125,000 x 3 = 375,000

Total $500,000

Some courts discourage punitive damages because they provide extreme windfall profits for the injured party. For that reason some decisions by juries have been reversed, and punitive damages themselves have been viewed as an example of the need to reform our judicial system.

A CASE IN POINT

Do you trust a jury?

This is not a construction story, but your insurance agent will tell it to you to convince you to settle a claim before trial rather than go in front of a jury. One of the most famous punitive damages awards ever made involved a woman who had spilled McDonalds' coffee on her lap. The coffee had been so hot it caused severe burns and several thousand dollars in medical bills as well as other damages, including lost time. She had to spend eight days in a hospital undergoing skin grafts.

The attorney for the woman knew the chance of receiving triple the cost of the actual damages ($160,000 for actual costs and $480,00 for punitive damages at three times actual) would be considerable, but he thought another strategy might produce even more for his client.

In front of a jury he pointed out that McDonalds makes over $1.5 million every day worldwide selling coffee. It would be appropriate if the client, having suffered so much and having spent eight days in a hospital, received one or two days' worth of McDonalds' coffee sales. It would not be anywhere near McDonalds' total daily sales, just their coffee sales.

The jury bought the argument and awarded $2.7 million in punitive damages above the amount of the actual damages.

The case was appealed and was ultimately settled. As part of the settlement agreement, the sum cannot be disclosed, but it almost certainly was something between $640,000 and $2.7 million.

Almost always a jury's sympathy will go to the injured, similar to the time when we were going in front of a jury to defend ourselves against the woman who claimed she had tripped over our debris on a subway project. We were convinced she had not. "Settle!" the insurance agent and attorney kept telling us, or we will. But when a detective photographically proved the claimant had one day switched the cast from one leg to the other, rather than have to settle, the case was thrown out.

Most Contract Administrators will not be so lucky, so beware the jury in a tort case, and conduct oneself accordingly.

15.1.1.2 Libel and slander

It is not hard to imagine one party saying something negative about another party on a construction project. If such a statement is not only negative or defaming, but it is also false, it can lead to a civil wrong under tort law. If the wrong is done in writing it is **libel** (and probably easier to prosecute), and if it is done orally, it is **slander**.

Of course, when competing for a project, the Constructor should promote positives of their own company and not defame the competition.

Exercise 15.2

Take a moment
Think positive

Below are three bad things you know about your competition. Change them to positive expressions about your own company:

1. You understand that the competition's bonding company has said "no" to any more bonded work until they finish what they already have.
2. The insurance company for the competition might not renew because of the poor safety record they have.
3. No one likes working for the competition, but times are tough so they have to.

Discussion of Exercise 15.2

How you express yourself can be very important. If we are seen as being negative toward our competition, we generally will not be seen positively ourselves. While being positive, it is also possible to cast light on your own qualities that are better than others. The possible positive comments related to the above might be:

1. I know this is not a bonded project, and we appreciate your wanting to save the extra costs, but if you would like to consult with the bonding companies of all those who have given you a price to do this work, we would be more than happy to provide you with the phone number of our agent.
2. We are very proud of our safety record. We believe it is second to none. Please feel free to compare our experience modifier against that of anyone in the industry. I am sure you are checking everyone else's, so here is ours.
3. Please drop by our office anytime. I know our industry has a bad reputation for attitude, but come and see why people like working at our company. We love to have a good time, and we do, because keeping all our projects on schedule is one sure way to have fun in this business.

Contract Administrators must also be very careful they do not libel or slander another party in relation to the prosecution of the work. The biggest temptation to do this would be to excuse one's own failure to keep on schedule or deliver quality to the Owner. Just as importantly, Design Professionals and other agents of the Owner must be careful not to libel or slander the conduct of the Constructor or any of its Subcontractors and Suppliers. This might be additionally difficult when the Designers are responsible for inspections and certification of percentages of completion.

Example:

Quality Granite Construction vs. Hurst-Rosche

Hurst-Rosche engineering firm had certified that the Contractor had satisfactorily completed its work, but later wrote that Quality Granite (Quality) had failed to complete the project in a timely manner and had done so with substandard workmanship. This letter caused Transamerica (the bonding company for Quality) to reduce future bonding with Quality. In court, Quality submitted evidence that the letter by Hurst-Rosche was defamatory. Quality also provided evidence that it had complied with the terms and conditions of the contract. The jury agreed with Quality, awarding damages to Quality for reduced bonding capacity, as well as punitive damages.

This then becomes even more difficult for the Owner's agents in relation to the competence of the low bidder. Particularly on public work, the Owner's agents might firmly believe the low bidder is not capable of performing the work. This may be due to an extremely low price, or it may be due to the lack of experience of the Contractor. On a public project below $100,000, where no surety is required under the Miller Act, then the advice of the Owner's agent(s) might be very important, but it does put the Design Professionals and/or Construction Manager in a difficult position related to libel or slander. Whatever advice that is given should be based on facts rather than opinions.

In *Riblet Tramway* vs. *Eriksen Associates*, the Design Professional thought it unwise to award the contract to the low bidder, because they had no experience of installing triple chairlifts. The court agreed, because the Engineer did not misrepresent the experience of the Contractor.

* * * * * * * * * *

Finding a way to be positive is an important habit the Contract Administrator should develop. A favorite expression of Robert Cook's was "If you can't boost, don't knock." As a Contract Administrator it is always worth keeping that expression in mind. No one in this industry can survive with a Pollyanna or "everything's coming up roses" attitude. Everything will not turn out perfectly fine without considerable effort to make it happen. Contract Administrators have to recognize what is wrong and not be afraid to clearly define it or call someone and let them know the issues in an honest, un-sugar-coated appraisal, but even negative criticism should be administered for the good of the project. Find a way to boost the project in a positive direction, rather than dwell on the negatives that have put us where we do not want to be.

* * * * * * * * *

15.1.2 Negligence torts

Negligence tort cases in the construction industry are generally more common than intentional torts, and, unfortunately, they can often be more costly – including both financial damages and human injuries. In negligence tort cases no intent needs to be proven. An injured party is able to collect compensation simply because an injury and/or damages occurred that another party is responsible for, due to their actions, or often, lack of action.

From the beginning, the Contract Administrator must recognize the principle of ***respondeat superior*** **or "reputed negligence."** This is the legal principle that binds the actions of the employee to the employer. In other words, as long as an employee is working within the scope of the employment, his or her actions are binding upon the employer. The employer will then have to defend negligent actions of employees as though they were the actions of the employer.

Example: An employee goes out for lunch and is asked to pick up stamps for the company. On the way there is an accident, due to the employee's negligence. The company is responsible, because the employee was engaged in work for the company.

Negligence is a broad issue and it affects both the Constructor and the Design Professionals. In fact, negligence can become the means for a Constructor to collect for damages that would otherwise not be available due to a lack of contractual obligation between the Constructor and the Design Professionals. Usually, in order to initiate a contractual claim, a Constructor would have to make it through the Owner, because it is with the Owner that the contract exists – not with the Design Professionals.

When defending a negligence claim, both Constructors and Design Professionals need to recognize that their particular expertise will be the standard by which negligence is judged. A prudent and reasonable Constructor would be expected to know and execute the construction in ways a typical homeowner doing a weekend fix-it-up renovation or repair would not. Similarly the Design Professional will be expected to understand and communicate plainly and without ambiguity the physical and structural requirements for successfully turning the "vision" into reality.

One method to judge "due care" is based on how well something could have been foreseen. If a reasonable Constructor could have foreseen the issue and still it was ignored, then the Constructor will likely be held liable. If a Design Professional should have recognized the error or ambiguity, then the judgment will likely be based on negligence on the part of the Design Professional.

At the same time, courts will recognize that risk of some sort is inevitable, and that life, itself, contains some elements of risk. Unfortunately, where the line is drawn often becomes an issue judges or juries decide differently.

Figure 15.1 The dark side of bridges
Photo credit: Charles W. Cook

Designers of bridges recognize that these are often the site of suicides. The Golden Gate Bridge carries thousands of cars and hundreds of walkers daily, but it is also the site of many suicides. Signs for crisis counseling, stating "there is hope – make the call," have been posted on many bridges, but, regrettably, some do not make the call. That unfortunate fact will not stop the world from constructing bridges, but imagine what would happen if the Designers were to be held liable for each suicide under negligence tort law.

Generally, a design error will be held against the Design Professionals, and an error in execution will be held against the Constructor, but that does not mean cases will not be brought to trial to determine exact responsibility.

Example: Generally, trench safety is the responsibility of the Constructor. If an eight-foot unshored, unsloped trench collapses on a worker, the Engineer is generally not held liable, since the safe execution of the design is the responsibility of the Constructor. The Engineer's duty was to design, not to construct. However, what is the case if the Engineer on a site visit witnessed the trench and recognized it as an unsafe condition? In the past most courts have ruled that the Engineer is still not negligently responsible. That verdict, however, is no longer certain.

Taking the example one step further, what if the Engineer noticed an unsafe condition related to the crane on the project? Now the public, rather than just the Constructor's people, are in "harm's way." Does the jury then give a pass for the Engineer who said nothing?

The standards may be gradually shifting, and no member of the construction team should expect a "free pass" for safety, regardless of previous rulings on tort law.

Although negligence on the Constructor's part may be viewed from a safety perspective, the Contract Administrator must recognize that negligence can carry over into completed operations. If the structure has a latent defect, the remedy and recourse of the Owner can be through a negligence tort rather than the warranty provisions of the contract. This can also be a means during and after construction for an Owner to hold a Subcontractor or Supplier liable, even though there is no direct contractual relationship between the Owner and the Subcontractor or Supplier.

On the other hand, it is possible that the Owner misuses the structure or the equipment installed. In such a case the Design Professionals and/or the Constructor, Subcontractors, or Suppliers might be called upon to defend a negligence tort, even though the work was done properly.

Example: Security equipment was installed but never properly maintained by the Owner, who had been offered but rejected a maintenance agreement. Eventually the Owner suffers a loss. Some Owners will initially look to the Contractor and whoever installed the equipment, as well as the Supplier of the equipment.

Every Contract Administrator, therefore, must realize that tort law is an additional layer of potential liability, and prudence dictates that the whole team must remain vigilant. While everyone may be judged by doing things on time and within budget, the other two legs of meeting standards of quality and doing so safely must not be ignored. In fact, they must be persistently in the forefront of everyone's mind and execution.

15.1.2.1 Cause and effect

One important issue (and perhaps defense) Contract Administrators should remember related to negligence is that the injury or damage has to be the direct result of the negligence (sometimes referred to as **causation**). Neither Constructors nor Design Professionals should be held responsible for the "negligence" or incompetence of others. Again, courts and juries may have different opinions, but the general rule is that there can be no negligence responsibility unless the negligence was the cause of the injury or damage.

Examples:

1. An electrician installs 15-amp circuits in the panel box. An Owner loads 20 amps on the circuit, blows the circuit and loses all the data currently entered on his computer. The electrician would not be liable so long as the circuits were not labeled 20 amps or the original drawings/contract did not call for 20 amps.
2. The same electrician installed high-hat lights in the study where the Owner works on his computer. A high hat directly above where the Owner works on his computer creates a glare on the screen. The Designer is not liable for negligence in causing the glare, provided the Owner and the Designer had not discussed specifics related to placement of the computer screen and the lighting.
3. That same high hat falls down and smashes the computer screen. Now someone is probably liable, and it most likely will be the electrician who installed the high-hat fixture, if it was not properly secured per the manufacturer's recommendation. On the other hand, if it was properly installed, it could be a long road to compensation for the Owner if the manufacturer's design is in error. Such errors of faulty manufacturing often lead to class action suits, since more than one homeowner was probably affected, and unless the damage is significant (or perhaps a genuine injury was sustained), the cost of one individual pursuing compensation is usually prohibitive.

15.1.3 Strict liability

Throughout the entire process, the Contract Administrator must be careful of the concept of strict liability. In such instances the fault of any one party does not need to be proven. All that needs to be proven is who is responsible for the act that caused the injury. This is a standard for many product liability claims. The manufacturer is responsible for producing the product, and during the use of the product the injury took place. Manufacturers will of course rely upon some defenses such as the product was abused or misused, or the product did what it should, but the user did not.

In construction, not only product liability could be an issue, but the very activities required on the project could cause damages.

Example: Blasting during excavation is a major concern not only for those on the jobsite doing the work, but also for surrounding buildings and their occupants. Damage to buildings or their contents, due to shaking of the ground or the impact of any explosion, can be a cause for concern and requires considerable documentation and warning to the Owners and occupants of the surrounding buildings that might be affected. Still, the potential for a strict liability claim is possible, not because the Constructor did something wrong, but because damage occurred that would not have occurred had the blasting not taken place.

15.1.3.1 Attractive Nuisance

I hate to think back to my youth and the number of times I could have been hurt exploring homes and other structures, and the equipment I climbed all over, on sites under construction when no one was there. I was fortunate, but many children are not. A construction site is what is termed an **Attractive Nuisance**. Builders have to recognize that unprotected construction sites are a tragedy waiting to happen.

In most cases large construction sites are well protected against intrusion, for many reasons. Theft of tools, materials, and equipment is always a possibility if the site is not protected. Nonetheless, other sites are often in close proximity to the public. Perhaps the most dangerous site, however, is the home builder's project within an existing neighborhood. Unprotected over the weekend, the site becomes a temptation for children to exercise both their bodies and their imaginations, sometimes with terrible results.

Figure 15.2 A pile of dirt can be a legal liability
Photo credit: Charles W. Cook

Left unprotected over a weekend, a pile of excavated dirt for a new home can be the first of many attractive nuisances children can find to play on over a weekend.

Courts and juries have consistently held that the danger within a construction project site of any kind is the responsibility of the Constructor. There is no defense allowed based on "the individuals should not have been on the site – they were trespassing." Such a defense related to children trespassing (and sometimes their parents) is negated by the Attractive Nuisance standard that unprotected construction sites are temptations too great to resist unless properly protected.

Amazingly, some states have recognized the potential for some properties to be actively used for recreational purposes, and statutes (not tort laws) have been passed allowing landowners to open their properties for recreational purposes and be clear of all tort negligence. Although this may seem a good will gesture to the community, the Contract Administrator is strongly advised not to go anywhere near that issue. Keep the site protected and clear of anyone not trained in construction. Keep construction sites for Constructors.

15.2 Contributory and Comparative Negligence

The real and effectual discipline which is exercised over a workman is that of his customers. It is the fear of losing their employment which retrains his frauds and corrects his negligence.

Adam Smith

There has been a doctrine of **Contributory Negligence** that absolves one party of responsibility for a negligent act if the other party was also negligent. Neither party in such a case can collect damages from the other. For years, courts did not look favorably upon any party that "contributed" to the injury or damage.

Example: Two DUI (driving under the influence) drivers both run the same red light and crash into each other.

Of course, it is quite difficult to determine absolute blame for some accidents or negligent incidents, and so the doctrine of Contributory Negligence has often been viewed as unfair if someone was negligent in a very minor way, such as a worker who is texting while crossing the project's access road. A driver runs into him because he was distracted while adjusting the radio.

For that reason, many states are awarding cases based on **Comparative Negligence**. Under Comparative Negligence the courts try to apportion the compensation for damages based on the degree of negligence for each party. Of course, such awards are hardly an exact science, and under Comparative Negligence a lot depends upon the judge or jury's perception. Some states also define **Contribution Negligence**, where more than one party is responsible for the same damages.

Example: After hours, when everyone is going home, and in the case above of the texting worker struck by the radio-adjusting motorist, if the jury found the worker 10% liable and the motorist 90% responsible, then if the worker suffered $100,000 in hospital expenses and out-of-work expenses, the motorist (or insurance company) would pay the pedestrian $90,000. If the motorist suffered $2,000 in damages and repairs, the worker would be responsible for $200, and the net judgment would be $89,800 against the motorist.

Certainly, when dealing with the complexity of construction claims, such Comparative Negligence judgments will almost always be made by people who do not fully understand the intricacies of the project or perhaps construction itself. Such rulings, therefore, will often be arbitrary.

Many states recognize that such comparative rulings are subjective rather than objective. Additionally, it has been recognized that in some instances more than one party may be responsible for damages to another party, and for that reason a different form of award has been rendered.

15.2.1 Deep pockets

Knowing one party might be more solvent than another, an approach to compensation known as **deep pockets** has been used by prosecutors. In a case of more than one party being held responsible in a Comparative Negligence case, then the prevailing party has the right to collect from the party with the "deepest pockets." Presumably that party is more capable of paying the damages than the other parties, and the "deep pocket" party then has the means, if it so desires, to pursue proportional compensation from the other comparatively negligent parties.

Example: A City inspector on site to check plumbing fixtures trips when stepping on an unprotected hole that had been cored for the concrete Subcontractor, using a Sub-subcontractor to do the saw cutting. The coring had been made necessary when the Designer realized an error in the original drawings had not called for access for wiring through the second-floor slab. The work had been finished less than two hours before the City inspector arrived, and the safety Engineer for the General Contractor had already walked the site first thing in the morning for her morning inspection.

Any two juries might come up with different comparative percentages of negligence for three or four of the parties involved, but the City inspector might wish to direct the full compensation for injuries and/or lost wages to the General Contractor (if found to be even partially responsible), and the General or its insurance company might then pursue compensation from the other parties involved.

In the example above, insurance companies would almost certainly be involved. If the injury had occurred to a worker rather than an inspector, Workers' Compensation Insurance would be activated. If the worker wanted to pursue action against one of the other parties, not an employer, then other insurance would usually go into effect for those parties. However, through the maze of comparative percentages, did the Constructor waive subrogation of certain parties through the contract language? Thus, does the contract in some cases prohibit the "deep pocket" party from pursuing compensation from others?

The Contract Administrator should discuss such issues with both the insurance company and the Constructor's own attorney.

In the Occupational Safety and Health Administration (OSHA), almost one fifth of all work-related fatalities occur in construction. The OSHA website lists that the four leading causes of death or injuries to workers in construction are: (1) falls, (2) electrocution, (3) being struck by an object, and (4) being caught in between objects. The purpose of insurance, and particularly Workers' Compensation Insurance, is to cover the financial costs, but the purpose of a good safety program monitored by everyone is to avoid the human suffering in the first place.

Figure 15.3 Construction is a dangerous industry
Photo credit: Charles W. Cook

15.3 Workers' compensation

Let all thoughtful citizens sustain them, for the future of Labor is the future of America.

John L. Lewis

As mentioned, Workers' Compensation Insurance would have been activated if a worker rather than the City Inspector had been injured in the example above. Workers' Compensation Insurance was discussed previously, but the major point of such coverage is to allow an injured worker to be compensated for injuries and damages (such as lost wages), without facing the prospect of a long trial before compensation is received.

The major effect of Workers' Compensation Insurance is to protect the worker, but another important effect is also to put the price of the insurance into the product. Each Contract Administrator knows the wages and other payroll burdens need to be included not only in the estimate, but also in all change orders. Workers' Compensation Insurance is a direct cost to the project for all the workers on that project.

However, buried within all the good for both the worker, who receives prompt compensation for bills and lost wages, and the Contractor, who can charge the costs of the insurance to the purchaser of the project – the Owner – is the fact that negligence is not determined between the Contractor and the injured worker. In fact, it could be the worker who was negligent, but the Contractor still bears responsibility for the cost through workers' compensation.

Regardless of fault, the injured worker cannot sue the Contractor for negligence. This is what is known as **exclusive remedy**. The employer's workers' compensation is the only means for the injured worker to collect against the employer, provided the employer's fault was negligence and not intentional. However, an injured worker can sue others for negligence. A Subcontractor's employee could, therefore, sue a General Contractor.

For this reason any accident on the project needs to be quickly and fully documented. The Contract Administrator must recognize that responsibility may be broadly defined, particularly if the injured party worked for someone else. This is true for all parties on the project, from Constructor through Subcontractors and Sub-subcontractors and Suppliers to the Design Professionals and Construction Managers, as well as Owners. The thought that someone else's workers' compensation will take care of everything may be right in most cases, but it is that one other time that can be a major concern for anyone who did not pay attention and document the incident.

> **Example:** If we go back to the City inspector example, and change him to a painter on the project, then the painting Subcontractor's Workers' Compensation Insurance would be the first source of payment for the worker. However, there are now a list of potential other parties that could be sued for negligence, including the General Contractor, the Electrical Subcontractor, the Coring Sub-subcontractor, and maybe the Design Professional for the original flaw in the drawings. In fact, the injured worker might pursue the Owner under what will shortly be discussed as "Peculiar Risk."
>
> What prospect of additional compensation and success in litigation the injured painter might have is only part of the issue. The other part is that defenses will have to be arranged, and they cost money and inevitably form a distraction. It is also likely that the General Contractor may be obligated by contract to pay the cost of defense for the Owner and/or Design Professionals, based on an Indemnity Clause in the contract.

15.4 Peculiar Risk

Be wary of the man who urges an action in which he himself incurs no risk.

Seneca

Owners and others engaged in construction know that the work is dangerous. The **Peculiar Risk** doctrine is intended to protect other parties against damages that might occur from actions taken by Constructors working for Owners who understand the risk but perhaps contract to companies that do not take due care of their actions. Courts have occasionally found that the Owner bears financial responsibility for such risk, particularly if the culpable Constructor is not capable of compensating the inured party in full.

Among the areas most often recognized for Peculiar Risk in construction are electrocution, falls, being struck by a vehicle or object, and asphyxiation, often associated with excavation.

The essence of Peculiar Risk is that an Owner needs to be certain Constructors are doing all they can to minimize the constant risk and dangers of construction. This has given rise to Owners expecting more and more from their agents in regard to such risks, and it also is the reason many contracts carry such broad risk shifting clauses.

Still, courts hold Owners liable for the actions of others working on their premises. This, of course, could be done in a comparative manner, apportioning responsibility among Owner, Constructor, and Subcontractor, or even Designer if the incident so warranted.

Example: A person is hurt while walking out the entrance as he was talking on a cell phone. The jury might find the walk-and-talker 5% liable for being careless and foolish, the Subcontractor whose employee had parked the forklift in the entrance way 25% liable for leaving the forklift there, the General Contractor 50% liable for not holding site-safety training programs, and the Owner 20% liable for letting the work proceed without a better site safety plan.

What each party must recognize, particularly the Owner in a Peculiar Risk situation, is that negligence is relative. The Owner may not have been negligent at all, but some portion of a damages claim may be applied to various parties in various amounts. Combine this with the "deep pockets" principle, and any party only marginally involved in a negligence tort case might become responsible for the whole amount.

As with many facets of law, no one should assume that Peculiar Risk will govern over the obvious fact that Constructors know that construction is risky. There is a doctrine that balances Peculiar Risk, and that is **Assumption of Risk**. Since everyone knows how dangerous construction can be, the Contract Administrator must not rely upon any thought that a sharing of Comparative Negligence will diminish the amount of damages any one party will have to pay simply because it will be shared by others. In fact, indemnification clauses within the contract with the Owner will generally place the burden of defense and damages on the Constructor. In turn, the Constructor may pass these down to Subcontractors through indemnification clauses.

Most of the tools and equipment Constructors operate during a project carry with them some instructions and "fine print" declarations that the user (Constructor) must observe, including several and specific safety procedures when using the tool. In so doing, the manufacturer is stating that the user is assuming the risk, particularly if one or more of the procedures are not observed (and there is often at least one).

15.5 Computing damages

While money can't buy you happiness, it sure allows you to choose your own form of misery.

Groucho Marx

Of course, the most feared part of damages is the debilitating injury, or even the death of one or more people in an accident. The calculation of compensation can never replace such losses. Beyond such a tragic loss, the courts will take into consideration various forms of compensation, including repairs to property that are required due to the actions of others, lost income while out of work or while a property or equipment is not available (e.g. a delivery truck is being repaired, due to an accident caused by another driver).

The Contract Administrator called upon to compute damages must keep in mind not only the cost of rectifying the damages, but the potential change in value from before to after the incident. One rule to keep in mind is that the cost to repair a damaged object cannot exceed the fair market value of the object. For this reason many accidents involving automobiles compute the current value of the automobile and determine the cost of repairs. If the cost to repair exceeds what the car is currently worth (not what one originally paid for it), then the insurance company declares the car "totaled" and writes a check for the fair market value rather than the cost to repair. Sometimes this seems a good deal to the car owner, and other times it does not.

In some instances the damage might be permanent, and in those instances the amount of compensation is computed on a one-time basis for the total settlement of all claims. However, if the damage might be temporary, but the length of the "temporary" is not certain, then a series of compensation requests might be honored up until the time there is no longer a need to pay damages.

15.6 The Golden Rule

Do unto others as you would have others do unto you.

Golden Rule

Exercise 15.3

Take a moment
Do unto others

Many cultures and religions have adopted some form of the Golden Rule as a foundation for all other principles. List as many as you wish, and mark which one, if any, actually is better than the Golden Rule we learned as children from either our parents or, perhaps, our kindergarten teacher.

Discussion of Exercise 15.3

There may not be any more precisely and succinctly phrased "golden rule" than the one we are all familiar with, but this exercise forms an important start of our understanding of ethics to be discussed in the final chapter.

As we conclude this chapter on tort law and look forward toward the final chapter on ethics, it is appropriate to close with the universality of the Golden Rule. Just as all cultures and civilizations have memorialized and practiced some form of the Golden Rule, tort law is truly the embodiment of a legal system to enforce what others should do for others.

No Contract Administrator will ever be able to eliminate all negligence on a project or within a company. Similarly, there will always be people more intent on raising themselves at the expense of others, and in the process they may commit intentional wrongs to anyone who opposes them.

However, being fair to others, while expecting equally fair treatment from others, is the essence of ethics and good business. Strong ethics promote good business and good business attracts better team players – ones who are determined to follow the Golden Rule and lift others while they lift themselves.

For the Contract Administrator it is important to keep in mind that we are all human. When everything is going well it is easy to be ethical. When things are crashing down around us, it is much more of a challenge to remain ethical. The key task of the Contract Administrator, therefore, is to ensure for everyone on the project that things do not come crashing down.

A CASE IN POINT

They say nobody wins in litigation, and it is true, not just because it costs money to prosecute or defend. I can honestly say we never lost in pre-trial settlements, court, or arbitration. But we never won, either. It is not just about the money we had to pay the attorneys; it is also about the time it takes from everyone, and often the pain one or both sides have suffered.

Here is a final story for this chapter that ties in comparative negligence, deep pockets, and tort law.

I had been driving to another project one day when I heard of a terrible accident at a train station we had renovated a couple years prior. A commuter had crossed the tracks from one side to another, and he had been struck by a train and killed. I wondered for a moment how the tragedy could have happened, but I continued to my own meeting.

I did not really think about the incident again until several months later, when we found ourselves involved in a negligence lawsuit brought by the parents whose son had been killed. Our own attorney said we were part of the prosecution's attempt to find the deep pockets, but I suspect initially we were just part of the broad Comparative Negligence net that the prosecutors spread to determine actual fault or negligence. Along with the Designers and the transportation authority that operated the train, we were advised that a hole had been cut in the fencing that we had installed between the tracks. The fencing was supposed to prevent pedestrians from crossing the tracks. Unfortunately, the student who had been killed had attempted to run across the tracks through the hole to catch his train.

Our defense was obvious, since we had properly installed the fencing and left the site years before with a code-compliant and satisfied inspection of all the work. The same was true for the Designers. Still, we were requested to work with the transportation company to assist in its defense, including attending pre-trial testimony taken through depositions.

Eventually the case was settled, but not before the parents who had lost their son had to relive the horror of the loss through that very painful testimony.

So again, I say, there are a lot of ways nobody wins in litigation.

INSTANT RECALL

- In relation to contract law, our compliance with tort law is implied.
- Tort law has evolved from the basic concept that everyone deserves the right and freedom not to be injured or to have their property damaged by another person(s).
- Sometimes there is a gray line between the civil wrong of tort law and a potential criminal wrong.
- The best way to handle tort claims, therefore, is to refrain from intentional acts, inspect against negligence, and ensure as best as one can (and insure if possible) against strict liability.
- The most common type of intentional torts in construction is those related to cases of fraud – misrepresentation of facts.
- If a false statement about someone is made in writing it is libel (and probably easier to prosecute), and if it is made orally, it is slander.
- As an employee is working within the scope of the employment, his or her actions are binding upon the employer.
- When defending a negligence claim, both Constructors and Design Professionals need to recognize that their particular expertise will be the standard by which negligence is judged.

- One important issue (and perhaps defense) Contract Administrators should remember in relation to negligence is that the injury or damage has to be the direct result of the negligence (sometimes referred to as "causation").
- There is a doctrine of Contributory Negligence, which absolves one party of responsibility for a negligent act if the other party was also negligent. For that reason many states are awarding cases based on Comparative Negligence. Under Comparative Negligence the courts try to apportion the compensation for damages based on the degree of negligence of each party. In addition, in some states Contribution Negligence is enforced when two or more parties are responsible for the same damages.
- The major effect of Workers' Compensation Insurance is to protect the worker, but another important effect is also to put the price of the insurance into the product.
- Even though the injured worker cannot pursue negligence against his or her employer, this is not necessarily the case in relation to other parties that may have been deemed comparatively negligent.
- Courts and juries have consistently held that the danger within a construction project site of any kind is the responsibility of the Constructor, and although this has been challenged in recent years on many levels, when it comes to children playing in an Attractive Nuisance construction site, the Constructor will be in a very difficult position to defend.
- Keep the site protected and clear of anyone not trained in construction. Keep construction sites for Constructors.
- The essence of Peculiar Risk is that an Owner needs to be certain Constructors are doing all they can to minimize the constant risk and dangers of construction. This has given rise to Owners expecting more and more from their agents in regard to such risks, and it also is the reason why many contracts carry such broad risk shifting clauses.
- Some portion of a damages claim may be applied to various parties in various amounts. Combine this with the "deep pockets" principle, and any party only marginally involved in a negligence tort case might become responsible for the whole amount.

YOU BE THE JUDGE

Pacific Western Construction Company, Inc. (Pacific) was awarded a contract to build a road in the Sequoia National Park. It subcontracted the trenching and installation of the culvert and drainage piping to Batchelor. Pacific did not inquire about Batchelor's safety plan, and the on-site representation from Pacific, Superintendent Hayes, did not direct it to shore or slope back any of its trenching. Trenches over five feet deep need to be either shored or sloped back sufficiently to prevent cave-ins. Hayes claimed he had not seen any trenches deeper than five feet, but others testified that he had. Hayes also pointed out that he left safety of the trenching to the Subcontractor, Batchelor.

An employee of Batchelor, Gregorio Jimenez, was working in an unshored, unsloped trench nine feet deep when it collapsed on him and he was killed. Another employee was also injured. The family of Jimenez sued Pacific, but Pacific defended itself on the grounds that the subcontract with Batchelor had a general **Hold Harmless Clause**.

How do you rule:

IT'S A MATTER OF ETHICS

Think to yourself or discuss among friends whether it is fair that the entity with the "deep pockets" can become solely responsible for all the costs of something when several parties contributed to the mistake or error. If not, what can or should be done about it? If you do not think it is fair, why do you think the law and courts have evolved to this point?

Going the extra mile

1. Complete the "You be the judge" exercise above and then go to the Companion Website for a discussion of the actual case.
2. **Discussion exercise I** (with friends or as part of a classroom discussion): The distinction between contract law, tort law, and even criminal law can be quite difficult to ascertain. For instance, does a contract exist between someone who buys something from another? If a severe injury results, due to the exchange, is there criminal wrongdoing? What if the seller intentionally deceived the buyer about the quality or about the danger? Does it matter if the exchange involves an object or a service? Is there a warranty implied, even if not expressed? Invent and discuss different scenarios where the case might be criminal, or a tort claim, or a contract issue (oral or written).
3. **Discussion exercise II** (with friends or as part of a classroom discussion): Punitive damages are controversial within our legal system. Some feel it is unjust enrichment to the party that has won the costs related to the injury. Others feel it is both a just compensation for injury and a deterrent for others who understand the consequences.
   a What is your view on punitive damages?
   b How much is reasonable (two times, three times, four times)?
   c Since there has probably been some involvement of public officials, should the punitive costs go to the court and state rather than the victim, thereby not enriching the victim beyond actual costs, but possibly lowering taxes for everyone?

4. Go to the Companion Website for additional study aids, including flash cards and sample questions.
5. A URL you might use for more in-depth discussion of tort law: http://tort.laws.com/tort-law.

6. Go to the Companion Website and watch Part II of the Ethics video.
7. Effective use of the correct vocabulary improves communication and promotes respect among colleagues. The reader should become familiar with the key words listed below that have been used in the chapter. Definitions can be found in the Glossary. Additionally, the student is encouraged to use the flash card exercise in the chapter's Companion Website as practice for possible quiz or exam questions.

Key words

Assumption of Risk

Attractive Nuisance

Causation

Comparative Negligence

Contribution Negligence

Contributory Negligence

Deep pockets

Exclusive remedy

Hold Harmless Clause

Intentional torts

Libel

Negligence tort

Peculiar Risk

Punitive damages

Respondeat superior or "reputed negligence"

Slander

Tort law

Treble damages

16 Statutes and regulations

Often hidden behind a labyrinth of bureaucracy, statutes and regulations can have a tremendous negative impact on the time and cost of a project if they are not followed. Usually boring and seemingly overly burdensome, handling statutes and regulations is an important part of each Constructor's responsibility.

Student learning outcomes

Upon completion the student should be able to. . .

- **recognize the broad expanse of statutes and regulations**
- **understand the importance of proper licensing, permitting, and building codes**
- **identify and analyze the critical issues of statutes and regulations, from licenses to retirement benefits**
- **identify evolving issues related to the environment**
- **analyze issues and actions in relation to specific areas of regulations, including bankruptcy and dispute resolution.**

16.1 Statutes and regulations

You have to learn the rules of the game, and then you have to play better than anyone else.

Albert Einstein

Just as with tort law, the Contract Administrator should not expect to find the statutes and regulations within the contract documents. In the case of tort law, we recognized that it is implied. We are all expected to follow the laws that protect individual rights and property.

In the case of statutes and regulations, however, most contracts definitely make them a part of the contract by reference. These usually come in the form of one or more risk shifting clauses. The great challenge for the Contract Administrator of knowing, understanding, and following all the statutes and regulations that could impact the project is that such a task is impossible. There are too many rules and regulations for a single person to grasp or even be aware of during the course of a project, or even a lifetime.

Although some Contract Administrators for the Constructor may object to Design Professionals passing responsibility to them for the many statutes and regulations that could be required in a project, the reality is just the same for them as it is for the Constructor. There are just too many statutes and regulations for one party or entity to cover completely.

The entire team has to be involved. The Design Professionals are expected to do the initial conformance, and the best do this very well. The Constructor is expected to add additional expertise, and through flow down clauses passes on to Specialty Contractors additional responsibilities in their particular areas. In such a process, if all goes well, the statutes and regulations are well covered by the individuals and the entities that are best equipped for or familiar with the specific issues. The process is not perfect, but it succeeds far more often than it fails.

Still, the amount of statutes and regulations is a challenge. All governments, from the local township or municipality to the Federal level, pass statutory and regulatory laws, and the language and scope of such laws usually require dedicated concentration to understand. Sometimes, depending on the size of the Constructor, some regulations are not applicable, but expert advice should be sought before ignoring a statute or regulation. In addition to each individual statute and regulation, the massive scope of potential statutory and regulatory laws is intimidating. Such a scope includes for the construction Contract Administrator:

1. licensing
2. permits
3. building codes
4. safety
5. payroll
6. environment
7. hiring and firing
8. labor
9. workplace conduct
10. liens
11. bankruptcy
12. dispute resolution
13. healthcare
14. retirement.

It should also be noted that some Owners, due to the particular nature of their environment, such as a hospital, a nuclear power plant, or a refinery, will also require additional rules and regulations for anyone entering their premises to do construction work. Such rules and regulations may decrease productivity, but the Constructor must recognize that the savings in property and lives are well worth strict observation of all such regulations.

Figure 16.1 Rainbow at the Denver International Airport
Photo credit: Charles W. Cook

Even in black and white one can see the rainbow at the Denver International Airport after a particularly violent storm. The tower cranes visible in the background to the left of the tail of the plane withstood the heavy winds, as they would be expected to do on every jobsite, but airports are particularly challenging for Constructors, and not simply for access. No debris (zero!) can find its way onto any part of the airplane gate areas, taxi, or runway surfaces. Even a small nut or a fragment of drywall can cause damage or even result in a tragic accident. There are many special provisions in various construction locations that the Contract Administrator must recognize go above and beyond typical requirements, and the Contract Administrator will have to know that there is a plan in place for all workers to understand and follow specific provisions and regulations on each project site.

16.1.1 Licenses

Design Professionals are well aware of the lengthy process required to become licensed. In some states, builders are being required to have a certified or licensed "Constructor" within the company. Many specialty trades require licensed individuals who can oversee the installation process.

There are exceptions, but generally before one can do business in most townships, municipalities, or states, some license(s) or registration will have to be obtained, and in the process some form of competence will be expected to be shown, through either testing or resume. Such a process is intended to protect the public and the eventual user of any space being built. It may seem burdensome, but it is very necessary. One would want a brain surgeon to be trained to operate on one's head. It may seem absurd that one would not want a brain surgeon to hang a suspended ceiling or certify a bridge, but frankly, who would care to sit in such a room or cross such a bridge? Construction may not be brain surgery, but it is dangerous, and many lives depend on the simple but extraordinary concept that, once completed, the structure will defy gravity for the rest of its existence.

Design Professionals and Constructors, therefore, need to be able to demonstrate competence to ensure the public and the end user will be safe and satisfied. Often Owners as well as the licensing body, such as the municipality, will want the Design Professionals, and subsequently the Constructor, to be able to demonstrate experience in the particular type of design or building that is being undertaken. In other words, designing football stadiums is not the same as building a nuclear power plant. Also, a little league field is not the same as a baseball stadium.

Related to expertise, it is important that the license be obtained in the specialty involved. This is critical for Subcontractors. A plumber should not be expected to be licensed in electrical work. Often a single subcontract is written by the General Contractor for all the HVAC work to also include the plumbing. Although some companies are licensed in HVAC and plumbing, often the plumbing work goes to a separate Sub-subcontractor. The Contract Administrator should be sure the actual party doing each specialty trade work is in fact licensed to do that work in that township or state.

Although Federal work does not usually require licensed Contractors, there is still a review process that the Constructor must pass in order to be qualified for the work. Even if the work is being done in a particular state with stringent licensing requirements, the Federal Supremacy principle will supersede the requirement to obtain a state license. The overall purpose of this is to allow the Federal government to bring Contractors that have shown expertise in such projects from anywhere in the United States to do work without interference from the state or local governments (presumably loyal to in-state and local Constructors).

16.1.1.1 Certification for warranty

Some Suppliers of specialty products will require the Constructors who are installing the equipment to be certified. Almost always the contract from the Owner to the Constructor will reference such requirements either specifically or through a broad form clause requiring "*all necessary certifications to be obtained before installation*."

The Constructor will often pass such certification requirements down to a Specialty Subcontractor responsible for the actual installation.

The importance of obtaining this certification for the Contract Administrator should not be overlooked. Often such certifications will include a pre-installation or application inspection. If the inspection discloses issues that prevent the installation or application of the product, the schedule can be seriously affected. Since many of the specialty products requiring inspections come close to the completion of the project, failure to pass an inspection can jeopardize completion, and, if Liquidated Damages are involved, the costs can mount quickly.

> **Example:** Paint manufacturers require reduced levels of moisture in masonry block walls or the paint will not be warranted. Often masonry walls are erected in buildings or sections of buildings that are close to the outside, and if the roof is not installed until later in the project, weather may continue to affect the block wall moisture content. Once the roof is installed (and assuming it does not leak down the walls) the painter may have to wait for the heat or A/C to be turned on to fully dry the blocks before painting. That could take some time that was not figured into the schedule.

The Contract Administrator should make certain all such inspections are scheduled as far in advance of the installation as possible.

Once the product has been applied or put in place by a certified installer, another inspection should take place. The key for the Contract Administrator is that the installation of any product that requires a certified installer will lose its warranty if not installed by the certified installer, but it must also be installed properly, and a post-installation inspection will help to document that fact.

16.1.1.2 Failure to obtain a license

No doubt some work is done by parties that have not obtained a license to do that kind of work. Although the work may be done correctly, such a practice should be strictly avoided. Penalties, including fines and potential imprisonment for a misdemeanor, along with past due taxes, and a lack of warranty are major problems, but another is that if a dispute arises, unlicensed Contractors usually have no recourse through the court system.

Although some states have relaxed litigation rules, many state courts will not hear a case presented by an unlicensed Contractor that has done work and a dispute over payment has arisen. In short, if an Owner finds a Contractor or Subcontractor is unlicensed, there very well may be no recourse for the unlicensed Contractor to seek payment if the Owner decides to withhold funds. This could also include the inability to file a lien against the property. This may seem unfair to the Constructor. Even though it was unlicensed, some work was performed. And there might be a ***quantum meruit*** claim asserted (based on the fact that the work has enriched the Owner, so the Contractor is due something), but the Owner is also in a difficult place related to warranty and defects if something does go wrong with a Contractor that cannot repair defective work, so courts often avoid hearing such cases.

16.1.2 Permits

The permitting process can begin as soon as an Owner recognizes the need for construction to take place. Such construction does not have to be totally new. It can be an addition or just a revision or renovation to an already existing structure. In some cases, Owners have a blanket permit allowing certain revisions to take place to existing facilities provided they do not significantly change the use or function of the space. Most construction, however, does require securing a new permit, and, depending on the schedule, an Owner may seek or start that process even before a Constructor is chosen.

Further into the process, Design Professionals or the Construction Manager might be expected to apply for the permit. This again usually saves time over waiting for a Constructor to apply for the permit, and it sometimes raises issues that need to be addressed sooner than later about the plans and specifications.

If it is not stated at the pre-bid meeting, it is important that the Constructor ask about the permit process, to make sure that permits are or are not to be included in the bid proposal. Owners often wish to wait until the bid proposals are submitted, since permits are usually based on price and can therefore be costly. If the successful bidder is to obtain the permits, then not only is that a cost that should be included, but the Contract Administrator must be aware of the length of the process in order to be realistic about the actual start date of the project.

In some instances, such as when a historic structure might be affected or modified, or the aesthetics of a neighborhood affected, review by an additional body might be necessary for the permit to be approved. This could be a historic commission or an art review board. Such reviews add additional time to the permitting process and the Contract Administrator must be aware of such an additional step to include it in the pre-construction schedule if the Owner or the Design Professionals have not already taken care of such reviews.

In addition to the overall construction permit, there are usually specialty trade permits required, including electrical and plumbing permits. It is at this time that unlicensed Contractors, discussed previously, are discovered and advised that they will need to apply for a license. In such a case the Contract Administrator can be faced with an additional delay.

16.1.2.1 Transfer of liability

One important caveat for the Contract Administrator is the potential for the Constructor to assume liability for issues related to the design during the permitting process. When the Constructor assumes the role of obtaining permits, the permits are submitted in the name of the Constructor rather than the Design Professionals. Even though a Designer's seal will be required on the drawing, the permit will be issued to the Constructor. This obligates the Constructor to build in accordance with the codes and regulations of the authority issuing the permit. Constructors therefore must be familiar with the codes and regulations. However, the requirement to build in accordance with issues uncovered during the permitting process will also be an obligation the Constructor assumes once the permit is obtained.

An inspector walking on the project and inspecting the work will contact the Constructor if there are any ambiguities, discrepancies, or errors in the execution of the work from what was originally "permitted."

It is important, therefore, that significant issues between the permit-approved plans and specifications and the contract documents are resolved immediately for all parties so that any potential or significant cost issues do not linger until they become major problems later in the project.

This is often difficult for the Contract Administrator, since the prospect of getting started is something everyone is eager for, and the permitting process has probably dragged on a bit longer than anticipated. After a bit of a delay obtaining the permits, neither the Constructor nor the Owner and the Owner's agents want to spend more time in the "pre-construction" process, but, depending on the significance of the issues, that is exactly what has to happen.

At this point the Contract Administrator must rely upon the ambiguity or discrepancy being a latent rather than a patent defect that a prudent or experienced builder would have recognized during the bidding process. Of course, depending on the language of the contract, Owners and agents of the Owner might argue about what is a Latent and what is a Patent Ambiguity. In such a case, it will be important for the Contract Administrator to understand how the various parties will approach interpretation of the contract. This will become important as the "team" moves forward on future challenges.

16.1.3 Building codes

Building codes can be local or national, and even international, and sometimes they do not agree. This can become a major challenge for the Contract Administrator to comply with the demands of inspectors working with different standards. Sometimes Subcontractors unfamiliar with a specific locale will work under one set of codes, only to find a local inspector will demand compliance with another set of codes.

Figure 16.2 The Woolworth Building, New York City
Photo credit: Charles W. Cook

Setbacks for the old and new: The neo-gothic style building pictured in the background is the Woolworth Building, for a brief time in 1913 the tallest building in the world. Buildings in New York City, as well as other cities with tall structures, require "setbacks" in accordance with building codes as the towers rise to "scrape the sky." There are various reasons for this, including some structural ones, and the design function to ensure some light gets to find its way to street level at different times of the day. Additionally, the setbacks are intended to protect pedestrians from objects falling from the upper stories.

One of the chief reasons for establishing building codes is safety. For that reason many of the codes relate to either fire or structural issues. Other codes have developed for:

- environmental reasons
- site drainage
- establishing or improving traffic flow
- providing access and use for those with disabilities
- ensuring minimal standards of products used in construction
- protection against potential natural disasters (e.g. earthquakes, hurricanes, floods).

In some instances, codes prohibit construction that would change the appearance of neighborhoods or adjoining structures. Knowing all the codes is impossible, but the Contract Administrator must recognize that the contract with the Owner will almost certainly require compliance with "all applicable laws and codes." Working with qualified Subcontractors is critical. The electrical, plumbing, and HVAC Subcontractors will probably have to obtain their own permits for work to be done on the project. All Subcontractors, however, need to be familiar with the codes most associated with their work. For example:

- Excavators and landscapers need to be familiar with environmental standards, drainage issues, and inspection of such work as bottom of footings.
- Drywall Subcontractors need to understand regulations related to type and quality of material to be used. Even though "called out" incorrectly on the drawings, using the wrong drywall for a particular application could easily be construed as a patent defect, and the cost to change the material to be the responsibility of the Contractor and not the Owner or Design Professional.
- If fire protection Subcontractors have installed the sprinkler heads incorrectly in relation to the code requirements, reworking the system could be quite costly.
- The hardware Supplier may have provided exactly what the drawings called for, due to the Owner wanting a certain master keying, but after the installation is complete the fire marshal rejects the work based on fire code requirements.

Figure 16.3 Sidewalk memorial in front of the demolished remains of two buildings (see "A case in point" below)
Photo credit: Charles W. Cook

The demolition was being done in part by a backhoe and crane. In response to the tragedy that killed six people, the Mayor of Philadelphia issued an executive order that all buildings being demolished next to an occupied building must be demolished by hand. Such a regulation obviously greatly increased the cost of demolition, and Contract Administrators must recognize how such local regulations that appear suddenly can affect the budget and schedule in one particular municipality or locale, while not being a requirement in another.

Complying with the building codes requires a team effort. When a Contractor or Subcontractor moves into a new area, there might be ordinances and statutes that have not been applied or required in other locales. Moving into an earthquake-prone area, for example, will undoubtedly require different construction practices. Sometimes a regulation will be put in place by local authorities due to an incident or accident that brought attention to a particular need or issue.

A CASE IN POINT

Not all stories have happy endings. On June 5, 2013, at 10:43 AM, the corner of Market and 22nd Streets, Philadelphia literally became the scene of a "sky is falling" tragedy. Immediately, members of Philadelphia's finest responded, combing with bare hands through the rubble of two collapsed buildings to pull out survivors. And almost immediately there was a search for why this happened and who was to blame.

From the debris fourteen people walked away with injuries, but six were carried away in body bags. The last survivor was finally found thirteen hours after she had been partially crushed under the weight of the debris from the building that had been under demolition and had collapsed onto the Salvation Army Store next to it.

Under the then existing Philadelphia building codes, the permit had been obtained, licenses existed, and required inspections had presumably taken place. After the fact, many people commented on how they had seen the problem and would have said something earlier. Some even said they were about to call for an inspection because the work was not being done carefully. Many had seen the problem, but prior to the tragedy no one had been their "brother and sister's keeper." A Contract Administrator cannot rely upon the fact that a Contractor or Subcontractor has a license or that a permit exists. Safe processes must be maintained.

During the searching for blame, some solutions were suggested, but the tragedy did not end on June 5, 2013. A week later, wracked with the horrible images he saw daily, and after sleepless nights of wondering what he could have done, a city inspector took his life, leaving the message behind that "It was my fault."

A 52-year-old, stable, hardworking, family man needlessly committed suicide, adding to the senseless death toll. It can happen to anyone. No matter how stable or self-assured one might believe oneself to be, the potential for a horrific tragedy is as close to each Contract Administrator as it was to those unfortunate shoppers and storekeepers in the Salvation Army Store at 22nd and Market Streets in Philadelphia.

When it comes to safety, the time to find fault is well before the accident has the opportunity to happen. The time to find solutions – positive solutions without blame – is immediately after you have found the fault.

Discussion of codes is seldom a specific issue in project meetings, but it should always be part of the Contract Administrator's focus. Through flow down clauses, the Contractor's obligations to the Owner to comply with all building codes is passed on to the Subcontractors, and often to critical Suppliers, but the real difficulty is not always so much the assessing of responsibility as it is the time that it takes to change or rework non-compliance issues.

By the nature of the inspection process, non-compliance issues are usually discovered just before something else needs to be started. Whether it is pouring the concrete in the bottom of a footing that is rejected, or it is the final inspection by the fire marshal before the **Certificate of Occupancy (C/O)** is issued, the delay can be both exasperating and, even worse, costly.

16.1.4 Safety

Safety was mentioned above, related to codes, but there is an even larger set of safety standards every Constructor must recognize, and that is the standards established by the **Occupational Safety & Health Administration (OSHA)**. Compounding the issue for

Constructors is the fact that some local governments have adopted additional, more stringent standards, and even states are permitted to address additional safety issues above and beyond the OSHA regulations. In addition, some Owners, due to the particular danger of the project site, will require additional safety procedures to be followed.

Figure 16.4 Safety is not just a hardhat issue
Photo credit: Charles W. Cook

Safety is a major issue and a concern for every member of the project team. The most effective Contract Administrators will be those who can establish an atmosphere that safety is everyone's concern. One individual may be assigned the task of overseeing safety on a project, but two eyes are not sufficient when it comes to all the potential hazards as well as all the details related to specific safety compliance by each trade.

A CASE IN POINT

Some companies use hardhats as a way to recognize the newcomers on a project. White hats are for veterans and yellow hats (or some other distinguishing color) are for those who have less than nine months' (or maybe a year's) experience with the company. That alerts others to keep an eye out for the rookie or "new kid on the block."

In the middle of my career, a good friend told me he had found an interesting way to change the worker's perspective on safety. Similar to our own hardhats, on the front of his company's hardhats there was the corporate name, surrounded by the slogan "Safety pays." Depending on where the individuals worked, there might be additional decals regarding training they had received for permission to work in certain environments or companies.

All that sounds fine, but it is seldom really that important to the wearer. The idea that safety pays is more likely considered a gimmick that management passes down to improve its own bonuses if it has a good year. Of course there can be incentives and rewards for no lost time, but they usually do not amount to much in most companies.

To change that whole perspective, my friend had each of his workers (over 400 at the time) bring in a photograph of their family, and that was glued to the inside of the hardhat. In that way, the last thing each worker saw before putting his or her hardhat on was a picture of the family, and that got the message of the purpose of staying safe into some hard heads better than any slogan on the front of a hardhat could.

Figure 16.5 Inside the author's hardhat: families grow and change, but the reason for coming home safely never does

Photo credit: Charles W. Cook

As mentioned briefly in the previous chapter, a significant number of the accidents that occur on construction projects fall under four main headings:

1. electrical (including contact with power lines and equipment, and extension cords being misused)
2. falls (including from unprotected sides of buildings, ladders, and scaffolds)
3. being struck by vehicles, objects being lifted, or falling from above
4. asphyxiation (loss of oxygen in some form, including trench cave-ins, drowning, toxic fumes, insufficient oxygen in enclosed spaces).

However, the Contract Administrator must recognize that there are also significant numbers of hazards throughout the project site. When we think of accidents, it is not merely loss of life that should concern us. Although not as tragic or grievous as a death on a project site, loss-of-time accidents or any injury that might impair a worker's quality of life immediately or in the future need to be guarded against.

Figure 16.6 The Empire State Building, New York City
Photo credit: Charles W. Cook

Although not the tallest in New York City, and certainly not in the world, the Empire State Building remains a popular New York City tourist attraction. Yet during its construction during the Great Depression, with as many as 3,000 craft workers a day on the site, it was estimated that at least one person per floor would lose his life (all male craft workers). Perhaps in 1930 it was remarkable that only two laborers, one ironworker, and three carpenters lost their lives during the construction. In addition, one woman passing by was struck by an iron worker's plank and later died of blood poisoning. As "successful" as that safety record may have seemed at the height of the Depression, today it would be completely unacceptable, and zero lost time accidents is the goal for the Contract Administrator to set for all personnel.

16.1.4.1 Recordable injury

When in doubt, make a record of an incident, but generally a recordable injury is one in which the employee is injured doing something on the project that requires more than elementary first aid. A lost-time injury needs to be recorded, but going to the first aid box for a band aid is not considered lost time. Generally, a recordable injury is one which a qualified medical expert or physician has diagnosed or treated, or an injury that keeps an employee from working or requires a restriction in the work load. Of course, any action that results in a fatality is also recordable.

Since there are some gray areas related to recordable and simple first aid, the Contract Administrator should be alert to every incident. For example, if non-rigid bandaging is applied, the treatment is considered minor first aid, but if a rigid splint or some other form of support is required, then the injury should be considered recordable and more extensive medical treatment is recommended.

Even though the step between elementary first aid and recordable injury may be blurred by the worker's own "brush it off" attitude, the Contract Administrator must recognize that the main purpose of good medical attention at the proper level is to prevent an injury from escalating into a more serious condition. For instance, it is possible that a minor cut may seem insignificant, but caution should dictate that the worker is up to date for tetanus immunization. Getting an immunization shot does not constitute a recordable injury or lost time, but the precaution may be worthwhile.

16.1.4.2 OSHA inspections

OSHA was signed into law by President Richard Nixon on December 29, 1970. The actual start of enforcement was delayed until April of 1971. Since then, several changes and many additions have been added to the regulations, and they will continue to be revised. The Contract Administrator, therefore, cannot rely simply on what was acceptable in the past. Knowing what is required for a specific project under current conditions and in accordance with any additional needs set by local authorities or the Owner is quite important.

The law provides the Secretary of Labor with the right to institute on-site inspections for compliance with OSHA regulations. The Secretary of Labor has therefore regionally assigned inspectors who will perform inspections of project sites. The inspector, known as an OSHA **Compliance Officer**, is entrusted with the task of making sure projects are complying with the current safety standards and regulations. Such inspections can be part of routine OSHA operations, or they can be the result of a complaint made by a worker regarding failure to follow safety procedures. In the event of an unfortunate accident, OSHA will also perform an inspection.

An employer might choose not to allow OSHA onto the site, but OSHA can then obtain a warrant for the inspection. It is probably more effective to request that the inspector return at some agreed-upon time when a safety officer with the company (perhaps another "hat" of the Contract Administrator) will be present to walk the site with the inspector.

There are basically two approaches to the inspection. One is to consider that OSHA is there to punish the Constructor with fines based on safety violations. The other is in some manner to recognize that doing things safely will ultimately help everyone, and perhaps prevent a costly accident from happening. In that regard some Constructors have actually asked for OSHA inspections. Such inspections are performed without the threat of a fine, but they will be followed by an inspection that will carry fines if the cited safety violations are not eliminated.

Once again, the success of the Contract Administrator might be just as closely aligned to how well one can handle the diplomacy as to how well versed one is in the safety regulations. On larger projects it is unlikely one person can do both, so a strategic recognition of how to proceed related to the entire subject of safety and who will handle the eventual safety inspection should be made well before the first inspector steps through the project gate and knocks on the trailer door.

After an inspection, the Constructor who has been fined does have a period of time in which to make an appeal. This is usually best handled by someone experienced in the intricacies of OSHA appeals, and the Contract Administrator would be well advised not to try to handle an appeal by themselves. It is better to hire someone experienced in OSHA citations to challenge the legitimacy of the fine if the cost imposed for the violation(s) is high – and they often are.

16.1.5 Payroll

Unless the project is a small one, there will almost certainly be someone specifically assigned from accounting to handle payroll statutes and regulations. Still, the Contract Administrator should be familiar with certain basics of payroll regulations. Some statutes and regulations have an impact on how long an employee can work before they are entitled to certain additional benefits that might be provided to other employees. These can include health and retirement benefits. If the Contract Administrator is tasked with the responsibility of monitoring periods and amounts of compensation, a close working relationship with the person in charge of payroll will certainly have to be established.

Local and state regulations regarding payroll vary considerably. The Contract Administrator will want to be familiar with the particular area and state in which the project is to be constructed, but some Federal laws set standards that must be followed.

16.1.5.1 The Fair Labor Standards Act

Through the **Fair Labor Standards Act (FLSA)**, the Federal government not only established a minimum wage for most employees, but it also set certain guidelines related to overtime and hours of work. Minimum wage can be increased by an act of Congress at any time. Generally, union Constructors do not have any difficulty complying with the FLSA, and whether union or non-union, almost all construction compensation is above the minimum wage. One area that remains difficult to monitor is the workers who are classified as Subcontractors rather than employees. Payroll taxes are not subtracted from a Subcontractor as they are from an employee. This was discussed in a previous chapter, but the Contract Administrator should be reminded that how other parties on the project under contract to the Administrator's company treat their employees can have an effect on the overall project, particularly related to safety issues, and also the overall schedule.

Individuals who are not employees may not carry Workers' Compensation Insurance, and an accident under Peculiar Risk circumstances could affect another perceived responsible Contractor as much as or more than the actual party that "subcontracted" the workers. Similarly, if workers are truly Subcontractors, the control of their time is their own. They may not finish or complete specific sections on a timely basis, or they may work longer hours or weekends, requiring project supervision to also be present.

16.1.5.2 Portal to Portal

One issue that might become important for the Contract Administrator to resolve relates to when the payroll "clock" starts. Congress enacted a statute that requires compensation for work being within the project site, but not for walking or driving to or from the work. This is referred to as the **Portal to Portal Act**. Still, the Contract Administrator must recognize that there may need to be exceptions.

Examples:

1. Drivers bringing vehicles back to be refueled will expect they are still on the clock.
2. Although generally cleaning up before the drive home might or might not be considered, cleanup after hazardous waste material removal should definitely be compensated.

16.1.5.3 On call

Another gray area related to such employees is whether they are "on call." If an employee is waiting for material and cannot leave the project site, that employee is definitely still entitled to be paid, even though technically "not working." The lack of work is considered the employer's responsibility, not the employee's. On the other hand, an employer can tell the employee to, "Go home and wait for me to call you back when the material arrives." In such a case, if the employee can truly leave the site and do something for his or her own benefit, then the employee is no longer "on the clock." The key is whether the employee can leave the project and do something that will benefit the employee rather than the employer. If an employee cannot use off-duty time for his or her own benefit, no matter how minor or inconsequential, but has to remain "handy," then the employee could be considered to be still "on the clock," and the employer is expected to pay the employee.

16.1.5.4 Local payroll statutes and regulations

In addition to the Federal laws, the Contract Administrator should be aware that state and local laws may require compensation different than FLSA. Such regulations may require compensation or taxation specific to a state or local township or municipality, and these will be in addition to or added to Federal requirements.

16.1.5.5 Davis Bacon Act

On public works, the Contract Administrator should be familiar with the provisions of the **Davis Bacon Act**, passed by the United States Congress in 1931. It requires that all workers be paid a prevailing wage on a project that receives any Federal funds. The prevailing wages are often determined by the union wage for each trade within a given area. States have passed similar laws for projects receiving state funding. Although this would seem to level the cost differential between union and non-union workers, union jurisdictional rules, as well as additional benefits not included in the prevailing wage for union workers, increase the hourly cost of the union worker over the non-union worker

16.1.6 Environmental regulations

Buildings, and by inference construction, are often viewed negatively in relation to the environment. Landfills are believed to be growing at alarming rates from construction waste and demolition debris. Asphalt parking lots and highways are blamed for changing climate patterns. Buildings are consuming vast amounts of the world's energy and returning huge quantities of pollution.

To blame Constructors may be unfair, but whether we blame the Constructor who builds or the consumers who demand the building, the environment is an important issue. Concern for the environment is affecting construction practices both from a public policy perspective and often from individual private initiatives linked to Green Building programs.

Figure 16.7 The tower at Hoover Dam
Photo credit: Charles W. Cook

Figure 16.8 Lake Mead, created by Hoover Dam
Photo credit: Charles W. Cook

The tower at Hoover Dam shown in the top photograph was the site of the last person to die during the construction of the dam, when he fell off the tower and drowned. In an amazing twist of irony, he was the nephew of the first person who died, while surveying the site for the dam. In 2011, in the background, the white stain on the rock, as also seen in the second photograph below it, shows that a severe lack of rainfall and, particularly, snowfall in the mountains has lowered the level of the lake behind the dam. This has caused increasing environmental concern and the question of just how much energy such artificial waterfalls can continue to produce.

In construction, the Contract Administrator must be aware that there are existing environmental hazards that must be recognized, and there are actions that must be taken to ensure damage to the environment from current construction is minimized.

16.1.6.1 The Comprehensive Environmental Response, Compensation, and Liability Act

Many of the regulations regarding hazardous waste encountered in construction can be traced back to the 1980 **Comprehensive Environmental Response, Compensation, and Liability Act (CERCLA)**, also commonly referred to as **Superfund**. Although there were other regulations in place before CERCLA, such as the 1976 Act entitled **Resource Conservation and Recovery Act**, 1980 marks a significant date in recognizing the impact of hazardous waste and the removal of the same from our project sites. In fact, 1980 is a good gauge for prudent Constructors to use when determining during a bid proposal phase whether the structure might have hidden hazardous waste, such as lead paint or asbestos, that will need to be addressed specifically during the demolition phase.

The Contract Administrator must be aware what responsibility for hazardous substance removal is placed upon the Contractor, based on potential risk shifting clauses in the contract. Within the CERCLA regulations is an important concept known as **Potentially Responsible Parties (PRP)**. Under this part of the regulations, the government attempts to involve any "party" that may be responsible for the cleanup of the hazardous substance. Generally, this will go back to the original Owner, but can also involve the existing Owner of the space.

The Constructor must be careful, however, since careless handling by the Constructor after uncovering the hazardous substance can make the Constructor the most accessible PRP. This can lead to fines as well as the responsibility for total cleanup, especially if the original Owner is not available.

Going one step further, CERCLA actually provides for total imposition of the cost of cleanup on any party, regardless of how little involvement they may have had in the original hazardous substance. If several parties share responsibility, but one party is obviously more capable financially of bearing the costs, then that party can be responsible for 100% of the cleanup. That does not prevent that party from seeking compensation from the others involved, but it just increases the initial burden of the process.

Although it might not seem fair, a Constructor can be found liable for materials installed years ago by someone else. The Contract Administrator, therefore, must be sure all parties involved in the project understand the seriousness of hazardous waste. Although the attitude might be that it will slow us down, the project must actually come to a halt in relation to the hazardous materials until proper cleanup can be accomplished. In order to avoid becoming a PRP, the Contractor must stop removing or disturbing the hazardous substance as soon as it is uncovered. Here again, the field personnel must be relied upon to be able to recognize hazardous substances. Innocently removing a hazardous substance will not be recognized as an excuse, and the Constructor will become the PRP most likely to assume the burden of full removal, as well as any fines that might be involved.

Ideally, the Contract Administrator will want to have the Owner take full responsibility for hazardous substances and their removal. The potential costs might seem like an enticing change order, but the additional responsibility for someone unfamiliar with handling hazardous materials may be more than a Constructor is ready to undertake. More importantly, the Contract Administrator should be aware that insurance coverage for handling or subcontracting hazardous substance removal might be excluded from the Contractor's coverage.

Even more ideal, whenever possible in private work the Constructor should exclude responsibility for hazardous substances from the bid proposal.

If the contract documents do refer to hazardous substances or place responsibility for any hazardous substances encountered onto the Contractor, then the Contract Administrator will want to include in discussions with the estimating team what steps were taken in an environmental review prior to submission of the proposal. This should include:

- Have hazardous substances previously been encountered on the site, and if so were they successfully removed?
- What hazardous substances might be encountered?
- Do the Owner, Design Professionals, or Construction Manager believe hazardous substances might exist?
- Did the site visit suggest there could be hazardous substances? if so, where and what kind?
- What contingencies, if any, are built into the project for hazardous substances?
- If they are encountered, what steps did anyone anticipate taking in regard to notification, delay, redirecting work area, removal?
- Who else could be involved in relation to uncovering hazardous substances?
- Who else might be considered responsible for the hazardous substances?

Just as important as being prepared if hazardous substances are uncovered is the restart of work once the issue has been addressed. It is important that the Contract Administrator gets proper authorization to resume work. Even though the last material has left the site, there may be tests or certifications that are required. Both the Owner and the government agency that reviewed the cleanup should provide the Constructor with written authorization to resume work. This can be important on many levels for the Contract Administrator, including insurance coverage, future PRP liability, and the safety of the Constructor's own personnel.

A CASE IN POINT

There is a famous, or perhaps to some an infamous, case of *Kaiser Aluminum and Chemical Corporation* vs. *Catellus Development Corporation*, in which soil that had been contaminated decades earlier was moved by a grading and excavating Contractor who knew nothing about the contamination. The site was being prepared for the construction of a new housing development.

Once the contamination was discovered, however, the Owner was required to remove the contamination and bring the site up to EPA and CERCLA standards. That done, the Owner then sued the Developer for the costs, and the Developer in turned sued the Contractor that had done the grading.

The courts ruled under CERCLA that the act of moving the contaminated material was sufficient to make the Contractor responsible for the costs, even though initially the Contractor had no idea of the contamination that had taken place decades earlier.

16.1.6.2 Other environmental regulations

There are other major environmental regulations, many dealing with preventing the destruction of our natural resources. The **Clean Water Act**, passed in 1972, regulates the disposal of pollutants into fresh water. Although this may seem more targeted to manufacturers, Constructors must be extremely careful in recognizing the consequences of storm water discharged from any construction site larger than one acre into fresh water sources.

The Constructor will be responsible for applying for a permit from either the EPA or state authorities to ensure compliance with Clean Water Act standards, as well as any additional requirements based on the nature and location of the project site. The Contract Administrator must not take this responsibility lightly, since the Act provides for civil fines of up to $25,000 per day. Worse yet, criminal penalties can be imposed, which will include similar fine amounts, and also the potential for prison time of up to a year for each violation.

Similar to the Clean Water Act, the **Clean Air Act** regulates what can be emitted into the air. Of course, this is a much broader resource surrounding all of us, but more and more EPA regulations are being enacted to ensure we do everything possible to improve rather than destroy air quality. For example, some states and the EPA are regulating emissions from construction vehicles in an attempt to significantly reduce greenhouse gases.

16.1.6.3 Toxic mold

Toxic Mold has become an important environmental concern and it is usually a complicated issue for the Contract Administrator. To begin with, the Constructor may or may not have insurance coverage for mold abatement.

There are many causes of Toxic Mold build-up, including restricted air flow, selection of materials, and Owner occupancy practices. Since it is almost always discovered after completion, for the Contract Administrator the realistic assessment of who is responsible is also tied to how important is the future relationship with the Owner and the Design Professionals.

16.1.6.4 Positive environmental initiatives

Figure 16.9 Southern Manhattan at night
Photo credit: Charles W. Cook

Although attractive to look at, the built environment by night and day is consuming a vast amount of natural resources and contributing large amounts of pollution to our world. For example, according to the EPA (see "Going the extra mile"), buildings in the United States account for over a third of the energy used and emit over third of carbon dioxide emissions back into the air. For this reason, the Constructor is often criticized simply for delivering what the public demands.

Primarily through the initiatives of concerned Design Professionals, Constructors are increasingly being involved in projects that help to improve or at least reduce the environmental impact over previous practices. Owners are requiring in their contracts that Constructors comply with efforts to achieve improved "Green Building" standards.

One of the most popular movements is directed by the United States Green Building Council (USGBC). It certifies buildings on a **LEED** (Leadership in Energy and Environmental Design) rating system. Based on the type of construction and the level of achievement in meeting environmental standards, a structure can be LEED certified up through Platinum Certification.

The Contract Administrator must recognize that an Owner seeking LEED certification for a building will require additional services from the construction team. Documentation of specific processes, from demolition and recycling of debris to commissioning the building, must be kept. Water and energy efficiencies for the site and building will be important.

In order to properly monitor and execute the LEED requirements a member assigned to the project from the Constructor must have taken a course and passed an extensive exam provided by the Green Building Council. This person is then LEED certified and can monitor the work for the Constructor.

The final caution for the Contract Administrator is to recognize that Owners find prestige more than profit in the certification achieved by their building. The Owner and the Designers want the highest standards possible. If they are seeking Gold Certification and receive only Silver Certification, the Contract Administrator will want to have carefully documented what issues along the way that were not the responsibility of the Constructor led to the lower rating.

16.1.7 Hiring and firing statutes and regulations

Most of the statutes and regulations related to hiring and firing deal with discrimination of some sort. The one notable exception is in relation to employment of undocumented workers. The Contract Administrator must be sure all the personnel on the project have a right to work in the United States. To do this, all employers are required to have every new employee fill out an I-9 form, provided by the Department for Homeland Security. Based on the documentation that an employee provides, the Constructor is required to verify the status of the individual. If the employee does not appear to be eligible for employment, the employer must not allow or continue the employment. Otherwise the employer is subject to severe penalties.

On Federal work, the employer must also comply with regulations known as **E-Verify**. This began as a 2008 executive order from President Bush requiring Contractors and Subcontractors on government projects to electronically check the eligibility of the employee through data banks of the Federal government.

The Contract Administrator must also be aware of conditions regarding immigrant workers, initially regulated by the **Immigration Reform and Control Act** 499 of 1986. In addition to granting legal status to some immigrants, the law made it illegal to recruit or hire unauthorized immigrants except for certain seasonal, agricultural purposes.

16.1.7.1 Discrimination practices

There are several statutes and regulations governing employment discrimination and the **Equal Employment Opportunity Commission (EEOC):**

- The **Equal Pay Act**, passed in 1963 and signed by John F. Kennedy, legislated for equal pay for equal work and abolished disparity of wages between male and female workers.
- The **Civil Rights Act**, signed into law in 1964 and amended in 1991, prohibits discrimination on the basis of race, color, religion, gender, or national origin. It also prohibits sexual harassment in the workplace.
- The **Age Discrimination in Employment Act (ADEA)** was passed three years after the Civil Rights Act to include prohibiting discrimination on the basis of age for anyone aged 40 years or older.
- The **Americans with Disabilities Act (ADA)** was passed in 1990 to prevent employers from discriminating against a qualified employee based on a disability, but an employer cannot be required to hire someone who could not perform the normal functions of the position. Further, employers must provide reasonable accommodation for any employee with a handicap, provided that would not require undue hardship on the employer.
- The **Family Medical Leave Act (FMLA)** requires an employer to allow up to twelve weeks of unpaid absence from the job for either the birth or adoption of a child or for a serious medical condition.

There certainly may be extenuating circumstances related to some issues. In construction the employer must balance the burdens, challenges, and dangers of the work with the fairness to an individual applicant. Although we should not put people in positions to fail or that are unsafe to themselves or others, we have to recognize that the statutes and regulations are in place to see everyone is given a chance to enjoy the fruits of the labor they are capable of performing.

16.1.7.2 Affirmative Action

The Contract Administrator should be aware of what is termed **Affirmative Action**, which might be a significant issue in many public contracts. Affirmative Action policies seek to help groups of the population that have previously suffered from discrimination practices, perhaps due to race, gender, or national origin. Some have challenged the regulations on the basis of reverse discrimination, but the Contract Administrator must recognize that most public contracts have requirements that have been agreed upon, and it is then necessary to comply. Typically, a certain level of participation of minority-, disadvantaged-, and/or women-owned businesses will be required in the project. Also, certain percentages of races and gender will be represented in the trades working on the project. The Contract Administrator should not ignore these requirements or wait until part way through the project to ascertain compliance with the percentages. By then, any "fixes" necessary might be extremely difficult or even too late.

16.1.8 Labor statutes and regulations

Some might consider there is a natural tension between management and labor. At one extreme, employers want to get the most work for the least amount of pay. At the other extreme, workers want to get the most pay for doing the least. That, of course, will lead to a dysfunctional relationship. Interestingly, the first recorded construction strike in the United

States was not over wages, and it did not take place between union and management, but was when house carpenters in Philadelphia stopped work in order to get a shorter work day.

In looking at the issues involved, it should not be difficult for a Contract Administrator to recognize the pressures on both sides that can create tension. Employers will be paid eventually for work put in place at what is usually an agreed lump sum price. An employee is paid more regularly for daily or hourly work, whether the work is done within the budget or not. The employee most likely needs to work, and the employer always has the threat of firing employees. Finding an effective balance is a challenge for the Contract Administrator, but statutes and regulations have been enacted to help guide employers and employees through the process.

Many of the previous issues that have been discussed relate to how employers treat those who work for them. One very big issue in construction is labor in relation to union affiliation. If a Contract Administrator is involved in a strictly non-union project, many of the issues regarding labor have probably been handled through employer–employee agreements or understandings. The exception to this would be a project where some workers might be approached by union representatives and a vote to remain **open shop** or become union is required.

In regard to a union representative soliciting non-union workers, unless an employer prohibits all solicitation, then the employer cannot bar the union representative from doing the same. Of course, once off site and before or after work hours, contact between workers and union representatives cannot be stopped. Contract Administrators faced with issues related to union solicitation should be very careful, notify upper management, and seek competent counsel.

It is possible to have a dual project site, where union and non-union labor work on the same project (but enter through separate gates). The prudent Contract Administrator will have to be alert to increased tensions before, during, and just after work each day.

On a strictly union site, the issues can still be difficult. The Contract Administrator must recognize that a set of contractual agreements exists that will have to be observed as closely as the contract with the Owner. Union employers have signed a contract that governs the work rules and conditions of the union employees. Among the union crew, a **shop steward** will be appointed to deal with any violations in the practices or actions of the employer toward the union workers.

One of the most widely used tactics of unions is the stoppage of work, or picketing. Laws and regulations exist regarding under what circumstances and where picketing can take place. This will not prevent some work stoppages or picketing from taking place in a **wildcat strike** action. Although illegal, and supposedly not sanctioned by the union itself, these can cause disruptions to get the attention of the Contractor, or sometimes, even worse, the Owner.

A CASE IN POINT

I was too young to remember, since I was only a couple months old at the time, but my brother told me this story, and if it had not turned out well I suspect my mother might have been raising both of us as a single parent.

I suppose my father got his attitude of "I believe we can do whatever we want" early in life, and I know it pretty much continued throughout his career.

A wildcat strike took place in the coal country of Pennsylvania. He was managing the construction of a power plant, and one morning the workers decided something was bothering them more than usual, so they mounted a wildcat strike.

No one was supposed to enter the site through the angry protesters, but Robert Cook did just that. In those days that was a bit like running a gauntlet, but it was also one way to start a dialogue. It eventually worked, the protesters cooled down, the issue was resolved, and apparently no one got hurt.

Of course, as my brother tells the story, my mother did have considerable work to do on our father's clothing that night, removing the dirt and stains from everything that had been thrown at him in the morning.

Those were indeed different times, but perhaps not different people. The Contract Administrator must recognize that below the threshold of what others consider fair are reactions produced by taut nerves and disturbed emotions. Keeping lines of communication open is the first and best step to avoiding dangerous escalation in a conflict.

A major cause of wildcat strikes or work stoppages can be what is termed **jurisdictional disputes**. Traditional Contractors and Subcontractors have maintained the right to assign work to a particular trade union. Tradition on the part of trade unions has usually been the expected rule for assignments. Sometimes, however, assignments are challenged. This can be due to a new Constructor deciding differently in a particular area than has been the way assignments were given in the past, a better price from a particular Subcontractor using a different trade, or a new product that more than one trade union wants to claim.

Example: Solar panels on a roof at first seemed to be the work of electricians, since they are intended to reduce the need for the occupant to draw on the supply from the local electric company. However, roof work has traditionally been given to carpenters. When solar panels first appeared on the market both carpenters and electricians realized that solar panels could bring substantial work in the future for members of either union. This prompted both trades to insist that the work was within their "jurisdiction." The Contractor who assigned it to one or the other sometimes faced a work stoppage or wildcat strike as the response of the union that did not get the assignment. The Contract Administrator should be aware of or check regarding the traditional union assignments in their area, which might be different from the jurisdictional assignments on a previous project in a different area.

When it comes to any work stoppage or other labor disputes, the Contract Administrator, with the help of others, should carefully seek neutral spaces in which to discuss the issues with individuals from both sides who are more intent on solving the dispute than on proving who is right because they are stronger. Good people can have a major disagreement, and one should not expect cooler heads to prevail until words speak louder than actions. Beware, however, that compromise usually does not settle anything for long. Someone may still be upset over what they did not get. Left unaddressed, such an upset can fester and erupt again. True resolution will be achieved when both parties find a collaborative solution.

16.1.8.1 National Labor Relations Board

In 1935, the **National Labor Relations Act** (NLRA) established the **National Labor Relations Board** (NLRB) to oversee disputes related to labor in the workplace. Its primary intent was to ensure that private sector employees have the right to form a union. The Board hears what are considered violations of such rights, including threats by employers against employees who are thinking of or attempting to form a union.

Most of the provisions addressing the concerns of Constructors are contained in Section VII of the Act, but the NLRB is constantly changing, and rulings may differ from one administration

Figure 16.10 President Franklin Roosevelt signs the National Labor Relations Act into law, July 5, 1935. Included in the ceremony and standing behind him is Frances Perkins, Secretary of Labor from 1935 to 1945, the first woman ever appointed to a presidential cabinet
Photo credit: United States Government

to the next. In fact, although the Board is meant to be made up of five members, a number guaranteeing no tie votes, depending on the cooperation between the President, who appoints all members to the board, and the Senate, which needs to confirm the appointments, the Board may have less than five members at any one time. It is possible for the President to make interim appointments when the Senate is not in session, but this usually leads to some future disapproval and lack of cooperation between the two branches of the government.

Generally, the Board is viewed as a pro-labor institution, even favoring unions over open shop employment. There are, however, provisions within the NLRA that prohibit unions from engaging in certain activities, and other provisions that protect employers' rights. Again, if the Contract Administrator recognizes that union activity might be taking place on a non-union project, it is very important to seek competent advice from someone who understands the statutes and regulations in regard to the NLRA, as well as any other regulations that have been passed. Inappropriate actions could place a Contractor in violation of employees' rights. Similarly, knowing what can and cannot be done can either positively or negatively affect the outcome for either side on the issue. (For more information on the NLRA see "Going the extra mile" at the end of this chapter.)

16.1.9 Workplace conduct

The image of the construction industry related to gender is of a male-dominated industry that does not bring out the best in men. From wolf-whistling loiterers on lunch break to individuals harassing the "weaker sex" for not carrying their share of the load, the industry has let a few set the standard that we are a 4-D industry: Dull, Dirty, Dumb, and Dangerous.

Almost everyone working in the industry knows differently. The infractions of a minority of workers have targeted the whole workforce for scrutiny in regard to workplace conduct, particularly gender-related harassment, but it can extend to other forms of discrimination, including race, age, and ethnicity.

Civil rights legislation began in 1875, when the United States recognized that the aftermath of the Civil War had not established equal opportunity for all. Progress was slow and uneven for almost a century. Then Congress passed the 1964 Civil Rights Act, to be amended in 1991. It broadened the scope of the 1875 legislation and stated that no employer can discriminate on the basis of race, color, religion, sex, or national origin. Harassment of anyone on such a basis is strictly prohibited.

If a party can assert a claim for discrimination or harassment, they are entitled to a jury trial that may result in severe punitive damages.

The Contract Administrator's best defense against any co-worker or supervisor being charged with workplace misconduct is to actively monitor all conduct and expect everyone else to do the same. Turning the other way is not acceptable, and it will ultimately lead to worse problems later in the project. Policy statements of "zero tolerance" should be more than the norm. They need to be first and foremost.

16.1.9.1 Substance abuse

Another area of workplace conduct that needs "zero tolerance" is substance abuse. Whether from drugs or alcohol, the danger to the individual user is compounded by the potential danger to everyone else. Construction is too dangerous to be working without one's full mental and physical resources. Some Owners and Constructors conduct random checks for substance abuse. Unfortunately in some cases, including in some union agreements, random checks have been prohibited.

The prudent Contract Administrator needs to remain alert. A beer can in a dumpster or too many bathroom breaks for someone might indicate that something needs to be investigated. Finding someone abusing drugs or alcohol may be saving more than just the quality of that person's work. It could be preventing injury or worse to someone else.

16.1.10 Liens

Liens were discussed extensively in a previous chapter. Although Owners will place contract provisions on liens and include requirements regarding lien waivers prior to payment, the actual laws governing liens vary from state to state. It will be important to work within the parameters of each state's lien law statutes and regulations.

16.1.11 Bankruptcy

Within a Contract Administrator's career, the challenge of bankruptcy will almost certainly arise, and most likely more than once. Multiple experiences with bankruptcy, however, do not make them any easier.

By becoming familiar with a few basic terms and the parties that will be involved in a bankruptcy, the Contract Administrator can begin to recognize what hazards exist and what relief might be available.

- *Chapter 7* – established by the United States Bankruptcy Code. An organization under Chapter 7 is totally liquidated and ceases to do business. The parties to which money is owed receive *pro rata* portions based on established priority positions and the amount of assets available for distribution.
- *Chapter 11* – this form of bankruptcy established by the Bankruptcy Code entitles the bankrupt party to seek reorganization of its debts so it can continue to exist. A party under Chapter 11 continues to do business while protected from its creditors. If an acceptable method of working through payment of debts can be worked out with a bankruptcy judge and all creditors, the bankrupt party can emerge from Chapter 11 repositioned to do business as before.
- *Debtor* – any party filing for bankruptcy. On a construction team this could include the Owner, any of the Design Professionals, the Construction Manager, the Contractor, any of the Subcontractors or the Suppliers. Under Chapter 11, the Debtor is termed **Debtor in Possession**, meaning the bankrupt party will continue to do business.
- *Estate of the Bankrupt* – or also Bankruptcy Estate, is the accumulated assets of the bankrupt party. It is from these assets that the creditors will receive compensation for their outstanding claims.
- *Secured Creditors* – these are the parties that have a legal position to share in the funds of the estate before other parties without such a position will receive compensation.
- *Unsecured Creditors* – parties without an established legal position to receive funds from the bankrupt's estate.

Ideally, there may be warnings before the actual notice, but if there are no signs of a party being in trouble, then as soon as a Contract Administrator hears of a bankruptcy, an attorney should be contacted to discuss the options and actions available.

There are some initial strategies to consider.

16.1.11.1 Chapter 7 strategy

Timing is everything when it comes to Chapter 7 bankruptcy. The sooner one files, the more likely one will be in line to receive something from the bankruptcy. The major challenge, of course, is that the party may be so out of assets that there really is not much left to pay any debts.

For the Constructor, the line forms at the end, and before assets may be divided among Contractors there is a strong possibility that the bankrupt party owed debts for Federal, state, and local taxes, as well as to a bonding company and bank. Those debts will have to be paid prior to remaining assets being made available to others.

The Contract Administrator faced with a party in bankruptcy should first consider if any bond is in place that can help with the outcome. Additionally, the Contract Administrator should be looking for the most expedient method of replacing the Chapter 7 bankrupt party. Bankruptcies will almost always affect the budget for the project, and there is little doubt the schedule will be negatively impacted. Quality of work is almost certainly in jeopardy, particularly if a desperate Constructor attempts to replace a bankrupt Subcontractor with another Subcontractor based strictly on price rather than experience or competence.

16.1.11.2 Chapter 11 strategies

Knowing that someone can possibly work their way out of bankruptcy is no reason to delay action related to the situation. Chapter 11 bankruptcy parties can still negatively affect the schedule of the project. If Liquidated Damages are involved, this could be disastrous, since the bankrupt party is not in a position to assume the costs even though they are probably responsible for causing the Liquidated Damages.

Open communication with the Bankrupt's representatives on the project and in the office is important, but it is also important to recognize that many of their actions will be governed by their attorney and the bankruptcy judge. At this point the promises they make are not worth any more than the promises made on the contract. Before withholding any payments due the bankrupt party, the Contract Administrator should again discuss such actions with an attorney. There can be penalties if a creditor violates the intentions or orders of the bankruptcy court.

One strategy a Contract Administrator might use is to issue joint checks whenever possible. A joint check would be made out to the party with whom the Constructor originally signed a contract, but the party that is owed money by that entity would also be included jointly on the check. The party that owes the money endorses the check first, and then the company that is owed the money can deposit it in its own account. In that way payments flow through to the correct party, but this should also be something discussed with an attorney to make sure if one does pay jointly the issue will be resolved and one does not wind up having to pay twice.

Additionally, foreseeing mounting costs for correction or replacement, a Contract Administrator might consider holding back monies due to the Chapter 7 or 11 entity based on prior applications for payment, but this too is potentially dangerous. It would be important to get an attorney's advice again before doing something that would be contrary to the demands of the bankruptcy court.

If the debt is actually reversed, and the bankrupt party owes the Constructor or someone else, the courts allow those debts to be put on hold, and eventually they may be negotiated. It is quite possible a Debtor will pay to a party it owes only a portion of what is actually due. This is referred to as **"pennies on the dollar,"** but the realization for most parties that are owed money is that "pennies on the dollar" are better than nothing at all at the end of a Chapter 7 line.

Underlying the difficulties with a Chapter 11 bankruptcy is the principle established by law that the Debtor in Possession has a right to continue to perform what contracts it believes it can complete. Under such circumstances, and with the approval of the bankruptcy court, both parties to the contract are expected to fulfill their contractual obligations. This could become quite stressful for the Contract Administrator. Again, this is a time for honest and open communication between the parties.

The ultimate strategy, however, for a Contract Administrator is to remain vigilant and to watch for any signs that a party is in trouble. If a Subcontractor or Supplier might be having severe difficulties with deliveries or is falling behind on job progress, some investigation might be in order. Termination of a contract prior to the actual declaration of bankruptcy can put the Contract Administrator in a much stronger position. Bringing in another Subcontractor or Supplier could be difficult and have a negative effect on the schedule, but not doing so could be even worse, depending how the bankruptcy court wished to proceed.

16.1.11.3 Stored materials

If a Constructor or another party has purchased materials but allowed them to be stored off site, it is very important to act quickly to have the material released. We have all heard that "possession is nine-tenths of the law," and materials paid for but not in one's possession might be difficult to obtain.

16.1.11.4 Bonded work

Bonds have been discussed extensively in a previous chapter, but the Contract Administrator should recognize that if a bond exists protecting a creditor's position, this can be an important source of relief from a bankrupt entity. Notice given by the Contract Administrator and a demand for prompt action will be important. Advice from an attorney should also be sought immediately.

16.1.12 Dispute resolution

Dispute resolution will be the topic of a future chapter. At this point it should be enough that the Contract Administrator realizes that the statutes and regulations governing dispute resolution can be unnecessary if an atmosphere exists on the project in which disputes can be reasonably and amicably resolved among the parties without resorting to an external source for judgment.

16.1.13 Healthcare

Regulations regarding healthcare continually evolve as the costs for insurance keep escalating, making coverage by employers more and more difficult. Generally, healthcare is best left to professional providers, and upper management will almost always determine what will be provided based on government regulations as well as what can be afforded.

The overwhelming principle for a Contract Administrator to keep in mind is that "what is good for one is good for all." Government regulations basically set the standard that what is provided or offered in healthcare to one person through the company must be offered to others. One partial exception to that requirement is that union employees have their own insurance through the union plans. The employer contributes hourly fees to the union to offset the cost of the insurance, but the plan covering the union employees is different than what the employer might choose for the other, non-union employees.

The Contract Administrator should also remember that Workers' Compensation Insurance paid as a payroll burden by the employer covers only work-related injuries. Other health issues need separate health insurance coverage.

Examples:

1. A worker on a project site trips on debris left in the middle of the walkway and breaks his leg. The medical costs will be covered, as well as lost time compensation, by the Workers' Compensation coverage.
2. The worker trips at home while playing soccer with his children. He breaks his leg. The medical expenses will have to be paid by health insurance. It is very likely there will be no lost time compensation, based on most medical policies.

Note: In the soccer example above it is obvious that the condition could not be hidden until the employee reported for work on the next morning. However, there are many accidents, aches, pains, and dislocations that can be disguised for a period of time. If the worker does something at home, but comes to work and then complains of an injury, the company may very well be required to cover under Workers' Compensation what actually did not happen at work. For this reason many companies have beginning-of-the-day exercises for the crews to both stretch and loosen the workers' muscles and joints for the rigors of the day, as well as determine that they are physically capable of doing what they will be required to perform. This will obviously also expose some "at home" injuries.

16.1.14 Retirement

Of course, retirement may look good to a Contract Administrator in the middle of a difficult project, but it also is usually a long way off for most. The main set of regulations governing private retirement packages comes from the Department of Labor under the 1974 **Employee Retirement Income Security Act (ERISA)**. No private employer is required to have a pension plan, but if an employer does intend to have a pension plan, there are regulations the Department of Labor will enforce, especially regarding who will be included and how funds will be managed and distributed.

Once again, decisions about retirement benefits will almost certainly be handled by upper management. If the Contract Administrator is upper management, retirement benefits are best discussed with an experienced attorney or plan broker, but for basic information the Contract Administrator might want to contact the Department of Labor website (see "Going the extra mile").

16.2 Uniform commercial code

Everyone lives by selling something.

Robert Louis Stevenson

Before concluding this chapter on statutes and regulations, one particular regulatory act should be discussed separately. The **Uniform Commercial Code** is a means to make sure material being sold between and within states is done with a uniform set of standards.

As mentioned before, the primary difference between a Subcontractor and a Supplier is that the Supplier does not furnish labor on the project site. An excessive amount of labor provided off site in the creation of something rather unique and specific to the project may actually qualify such an entity as a Subcontractor.

Usually, however, the lack of actual personnel on the project site means the delivery of a product to the site gives such a party the rights and responsibilities of a Supplier. Purchase orders may have been written containing very specific details of the rights and responsibilities of both parties, but more often than not, such purchase orders are minimal and often non-existent. A signed delivery slip will be the basic acknowledgement that something was supplied. Fortunately, every state except Louisiana has adopted a set of rules proposed by the Federal government known as the Uniform Commercial Code (UCC), which defines the basic rights and responsibilities of parties doing commercial business together. Although Louisiana has not adopted the UCC, it has adapted its own state laws to reflect the same rights and responsibilities.

Realizing a contract can be verbal, the rules of the UCC are intended to apply to any sale of goods where the labor is incidental to the transaction.

> **Example:** The company that delivers concrete in a truck is a Supplier, even though there is a truck driver involved. The company that provides workers to pull the concrete down the shoot, screed it, and finish the slab is a Subcontractor.

In the example above, by the end of the pour an additional few yards are sometimes ordered. Due to time, these additional yards are seldom put on a purchase order, but the contract to pay for them still exists. Even if there is no actual purchase order for what has been delivered the basic Contract principle that we learned earlier still exists, since a *quid pro quo* arrangement exists – where an offer is accepted, and some form of compensation (or consideration) will be given. This is true for "bill me later" transactions, credit card orders, or COD purchases. Through the many and varied commercial transactions that take place, from store pickups to internet sales through all the products delivered to a construction site, the rules of the UCC apply.

However, the Contract Administrator needs to recognize that the more significant the purchase, the more the purchase needs to be documented or confirmed in writing. More than the price, delivery of purchased items can be very important. Items that are specifically manufactured for a project or have required shop drawing submittals and approval need to be accurately detailed in the important aspects of the purchase, submissions, and delivery, particularly if the items being ordered are involved in the critical path or even if they have very little **float** available between the delivery and installation of the material.

Obviously cost is an important item related to purchase orders, but even more than cost is the requirement to detail quantity. Whether dealing with lineal feet, cubic yards, tons, or some other unit, the quantity of the delivery often needs to be defined before the cost of each unit and the total cost can be determined. Without specific quantities a "contract" under the UCC is difficult to enforce.

16.2.1 Shipping under UCC

The Contract Administrator will want to be very careful in relation to when the Contractor takes possession of the material. The term the UCC uses is **Free on Board (FOB)**. The best term would be "FOB jobsite." If the Supplier negotiates the FOB terms to be the warehouse where the goods are manufactured or stored, then the cost of insuring the product until it gets to the project will be on the Contractor. If the project site is the FOB destination, then the Contractor does not take responsibility for the goods until they actually arrive at the project.

Another term the Contract Administrator must be careful of is **Cash on Delivery (COD)**. Paying for a product when it is delivered will usually eliminate the Contractor's opportunity to inspect the merchandise. This can be extremely critical if the product has some "performance" requirement (e.g. heating or air conditioning) once it is incorporated into the project. Obvious structural damage from shipping may be readily apparent, but the performance ability of equipment is not usually detected until it is hooked up and running.

The Contractor has the right to reject material that is delivered, but there has to be a reasonable cause for such a rejection, and the situation may become a major difficulty for the Contract Administrator because the schedule may be adversely affected by the rejection of material.

The Contract Administrator, however, must promptly notify the Supplier of the rejection of the material or equipment. Whether the material or equipment is rejected based on damage immediately apparent or after the material or equipment was put in place and is running could jeopardize the Contractor's right to have the material or equipment replaced at no cost without a timely notice of rejection.

16.3 No end in sight

It's not wise to violate rules until you know how to observe them.

T. S. Eliot

By now, the Contract Administrator must realize that learning all the rules will be impossible. Yet not observing one or more could be disastrous. There can be no better reason, therefore, to find competent help and experienced "team mates" than the necessity of observing the vast array of statutes and regulations.

Constructors often are upset by some risk shifting clauses placed into the contract documents by the Owner or the Design Professionals requiring the Contractor to be responsible for the vast number of statutes and regulations that might affect the progress and final construction. At the same time, most if not all of these risk shifting clauses are passed on to the Subcontractors, Sub-subcontractors, and Suppliers. The theory has become that those who should know best about particular issues need to be responsible.

As members of the industry extend that concept a bit farther, they arrive at the conclusion that as responsibility increases, so should involvement earlier in the process increase. Design-Build and Integrated Project Delivery methods are moving closer to that "Master Builder" concept where a single entity can be responsible for "doing it all." Until then, the Contract Administrator will continue to be challenged to observe not just the contract requirements, but the statutes and regulations that are all part of the process of doing business.

Exercise 16.1

Take a moment
Where do you want to grow?

Before proceeding to "Instant recall," work through this exercise. It could be a career changer or even save a life.

Grade yourself on the three major topics of statutes and regulations that have been discussed in this chapter; then determine which one(s) you really need to work on and how soon you need to improve, for the sake of your future career and the success of the construction team you are or will be a part of. Grade 1 (low) to 5 (highly skilled and knowledgeable):

Area	*Grade*	*When to work on*
Safety		
Environment		
Employment practices		

Discussion of Exercise 16.1

The key to the above exercise is to recognize the need for continuous improvement beyond this textbook. You the student will be entering a dynamic and ever-changing environment, and the Contract Administrator cannot rely upon what he or she has learned. One must keep learning, keep improving. The best way to do that is to recognize areas for improvement and then set goals to address your needs.

Best wishes! It's a journey, not a destination.

INSTANT RECALL

- In the case of statutes and regulations, most contracts definitely make them a part of the contract by reference.
- Design Professionals and Constructors need to be able to demonstrate competence, so as to ensure the public and the end user will be safe and satisfied.
- Although some states have relaxed litigation rules, many state courts will not hear a case presented by an unlicensed Contractor that has done work and a dispute over payment has arisen.
- The requirement to build in accordance with issues uncovered during the permitting process will also be an obligation the Constructor assumes once the permit is obtained.
- It is important, therefore, that significant issues between the permit-approved plans and specifications and the contract documents are resolved immediately for all parties so that any potential or significant cost issues do not linger until they become major problems later in the project.
- One of the chief reasons for establishing building codes is safety. For that reason many of the codes relate to either fire or structural issues. Other codes have developed for:
  - environmental reasons
  - site drainage
  - establishing or improving traffic flow
  - providing access and use for the handicapped
  - ensuring minimal standards of products used in construction
  - protection against potential natural disasters (e.g. earthquake, hurricane, flood).
- The real difficulty in complying with building codes is not always so much the assessing of responsibility as it is the time it takes to change or rework non-compliance issues.
- Compounding the issue for Constructors is the fact that some local governments have adopted additional, usually more stringent, standards, and even states are permitted to address additional safety issues above and beyond the OSHA regulations.
- The most effective Contract Administrators will be those who can establish an atmosphere that safety is everyone's concern.
- The success of the Contract Administrator might be just as closely aligned to how well they can handle the diplomacy as to how well versed they are on the safety regulations.
- Who will handle the eventual safety inspection should be decided well before the first inspector steps through the project gate and knocks on the trailer door.
- Through the Fair Labor Standards Act (FLSA), the Federal government not only established a minimum wage for most employees, but also set certain guidelines related to overtime and hours of work.
- One issue that might become important for the Contract Administrator to resolve relates to when the payroll "clock" starts.
- The Contract Administrator must be aware there are existing environmental hazards that

must be recognized, and there are actions that must be taken to ensure that damage to the environment from current construction is minimized.
- Many of the regulations regarding hazardous waste encountered in construction can be traced back to the 1980 Comprehensive Environmental Response, Compensation, and Liability Act (CERCLA).
- Within the CERCLA regulations is an important concept known as Potentially Responsible Parties (PRP). Under this part of the regulations, the government attempts to involve any "party" that may be responsible for the cleanup of the hazardous substance.
- Careless handling by the Constructor after uncovering the hazardous substance can make the Constructor the most accessible PRP.
- The Contract Administrator, therefore, must be sure all parties involved in the project understand the seriousness of hazardous waste.
- Ideally, the Contract Administrator will want to have the Owner take full responsibility for hazardous substances and their removal.
- After hazardous substances have been removed, it is important that the Contract Administrator gets proper authorization to resume work.
- Constructors must be extremely careful in recognizing the consequences of storm water discharged from any construction site larger than one acre into fresh water sources.
- The Constructor will be responsible for applying for a permit from either the EPA or state authorities to ensure compliance with Clean Water Act standards, as well as any additional requirements based on the nature and location of the project site.
- Toxic Mold has become an important environmental concern and it is usually a complicated issue for the Contract Administrator.
- The Contract Administrator must recognize that an Owner seeking LEED certification for a building will require additional services from the Construction team.
- Most of the statutes and regulations related to hiring and firing deal with discrimination of some sort.
- All employers are required to have every new employee fill out an I-9 form provided by the Department for Homeland Security.
- On Federal work, the employer is subject to further regulations known as E-Verify.
- In 1935, the National Labor Relations Act (NLRA) established the National Labor Relations Board (NLRB) to oversee disputes related to labor in the workplace.
- The Contract Administrator's best defense against any co-worker or supervisor being charged with workplace misconduct is to actively monitor all conduct and expect everyone else to do the same.
- Another area of workplace conduct that needs "zero tolerance" is substance abuse.
- Under Chapter 7 bankruptcy an organization is totally liquidated and ceases to do business.
- A party under Chapter 11 bankruptcy continues to do business while protected from its creditors.
- Related to healthcare, upper management will almost always determine what will be provided, based on government regulations as well as what can be afforded.
- The Contract Administrator should also remember that Workers' Compensation Insurance paid as a payroll burden by the employer covers only work-related injuries.
- There can be no better reason to find competent help and experienced "team mates" than the necessity of observing the vast array of statutes and regulations.

YOU BE THE JUDGE

The plaintiff (Fraser) is a resident among many in townhouses owned and operated by Townhouse Corp. (Townhouse). In a lower court decision, the plaintiff's claim of severe health issues caused by Toxic Mold was denied based on the fact that the body of science and testimony in the 2008 case did not find that the mold was the actual cause of health issues, including respiratory difficulties. Fraser, along with several other residents of the townhouse complex, appealed to the New York Supreme Court.

In reviewing the case, the Supreme Court recognized that the lower court had ignored expert testimony supporting the plaintiff's claims that the dampness produced mold that caused health issues, but the lower court had requested a Frye test, which is a hearing to determine whether the claim is generally accepted theory within the existing scientific studies. The court, in reviewing the Frye test hearing evidence, concluded that mold could not be determined as the actual cause of the problems the plaintiffs experienced.

Before the New York Supreme Court the plaintiffs were unified in their complaint that the dampness had caused mold and the mold had caused serious health issues for them all.

How do you rule:
If you rule in favor of the plaintiffs, what would be your remedy:

IT'S A MATTER OF ETHICS

The Americans with Disabilities Act (ADA) recognizes that some work cannot be performed by some individuals with certain disabilities. Nonetheless, we now have soldiers crippled by IEDs returning to duty with an artificial limb. With such advancements in bio-medicine and mechanical engineering for the human body, where do you see the limits of employment for different disabilities within the construction industry?

Going the extra mile

1. Complete the "You be the judge" exercise above and then go to the Companion Website for a discussion of the actual case.
2. **Discussion exercise I** (with friends or as part of a classroom discussion): There is an expression, "ignorance of the law is no excuse." Most people are familiar with the intent of such a statement. Still, with all the different statutes, regulations, and codes, what is truly fair if some issue related to an unanticipated regulation arises? Who on the construction team (Owner, Designer, Constructor, Subcontractor) should be responsible if something is done the way it has always been done, and it turns out to be against the code? Who pays if it needs to be changed, and under what circumstances? What if the regulation is not practical (e.g. did you know that it is illegal to take a bath in Boston on a Sunday? Has the regulation not been repealed because people take showers instead? Or it is just a silly law no one observes?) What if a regulation is not as good or safe as the way a master craftsman now does it?

3. **Discussion exercise II** (with friends or as part of a classroom discussion): Discuss the issue of the poor grades the construction industry receives when it comes to environmental impact:

    a. Is it fair?
    b. Who is chiefly responsible?
    c. What is the cause of this impression?
    d. What can be done about it now?
    e. What should be done in the future?

4. Go to the Companion Website for additional study aids, including flash cards and sample questions.
5. Each of the Federal statutes, regulations, and government agencies mentioned in the "key words" has excellent internet domains. You might wish to consult any that interest you. For instance:
   www.nlrb.gov/national-labor-relations-act, an excellent site if you are interested in or working on a project with union employees.
   www.osha.gov, a very comprehensive site that is filled with important information related to safety in all industries, including construction.
   www.dol.gov/compliance/laws/comp-erisa.htm, Department of Labor website.
6. For those interested in LEED and Green Building, be sure to check out www.usgbc.org and www.epa.gov and search "green building."

7. Go to the Companion Website and watch Part III of the Ethics video.
8. Effective use of the correct vocabulary improves communication and promotes respect among colleagues. The reader should become familiar with the key words listed below that have been used in the chapter. Definitions can be found in the Glossary. Additionally, the student is encouraged to use the flash card exercise in the chapter's Companion Website as practice for possible quiz or exam questions.

Key words

Affirmative Action

Age Discrimination in Employment Act (ADEA)

Americans with Disabilities Act (ADA)

Cash on Delivery (COD)

Certificate of Occupancy (C/O)

Civil Rights Act

Clean Air Act

Clean Water Act

Compliance Officer

Comprehensive Environmental Response, Compensation, and Liability Act (CERCLA)

Free on Board (FOB)

Immigration Reform and Control Act

Jurisdictional disputes

LEED

National Labor Relations Act

National Labor Relations Board

Occupational Safety & Health Administration (OSHA)

Open shop

"Pennies on the dollar"

Portal to Portal Act

Davis Bacon Act

Employee Retirement Income Security Act (ERISA)

Equal Employment Opportunity Commission (EEOC)

Equal Pay Act

E-Verify

Fair Labor Standards Act (FLSA)

Family Medical Leave Act (FMLA)

Float

Potentially Responsible Party (PRP)

Quantum meruit

Resource Conservation and Recovery Act

Shop steward

Superfund

Toxic Mold

Uniform Commercial Code

Wildcat strike

CHAPTER 17

Project closeout

Project closeout begins immediately after signing the contract. Most Contract Administrators wait until the end of the project. That puts unnecessary pressure on every element of the construction team, usually delays the final payment of the project, and often has a final negative impact on client–Contractor relations. Knowing how to get through project closeout is as important for the future of the company as putting the "bricks and mortar" in place originally.

Student learning outcomes

Upon completion the student should be able to. . .

- **analyze priorities for what to do first and last**
- **describe important project closeout requirements**
- **recognize the importance of the whole project team in relation to project closeout**
- **describe the effective use of the punch list**
- **evaluate post-project efforts for greater success.**

17.1 Last things first

Don't wait for the Last Judgment. It takes place every day.

Albert Camus

We briefly described project closeout in a previous chapter. From the beginning, the Contract Administrator will need to put every activity in context with the final project completion. The most fundamental reason for this is, of course, financial. Buried within the retainage that has been kept from each application for payment is the profit, and sometimes a portion of the overhead or general and administrative costs the Constructor has expended during the project. Compounding that dollar amount is the fact that the Constructor and all who have still not been paid in full may have borrowed money to meet the immediate payments, particularly labor and the labor burdens of insurances and taxes. The borrowed money requires interest payments to be met that continue until the principal is paid back.

There are, however, other reasons to quickly finish a project once the actual construction has been completed. Owners, as well as Design Professionals, do not want to linger over past issues. The project has undoubtedly been challenging, but the small details of training, operation manuals, and final cleanup hold little appeal for anyone. The thrill is gone from what has already been accomplished and the attention of most everyone is on some future endeavors that are beginning elsewhere.

The Contract Administrator should closely analyze the contract itself for what items and issues will need to be addressed before final payment is made and the project completed. Here is just a partial list of what the Contract Administrator might typically be looking for from the beginning of contract administration:

1. "as-builts"
2. training and instructional sessions
3. operation and maintenance manuals
4. guarantees and warranties
5. waivers
6. punch list items
7. completion notices
8. archival records
9. project review and cost documentation
10. follow-up with Owner and Design Professionals.

Exercise 17.1

Take a moment
What hat do you wear?

Managing a project and administering a contract requires many different functions. In a large corporation there is a Chief Executive Officer (CEO), which is usually the top position, a Chief Financial Officer (CFO), who takes care of the major accounting and money issues, and a Chief Operations Officer (COO), who has overall control of the work in progress. On any given project, the functions of all of those "chiefs" might become yours. In the space provided, write what functions within the contract and the project itself fall under each "chief" hardhat you have to wear:

CEO:

CFO:

COO:

Discussion of Exercise 17.1

Contract administration requires many talents and working with different individuals. Although the final goal might be the same, the reason and path to get there might be very different.

The project CEO has to be able to deal with the Owners and Design Professionals, as the point person for the company. In addition the CEO needs to make sure all Subcontractors and Suppliers are moving forward in a positive fashion toward completion. As mentioned, this can be a challenge, since the Owners may be driving for emphasis on schedule, and the Subcontractors might put an emphasis on efficiency and productivity based on personnel, equipment, and space available. Additionally, the Suppliers may have issues related to what is possible regarding specified material and delivery. Coordinating all the subcontracts and purchase orders with the prime contract with the Owner can be a very time consuming, but also very necessary task at the beginning of each project.

The project CFO functions are probably how you will be judged by your own upper management. Bringing the project in on time will certainly please the Owner, and in all likelihood, that will preserve the profit expected in the original estimate, but upper management will be looking to how much money the contract produced. Ultimately, how much money one makes is how one will be appreciated, and how much money one loses will not be. Cost reports, change orders, payment applications, and related documents will be very important for you to make sure they flow quickly and efficiently each time.

The project COO will have to be very concerned about the schedule and the field forces. In addition, an effective safety program is mandatory. The COO needs to be working on the project closeout from the beginning, and also coordinating all efforts in the field with the requirements of all the statutes and regulations. If there is a problem with some aspect of the project the COO might need to pass it to the CEO for resolution, but that should be the easy part, because, as mentioned above, you are also the CEO.

A CASE IN POINT

A good friend of mine, and a great contributor to the construction industry, taught me the above CEO/CFO/COO functions of project management. You have already "met" Pete Filanc if you watched the "Moving the Gruber Wagon Works" video on the *Constructing Leaders* website. He was one of the founding Constructors of the Design-Build Institute. At the time I recorded Pete's comments he was suffering from cancer. He battled it valiantly, and for a while it was in remission. Through all the medical challenges Pete was smiling, and I suspect his positive attitude added months, maybe even years, to his life. Just as importantly, he was always willing to share stories, often related to mistakes he had made, so others could benefit from them and not make them also.

I remember well how he emphasized that he had almost lost his company by not paying enough attention to what was happening – by not wearing the right hat at the right time. We need to let others do what they do best, but we still need to wear our own hats of responsibility on the project and within the company.

I would like to think the lesson for all Contract Administrators is that we have to rely upon others, but we do not give up the varying responsibilities of the different hats we wear.

Pete was taken from us too early, his life cut short by the cancer that consumed his body but not his spirit. In the last week of his life he had the pleasure of walking his second daughter down the aisle at a wonderful wedding ceremony. He was exceedingly proud and, as always, happy.

And I believe he would be very happy to think that in the future there will be Contract Administrators who might learn from his lesson of knowing we all wear many hats, and each brings with it different responsibilities at different times. We all have to pay attention to all the challenges and to continue to do our best.

17.1.1 "As-builts"

The Contract Administrator will not be able to detail field changes or complete what are known as the **"as-built" drawings**. It will be important, therefore, that the Contract Administrator assigns someone from the beginning of the project to maintain and document all conditions and changes made in the construction that differ from what was shown on the drawings. This includes any changes made through a change order process or RFIs. In fact, some Contract Administrators ask that whoever is keeping the "as-built" record attach pertinent RFIs to the plans that are being modified.

* * * * * * * * * *

An Owner can require "as-built" drawings by contract, but even if "as-built" drawings are not a requirement, the Contract Administrator might want to be sure to keep them up to date during the progress of the project. They can become important records for many reasons, including dispute resolution and, even more importantly, they will be a significant source of information for certain future projects, giving the Contract Administrator who thought enough to keep the "as-built" record a valuable advantage to negotiate the work.

* * * * * * * * *

Today, the process of "as-builts" is complicated when using electronic media such as BIM. Of course, there is the immediate challenge of having qualified personnel who can in fact make a change to the BIM electronic renderings. Another crucial factor related to BIM documents is a matter of "who owns the documents." From the Design Professional's perspective there might be issues of liability based on changes made to existing documents. Changes made to the original BIM documents can affect future work as well as warranties. If electronic modeling is involved in the construction process, the Contract Administrator will certainly want to have discussion and agreement from the Owner and Design Professionals about how to proceed with "as-built" documentation.

The most important step a Contract Administrator can make after assigning someone to maintain the "as-built" drawings is to monitor progress on the documentation on a weekly basis. In the beginning, maintaining the "as-builts" will be easier than throughout the rest of the project, but once an individual understands that the "as-builts" are a critical part of the project's deliverables, the more complete and up to date the "as-builts" will be.

Going back to document field conditions and changes to the original plans after the project is completed is a nearly impossible task and the quality of the records will deteriorate. Owners and/or Design Professionals may even reject such work until it is more detailed and accurate, and this will be a contractually justified reason to withhold final payment. Therefore, a Contract Administrator might want the Design Professional or other Owner representatives to do an early review of the "as-builts" during the project.

Keep in mind that the more carefully the "as-built" documentation is maintained and the more detail that is shown, the better future issues can be easily resolved. The "as-builts" may be used to resolve issues of latent defects. They might also be needed to locate critical components of the structure for future installations or renovations.

A good set of "as-builts" will not only be appreciated, they could mean moving to the front of the line in consideration for future work. For that reason, the Contract Administrator should think of the "as-builts" as a marketing tool and the quality should compare to the original drawings or electronic renderings in detail and accuracy.

Changes and differing site conditions are inevitable. The "as-builts" should not be used to detail incompetence or ambiguities related to the efforts of the Design Professionals. The Constructor's field forces are not competing with the Design Professionals, but supplementing the Design Professionals' efforts with quality "as-builts," and, if done and delivered with respect, that can be a genuine compliment to the expertise of the Design Professionals by the Constructor's field personnel.

17.1.1.1 Delivery of "as-builts"

Once completed, the Contract Administrator must be certain that the "as-builts" are delivered to the proper person. If the contract does not state who should receive the drawings, the Contract Administrator should have the project minutes document who is to receive them.

Additionally, the Contract Administrator will want to have a confirmation in writing that the documents were received, and by whom.

When working with electronic media, sending multiple copies may be easier and storing on a server may be more accessible than to have a single set of physical "as-builts" done and delivered, but the Contract Administrator should recognize the need for back-up of the documents stored electronically. An Owner's own reception and storage may be compromised and require another transmission of the "as-builts" sometime in the future.

17.1.2 Training and instructional sessions

In recent years the complexity of equipment and other installations in buildings has required more than just a pass-off of operational manuals. Full training sessions are conducted with the people who will be in charge of operations and maintenance for the structure.

A CASE IN POINT

Remember the PSFS story and Frank Cook – the "clerk of the works"? Once a building is completed there are new challenges and opportunities. After the completion of what is now the historic landmark PSFS building, my father's Uncle Frank recognized a great need for the building Owners to maintain the structure that contained several very different innovations in electrical lighting, push-button elevators, and central air conditioning. Many of the features were "firsts" for office towers in the United States, and keeping them running would become essential for the retail stores on the first floor, the bank on the second floor, and all the offices in the rest of the building.

Frank Cook enjoyed several months after the completion of the building coordinating the maintenance as the building's first Superintendent. Eventually other challenges and projects would draw him away from the sedentary building, but the lesson about the need for continued maintenance is important.

Contract Administrators or other personnel with the Constructor should not turn away once the building is completed, just to rush to the next job. Servicing a client after "project closeout" can be and often is more profitable than the project itself. Maintenance can be an important revenue source for Constructors. Commissioning a building, maintaining it through a warranty period, and routine drop-by visits are one way to keep in front of a client and an excellent source of future work for one or two workers. It is often a very profitable source of work for the maintenance staff in the mechanical and electrical service departments of the Subcontractors who were involved in the original construction of the structure.

Figure 17.1 For the Contract Administrator, the cityscape can be a forest of finished buildings or a landscape of opportunity. From basement equipment to new green rooftops, there are possible maintenance agreements that can be very worthwhile for constructors interested in pursuing post-construction services
Photo credit: Charles W. Cook

The key factors for the Contract Administrator in regard to training and instructional sessions are:

- review the contract with the Owner immediately for all training requirements
- be sure training and instructional sessions are incorporated into the subcontracts and purchase orders of the specific Subcontractors and Suppliers that will be needed to give the training
- identify as early as possible the representative from the Owner who will be requiring training
- schedule training sessions as part of the project schedule and, above all, do not wait until the end of the project to begin training
- consider videotaping the training sessions for future reference by the operators of the equipment.

When Subcontractors and Suppliers submit breakdowns of their payment application, it is often easy to overlook the need for instructional sessions, and they may be minimized or brushed aside as merely part of the retainage, but retainage is meant for other issues. Instructional training should be considered a line item. If it is not, the Contract Administrator may have a more difficult time scheduling training on an "as soon as possible" basis. If that happens, then the individuals who are to receive the training may be "jammed" with sessions at the end of the project, and the training, itself, may be less than effective. Call-backs by the Owner's operators to be reshown how the system works will not be enjoyed by anyone. It becomes an unnecessary expense for the party coming back, and it will probably produce some discomfort or even displeasure from the Owner.

17.1.3 Operation and maintenance manuals

Closely associated with training and instructional sessions are the operation and maintenance manuals. Years ago, the equipment and materials were not as sophisticated or complicated as they are today. In the past the Constructor usually handed to the Owner a stack of papers describing the equipment and how to operate and maintain what had been installed. Today, the training session will include a complete run-through of the operation and maintenance manuals.

One thing that has not changed, however, is the importance of securing these written documents as early as possible. Often operation manuals will be forwarded with the equipment when it is delivered from the Supplier. In most instances, the crews assembling the equipment are quite familiar with the installation, and those manuals get misplaced or overlooked as not necessary for the installer. But they are very necessary for the Contract Administrator. Depending on the size of the project, it is always helpful if someone else takes responsibility for collecting and assembling all the operation and maintenance manuals. The Contract Administrator needs to stress to everyone that this is far easier to do as the equipment manuals arrive than it will be after the installers have left the project site.

Doing paperwork is never enjoyable, but chasing paperwork that is somewhere off site is usually frustrating and often immensely time consuming. It is important for the Contract Administrator from the beginning to understand both the number and quality of the operation and maintenance manuals. Three ring binders are often adequate, but sometimes the manuals are expected to be more permanently bound, and as technology continues to add possibilities, electronic operation and maintenance "manuals" are becoming an effective way to stay current and provide the Owner with instructions that can also include video demonstrations.

17.1.4 Guarantees and warranties

Warranties can be expressed (clearly described in the contract documents) or implied. Implied warranties relate to what any reasonable Owner should expect from a competent Constructor.

The biggest issue related to guarantees and warranties for the Contract Administrator might arise during the course of the project. Perhaps in order to maintain the schedule some shortcut(s) was taken related to installation procedures. This may have been noted by the Owner or the Design Professionals, and an issue of whether or not to accept the installation may rest on whether a manufacturer would stand by the warranty if an item was not installed in accordance with the manufacturer's specifications.

Example:

1. Exterior paint may have been applied when the temperature was below what the manufacturer suggests is acceptable.
2. A substitute was made for mechanical equipment, due to availability, requiring a different configuration and hook-up from the original design.

When a warranty starts is another issue the Contract Administrator must deal with on every project. Often equipment will have to be hooked up and running in order to move forward with the project. The air conditioning and heating will be required to bring interior rooms up to proper temperature and humidity. This may happen several months before the end of the project, but the Owner may not accept the equipment until project completion. If the Contract with the Subcontractor or Supplier of the equipment did not include an extended warranty, the Contract Administrator may have to buy additional warranty coverage in order to satisfy contractual obligations with the Owner.

If the Contract Administrator has warned everyone to be on the lookout for and collect operation manuals, then it is just as important to make sure warranty periods do not start at the time the equipment is received on site.

One other issue that can arise for the Contract Administrator is after the project is complete and something goes wrong. If it is repaired, what happens to the warranty? Conditions vary from state to state, and even case by case. In some instances the warranty stops during repairs and then starts back up again. In other cases the Owner will want the warranty to resume anew from the time the equipment is repaired. Nothing is easy, but the Contract Administrator will have to find both a reasonable and a financially viable solution involving all the parties.

Perhaps the best defense regarding the guarantee and warranty issues is a good offense from the beginning. The Contract Administrator should make sure the purchase orders and the subcontracts clearly define the beginning and ending as well as extent of the warranty in accordance with the prime contract as well as what the Owner expects.

17.1.5 Waivers

Almost certainly the contract will require **waivers** from the Constructor to the Owner. Some waivers might already have been implemented, such as when notice of certain conditions are required and the failure to do so waives the Constructor's right to time and/or money, due to the unanticipated condition.

If there was a notice requirement regarding time extensions, and the Constructor failed to notify the Owner within the time limits of the need for additional time, due to some conditions encountered on the project, then the Owner can hold the Constructor responsible for the original completion date without an extension of time. The same would be true regarding notice requirements for a change in cost.

Examples:

1. Due to many different change orders accumulated over the length of the project, the project was delayed not by a single change order, but by the total amount. Failure to request an extension of time will result in the Owner's right to expect no extension of time exists.
2. On a Multi-prime Contract another Constructor causes a delay, but the party that is delayed never files for an extension of time. The delayed Constructor has waived any right to an extension of time.
3. The Owner's contract requires notice of an increase in cost before any extra work is performed. The Constructor that proceeds with extra work without notice to the Owner waives the right to receive compensation.

The Contract Administrator must recognize and balance the rapport and communication between the parties with the contractual obligations for notification. Doing business as it has always been done may be protected by an estoppel argument, but formal notification done in a polite manner will negate the need for such arguments as the Contract Administrator approaches the completion of the project.

At the end of a project, the most common form of waiver is the waiving of rights to sue the Owner or lien the property for payments that have already been made. The Owner does not want to pay twice for the work, and so the Contract Administrator will be required to not only provide a waiver for their own work, but also waivers for everyone else on the project. In exchange for final payment, the Constructor is basically expected to guarantee no liens will be placed on the property by anyone the Constructor has involved in the project.

Again, expecting to collect a waiver from the excavator or concrete Contractor months or even years after the work has been completed is an additional burden. The best way to handle such a requirement is to collect waivers during the entire progress of the project.

17.1.5.1 Consent of surety for final payment

Although not strictly a waiver, on bonded work, the Owner will often require a **Consent of Surety for Final Payment** form. This might usually be considered a formality, since everyone should be happy to finish the project, but the Owner is making sure of two important issues: (1) the Surety has no reason to believe any outstanding financial issues exist between the Contractor and any Subcontractors or Suppliers, and (2) the Surety has to agree that although the project is completed the bonding company is still required to uphold any obligations that may arise after final payment if the Constructor fails to do so in some way.

17.1.6 Punch list items

From the beginning, the Contract Administrator should request, and insist as strongly as possible, that there will only be one punch list kept for the project. The Constructor will not be able to dictate who assembles the punch list, but it is important that multiple punch lists will not be circulating around the project. It can become extremely frustrating for everyone when one punch list is completed, only to find one or more items never identified before on another punch list that need to be fixed. Often these items could have been taken care of quickly and easily if they had been known while another set of punch list items were being addressed.

In addition, the Contract Administrator should recognize and promote that a punch list should be a continuous process. It is far better to have trades that are leaving a project complete all or as many items as possible before they are no longer on the project, than to have a large number of items that they must return to address long after they have gone on to other and possibly bigger and better things.

One issue the Contract Administrator should also be clear on is that change order work is not a punch list item. Change order work is beyond the original contract scope. Only if change order work needs to be repaired should it appear as a punch list item.

17.1.6.1 Pride of workmanship

A "zero punch list" project is possible, but very seldom achieved. It can be achieved only if all the team partners and personnel involved in the project recognize that the "zero punch list" project depends on everyone working together. The point is not to find fault but to find solutions. From the Superintendent who walks the project on a daily basis to every foreman and craft worker, the view should always be toward quality execution.

The Contract Administrator can instill the challenge that addresses the intrinsic motivator of "pride of workmanship." If that motivation permeates the project, then the attitude of "let's do what we have to do to get out of here," or "It's good enough for government work" will be replaced by crews intent on doing their best and delivering quality.

For that reason someone on the Constructor's staff who walks the project daily should set the standard for what is expected. Walking past deficient work and ignoring it will only add to the length of the Architect's punch list in the end. In this way, although it is not an actual punch list, the Constructor will be able to point out items that need to be fixed when the workers are present, and it will set the standard for future installations on the project.

If some items are so significant that the Constructor actually details them for some trades or Subcontractors, it is important for the Constructor to emphasize that this is not the Architect's punch list. These are some obvious issues that need to be addressed, but the Designers will almost certainly have others. Addressing these issues will shorten but not eliminate the obligation to complete the Architect's punch list.

Exercise 17.2

Take a moment
Stop procrastinating

Putting things off can be a major problem when it comes to project closeout. Below are listed the three major causes of procrastination. Using a total score of 100, give a numerical value to each cause, based on how much it generally is the reason you put things off.

1. Things are not pleasant ______
2. It is going to be difficult to do ______
3. There are other things I want to do ______

TOTAL 100

Assuming that you have placed one higher than the other two, write a brief action plan to help you stop procrastinating related to that reason.

Discussion of Exercise 17.2

Procrastination is universal. We all fall victim to it at one time or another, but the consequences for a Contract Administrator can be enormous. Putting off a punch list can seriously delay final payment, and that can be disastrous, particularly for relations with Subcontractors who have completed their work and are awaiting retainage. Not collecting operation manuals and conducting training sessions is just as serious. Getting in and getting out is the key to good business, and the Contract Administrator is at the point of that attack.

Here are a few ideas to help eliminate or conquer procrastination:

1. Admit when you are procrastinating.
2. Recognize the reason why you are procrastinating (start with the list of three in the exercise above).
3. Analyze what needs to be done.
4. Create a schedule, putting the tasks in manageable and sometimes "one bite at a time" chunks, to eat the elephant.
5. Delegate if it is appropriate and you have the personnel who can do it.
6. Ask others to hold you responsible and monitor your progress.

17.1.6.2 Punch list as a sales tool

As in most relationships, our final impression is normally the lasting one. Although we may strive toward a "zero punch list" project, it is almost certain some items will need to be addressed. How the Contract Administrator sees to the quick and effective resolution of any items the Design Professionals or the Owner needs to have dealt with will be important in relation to future work with them.

Generally, punch list work is what remains after Substantial Completion and the Owner has moved into the space and is using it for its intended purpose. Continually interrupting the Owner's employees for repair or touch-ups can become annoying, so the sooner the space is occupied by only those who are meant to be there, the better it is for everyone.

No one expects perfection, but addressing punch list items in a timely manner to everyone's satisfaction will be a positive step toward future consideration and perhaps even negotiation of private work.

17.1.7 Completion notices

Although the Owner and the Design Professionals may recognize when the work is completed, and the Owner may have even established Substantial Completion by moving into the structure, the Contract Administrator must recognize that the contract may call for specific notifications of Substantial Completion and final completion. This may seem unnecessary, but the notice becomes a legal document. Hopefully, no issue will arise related to the actual notice, but a day one way or the other could affect a warranty, perhaps a bond issue, and certainly would have an effect on any Liquidated Damages that might be due.

Many contracts will call for the Designer to review and verify the date of Substantial Completion. This might even entail the Designer signing a **Certificate of Substantial Completion** and including an estimate of the cost of work that remains uncompleted or deficient.

In addition to filing such notices based on contractual obligations, the Constructor should also be sure to notify any parties financially involved in the completion of the project, including the bonding and insurance companies. If carried by the Constructor rather than the Owner, Builder's Risk Insurance, for example, should be cancelled. The fee for that insurance is based on duration as well as amount.

17.1.8 Archival records

Even though the Contract Administrator has confirmed that "as-builts" were delivered to the appropriate individual, it is best to maintain an archival record of the project for the Constructor. With electronic media becoming an effective storage method, the huge storeroom spaces of the past are no longer necessary.

The archived project may never be referred to again, but it could become both an effective sales tool when a client wants to revisit some element in the past project as well as a great learning tool for future estimates and construction methodologies.

Assembling the material is often a thankless task, so the Contract Administrator might want to make this a priority from the beginning of the project rather than something to be done on a last-days basis when individuals are usually more excited about getting on to the next project than reliving the past.

A CASE IN POINT

This is another true story told to me by a good friend, but I will just keep it rather general for reasons I think you will understand. Pete is an avid reader, and while starting to read a book written by a distinguished university professor, he found a statement about how a major Owner had developed a seminal work for construction productivity back in 1954. What amazed Pete was that the project that was mentioned had been constructed by the company Pete now worked for. Pete's company had been involved in a major change in how construction can be done.

Wanting to know more, he asked his boss, who informed him that it was true. For six weeks the Owner and the Constructor's project team met at an out-of-state location and developed an extraordinarily effective model for improving productivity.

Pete's driving question became, "What happened? Why aren't we still doing it, and, in fact, why are we not working for that client right now?"

Once the project was completed, the Owner continued with the construction process, but the then head of the company that Pete now worked for was not in favor of it. The Owner dropped the Constructor, and it took two decades for Pete's company to regain the Owner's confidence and the right to work on the Owner's projects again.

Learning from the past can be a valuable tool on many levels, including estimating advantages, productivity improvement, and marketing opportunities. The Contract Administrator can give a company an incredible advantage by making sure lessons learned are disseminated.

17.1.9 Project review and cost documentation

Included in the archival records will be cost documentation. Although seldom a part of the Owner's contractual requirements, the Contract Administrator should retrieve significant elements of the project execution and costs to disseminate among the estimators and Project Managers of the company for future consideration. Winning the next competitively bid project may depend on information learned on an earlier project. Additionally, by letting others know what worked well in addition to what needed better execution, future projects will benefit and the whole company can be more effective.

17.1.10 Follow-up with Owner and Design Professionals

Sometimes it is necessary to "divorce" an Owner or Design Professional after a very bad experience. Hopefully this is the exception rather than the rule. The greater end result is to want to do more work with both the Owner and the Design Professionals. Working well together and leaving a good final impression are two critical steps toward that goal.

The next step, however, is to come back. By revisiting the project you may encounter some warranty issues, but you might also walk into some additional work. After an Owner occupies a space, better ideas emerge on how to function within that space. Sometimes those occupying the space who never had input into the final product have ideas that are worth pursuing. If the Owner has the funds, it is quite probable the revisit could result in some work being negotiated.

Also, it is often in these follow-up visits that the Contract Administrator becomes the first one notified of other projects being considered by the Owner in the future. Being on the inside track can help sales. It might lead to negotiated work, and if the whole team had worked well on the previous project, bringing the Design Professionals into the conversation might even lead to a Design-Build project delivery method.

This is the true culmination of contract administration. Fulfilling the contract requirements is the obligation of the Constructor, and the Contract Administrator must oversee that. However, from the moment the contract is signed, the ultimate goal is excellence in execution so that even more work and better projects will be the future result.

Figure 17.2 Woolworth Building at Night
Photo credit: Charles W. Cook

Frank Woolworth, the Five and Dime store merchant, was so proud of his office building that he had it printed on his store merchandise, from playing cards and dominoes to boxes of a dozen sewing needles selling for ten cents. Today, Owners still justifiably take pride in their structures, and Design Professionals look forward to the completion of their vision. In some instances Owners and Designers are seeking significant Green Building recognition and certification. It may not be a Woolworth skyscraper, but the Contract Administrator wants to keep the goal in mind that the final product is meant to be a team accomplishment worthy of the pride of all who made it possible.

A CASE IN POINT

The story goes that a hospital administrator had finally let a Contractor back in to do work on a trial basis after a twenty-year absence due to a dispute that had arisen over a previous contract, once again affirming that, regardless of who wins a dispute, the Constructor can still lose.

The Contractor was happy to be back with a good client, but then the CEO of the construction firm (Pete, again) got a call from the Vice President (VP) of facilities that someone in the company had gone around him. That was not an encouraging start to the conversation.

What had happened was that the President of the hospital had repeatedly asked the VP of facilities to get the City to fix the pothole in the middle of the street at the entrance to the hospital. The VP had gotten nowhere with the City for months, but that morning the President had seen a worker with the Contractor's sweat shirt on patching the pothole. When the President went out into the street and asked the worker how the City had finally approved fixing the pothole, the worker said, "Oh, I didn't get permission from the City. I just saw this pothole, and I didn't think it should be here in front of your hospital where people enter, so I decided to fix it myself."

The VP of sales concluded the conversation with the CEO of the construction firm with the comment, "The President figures you guys care enough to fix any problem, so now I will never be able to get rid of you."

Sometimes it is what you finish, not just when you finish.

17.2 Pareto rule

How did it get so late so soon?

Dr. Seuss

Also known as the 80–20 rule and named after the Italian economist Vilfredo Pareto, the **Pareto Principle** in the construction industry has been adapted to suggest that we spend 80% of our effort on the last 20% of the project. The Contract Administrator can circumvent the drain of time and energy by working early on the final contract issues and obligations. The more some requirements can be communicated and delegated to other parties, the easier it will be for the Contract Administrator, and almost always the final completion will move more quickly and smoothly than a scramble at the end to gather materials and complete the finishing touches.

Project closeout is indeed the ultimate responsibility of the Contract Administrator. If a "stitch in time saves nine," then addressing the ten issues listed in this chapter early and often will save 80% of the effort of the Contract Administrator being expended at the closeout of the project.

In order to do that, the Contract Administrator should consider:

- assembling a detailed list of project closeout requirements based on the contractual obligations as well as early conversations with the Owner and Design Professionals
- assigning contractual obligations to appropriate Subcontractors and Suppliers to fulfill all requirements relative to their scope and well before final completion

- whenever possible, delegating to other individuals the accumulation and assembly of materials and other contractual obligations such as training
- monitoring and confirming delivery or completion of items for project closeout on a timely basis with the Owner and the Design Professionals.

On large projects, the Contract Administrator might have a distinct advantage, since sections of the project may close out prior to the whole project. If this happens, the Contract Administrator has a mini-preview of the overall project closeout, and this can be used to train and prepare others on the project for final completion.

INSTANT RECALL

- From the beginning, the Contract Administrator will need to put every activity into context with the final project completion.
- Here is just a partial list of what the Contract Administrator might typically be looking for from the beginning of contract administration:
  - as-builts
  - training and instructional sessions
  - operation and maintenance manuals
  - guarantees and warranties
  - waivers
  - punch list items
  - completion notices
  - archival records
  - project review and cost documentation
  - follow-up with Owner and Design Professionals.
- It will be important that the Contract Administrator assigns someone from the beginning of the project to maintain and document all conditions and changes made in the construction that differ from what was shown on the drawings.
- The most important step a Contract Administrator can make after assigning someone to maintain the "as-built" drawings is to monitor progress on the documentation on a weekly basis.
- Once completed, the Contract Administrator must be certain the "as-builts" are delivered to the proper person.
- The key factors for the Contract Administrator in regard to training and instructional sessions are:
  - Review the contract with the Owner immediately for all training requirements.
  - Be sure training and instructional sessions are incorporated into the subcontracts and purchase orders of the specific Subcontractors and Suppliers that will be needed to give the training.
  - Identify as early as possible the representatives from the Owner who will be requiring training.
  - Schedule training sessions as part of the project schedule and, above all, do not wait until the end of the project to begin training.
- It is the important to secure all written documents, such as operational manuals, as early as possible.
- At the end of a project, the most common form of waiver is the waiving of rights to sue the Owner or lien the property for payments that have already been made.

- From the beginning, the Contract Administrator should request, and insist as strongly as possible, that there will only be one punch list kept for the project.
- One issue the Contract Administrator should also be clear on is that change order work is not a punch list item.
- No one expects perfection, but addressing punch list items in a timely manner to everyone's satisfaction will be a positive step toward future consideration and perhaps even negotiation of private work.
- Although the Owner and the Design Professionals may recognize when the work is completed, and the Owner may have even established Substantial Completion by moving into the structure, the Contract Administrator must recognize that the contract may call for specific notification of completion.
- Even though the Contract Administrator has confirmed that "as-builts" were delivered to the appropriate individual, it is best to maintain an archival record of the project for the Constructor.
- The Contract Administrator should consider:
  - assembling a detailed list of project closeout requirements based on the contractual obligations as well as early conversations with the Owner and Design Professionals
  - assigning contractual obligations to appropriate Subcontractors and Suppliers to fulfill all requirements relative to their scope and well before final completion
  - whenever possible, delegating to other individuals the accumulation and assembly of materials and other contractual obligations such as training
  - monitoring and confirming delivery or completion of items for project closeout on a timely basis with the Owner and the Design Professionals.

YOU BE THE JUDGE

A contract was awarded by a Developer (Burzynski) to a utility Contractor (Keith), containing Liquidated Damages at $150/day. By contract both parties agreed the damages were not a penalty. Keith was to perform all sanitary, sewer, water main services, earth work, utilities trenching, and manhole construction. The project was delayed a total of 83 days. The lower court divided the cost and assessed damages against the Contractor for 65 days, since Keith had been delayed by Burzynksi for 18 days while waiting for paving to be accomplished under a separate contract. After the paving was completed the final work on the manholes under Keith's contract was done.

On appeal, Keith asserted that it had stopped work because it had not been paid for work it had done and in addition it could not proceed with the manholes. Keith further asserted that dividing the Liquidated Damages was inappropriate on a basis of proportionate fault. Keith did admit that it had delayed the project due to lack of some materials, but Keith counter-asserted that delays were also due to Burzynski, and there was no need to prove such delays beyond the one already shown, since, once it is shown that one delay is caused by the other party, there is no longer cause for any Liquidated Damages to be assessed.

How do you rule:

In what amount:

IT'S A MATTER OF ETHICS

Imagine that you have been trying to get a project closed out for several weeks (sometimes it takes months), and you have finally completed everything you need to in order to meet the contract requirements. All inspections are completed, all punch list items are finished, the Design Professionals are satisfied and ready to sign off for the Owner's final acceptance. At this time you learn from a Subcontractor that the manufacturer of a critical piece of mechanical equipment is about to recall the equipment. This will not become public knowledge for at least two weeks, and by then you will have final payment and be clear of the project, which has been a bit of a nightmare for you. You are convinced you have spent 80% of your life on the last few weeks. Even worse, however, is that the piece of equipment was so large that you had to build part of the structure around it, and removing it will require considerable work, including some structural issues. You have checked the contract and specifications and it looks as though someone might even construe the equipment to be a "performance specification," and that could place the entire burden on you and/or the Subcontractor.

You are on your way to receive a check from the Owner as final payment. What do you do?

Going the extra mile

1. Complete the "You be the judge" exercise above and then go to the Companion Website for a discussion of the actual case.
2. **Discussion exercise I** (with friends or as part of a classroom discussion): The Pareto Principle was originally set by Joseph Juran, a business management consultant, to express the concept that 80% of the effects of our efforts come from 20% of our efforts. It has subsequently been revised, altered, and even reversed to suggest just about any 80–20 ratio. How true do you believe it to be? Do you accomplish more from less? Or do you sometimes feel you are spending more and accomplishing less? Is Juran right that we can rely on just 20% of our effort and accomplish more? What happens if we let things go until the last minute? Does that affect how much effort we have to put out in order to accomplish more? The same? Less?
3. **Discussion exercise II:** I know of one company that did in fact achieve "zero punch list" on several of its projects during one year. Unless you create a corporate and project environment, that cannot happen. Discuss among friends or within the class what it would take for a Contract Administrator to create a project environment to achieve a goal of zero punch list items throughout the project or by project closeout.

4. Go to the Companion Website for additional study aids, including flash cards and sample questions.
5. The Pareto Principle as first put forth by Joseph Juran was meant to have positive potential for management of projects and companies. You can consult many different management experts through searching the web. You might begin with searching the Pareto Principle at various sites on the internet.

6. Go to the Companion Website and watch Part IV of the Ethics video.
7. Effective use of the correct vocabulary improves communication and promotes respect among colleagues. The reader should become familiar with the key words listed below that have been used in the chapter. Definitions can be found in the Glossary. Additionally, the student is encouraged to use the flash card exercise in the chapter's Companion Website as practice for possible quiz or exam questions.

Key words

"As-built" drawings

Certificate of Substantial Completion

Consent of Surety for Final Payment

Pareto Principle

Waivers

CHAPTER 18

Dispute resolution

Avoiding disputes resolution is the purpose of all the previous chapters of this text. However, sometimes disputes will have to be resolved by a judge, jury, or some other neutral entity. This chapter will describe how the construction professional needs to prepare for and support the legal process.

Student learning outcomes

Upon completion the student should be able to. . .

- **recognize and understand different forms of dispute resolution**
- **apply principles of dispute resolution to best practices for different situations**
- **analyze and discuss the advantages and disadvantages of court, mediation, arbitration**
- **describe the role and responsibilities of contract administration in the dispute resolution process**
- **evaluate how to prepare for different dispute resolution processes.**

18.1 Dispute resolution or dispute management

Disagreement is something normal.

Dalai Lama

Many have come to consider the term "dispute resolution" to be inaccurate. They believe disputes are inevitable, a part of our lives. They are so much a part of our lives that disputes are continuous rather than resolved. For them, "dispute management" is the more accurate way to view or approach the process of handling disputes. However, for the Contract Administrator, although disagreeing might be inevitable, continuing to disagree over an issue is not at all beneficial.

The Contract Administrator should always look for ways to resolve conflict so that the team is stronger after the resolution than it was before. This is challenging, and success is never a certain outcome.

18.1.1 Conflict resolution modes

In the latter half of the twentieth century, Kenneth Thomas and Ralph Kilmann graphed an approach to conflict resolution that set a standard of five basic modes for resolving conflicts.

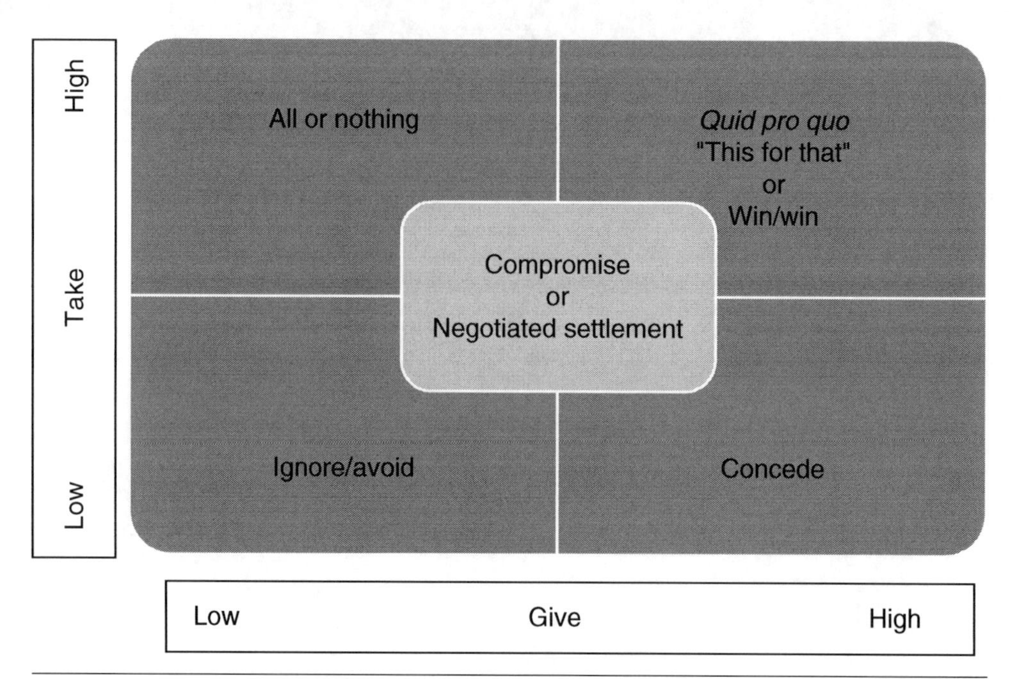

Figure 18.1 Five modes for resolving conflicts
Source: Adapted from Kenneth Thomas and Ralph Kilmann's "Five common modes of conflict resolution"

Since then, others have refined and redefined the five modes, but overall they remain five in number, based on how much or how little each side intends to give or take from the conflict.

18.1.1.1 Ignore or avoid the conflict

This is the territory of procrastinators, but it is also the realm of self-preservation. The first rule of self-defense for a master black belt is to avoid the confrontation. Walking on the wild side or "going where angels fear to tread" is for the foolhardy. Or, if the other side has a cannon and you have a pea-shooter, unless you are David versus Goliath, it is best to avoid that conflict.

Unfortunately, a Contract Administrator cannot ignore some conflicts indefinitely, and so a fair appraisal between what needs to be done and the best time to do it is necessary. Often the notice clauses will dictate certain deadlines for bringing potential conflicts forward. However, if "timing is everything," then sometimes "sooner is better than later."

Exercise 18.1

Take a moment
"Go" or "No go"

Write below three reasons you would wait to either inform someone you know well about a conflict or discuss it with them, and then three reasons you would immediately let someone know of a conflict.

Delay:

1.
2.
3.

Inform immediately:

1.

2.

3.

Discussion of Exercise 18.1

There are two conflicting aphorisms that are basically clichés I always wonder about when confronted with a challenge: (1) “He who hesitates is lost,” and (2) “Look before you leap.”

How each of us responds to a challenge is often deeply rooted in our own emotional makeup. At one extreme our reaction might be instinctive and even self-preserving, not to bring up a conflict. At the other extreme we might be aggressive to push forward for an immediate resolution. Whatever our natural response is, we need to consider if this is actually the best response, based on the circumstances.

The above exercise is not intended to give answers. Each situation will require a different “best response.” Factors that will figure into that response include:

- How strong or weak is your own position currently, and will it change in the future?
- What is your knowledge of the facts involved?
- How important is the resolution to you?
- When would the other party think you should have brought the conflict forward?

18.1.1.2 Concede

It is sometimes beneficial to give in on some matters. The expression “it is more blessed to give than to receive” has merit in many instances. The Contract Administrator might concede an issue because it makes little difference one way or another. Similarly, the Contract Administrator might concede an issue because there will be something more to gain later.

The real challenge for the Contract Administrator is to recognize and balance potential gain and loss, especially if the gain is only a possibility somewhere in the future.

The other possible drawback to conceding early is that it may be expected often. If a Contract Administrator concedes on one point, an Owner and/or the Design Professionals may expect that to be the standard for the rest of the project. It is important, therefore, that when conceding the Contract Administrator does fully express the justification of the Constructor’s position. Although this may sound overly cautious, it is possible that what was conceded in the past could come back in another ambiguity or latent defect later in the project, with far more consequences.

Example: A delay by the Owner may have been covered under the No Damages for Delay Clause, and the Owner denied an extra. Later it is necessary for the Contract Administrator to begin Constructive Acceleration without authorization for an extra from the Owner. The documentation that the delay took place will be important, even though at the time the Contract Administrator had moved on immediately after the delay without compensation. In essence, what the Contract Administrator initially conceded became a bigger issue once the full effects of the delay on the project were recognized.

18.1.1.3 *Quid pro quo* ("this for that," or win/win)

As we have already learned, this is Latin for "this for that." In the Thomas–Kilmann conflict resolution model this mode is termed "collaboration," based on the concept that everyone works together to find a solution. I prefer *quid pro quo*, on the basis that everyone gets something out of the resolution, but it is usually not the same thing. Sometimes it is an exchange of product for money. Sometimes it is a better product for additional cost. Sometimes it is a swap of something less important for something more important. It could also be just being appreciated or the pleasure of being in on something worthwhile and helping others achieve their goal – making dreams come true. In any collaboration, whether for money or for something else, if there is not a *"quid"* for the *"quo,"* then the success of the collaboration will quickly fade.

Whether money or pride in workmanship, each party to the collaboration needs a *"quo"* if some *"quid"* is to be invested, and vice versa.

Under most circumstances, this is the mode everyone would wish a team would function under, and in such an atmosphere conflicts would not pass the discussion phase. This is "construction in heaven," however. There will always be conflicts and the Owner, Design Professionals, Constructors and others involved in the process will need to find the best resolution possible. Final payment may depend upon it. Future work may depend upon it. Even one's health may depend upon it.

When two people think alike, the chance to improve is greatly diminished, if not eliminated. Change begins when one person sees something differently. The Contract Administrator should therefore encourage disagreement, at least in the form of free expression of ideas or different ways to do things. The key is to emerge from the disagreement better than before. A change in plan or schedule, an improved installation method, a better-specified product can all emerge if someone first asks "Why are we doing it this way?"

Example: For years, Constructors planned for large staging areas to store materials and the equipment to handle them. With the advent of Lean Construction practices, the concept of "just in time delivery" has returned to the construction process. To have effective "just in time" delivery, however, the entire team needs to coordinate, stay current on RFIs, and work together to maintain the tight schedule with all the other trades involved in the project. This is collaboration at its best and it certainly began with a *quid pro quo.* In most cases for the Constructor and Subcontractors that *"quid"* was if we do this right, we all get done sooner, and that means protection of profits, or even more profits – quite a *quid.*

18.1.1.4 Compromise or negotiated settlement

Unfortunately, this mode is usually considered by many Constructors, Owners, and Design Professionals to be the most useful and easily reached resolution in a conflict. Although that may be true, it is also true that compromise is usually the least satisfying and effective resolution mode. In almost all cases of compromise someone is not happy or there is a latent problem smoldering below the surface that will hinder progress in the future. In fact, it is often possible that neither party is happy with the compromise, and so the "battle-lines" are drawn for future conflicts.

A CASE IN POINT

Figure 18.2 Constitutional Convention of the United States, Philadelphia
Photo credit: United States Government

The United States Constitution is considered a remarkable document, but without what is called "the great compromise," no Constitution among the thirteen colonies would have been possible. "The great compromise" was necessary for the short-term gain of forming "a more perfect union," but buried within "the great compromise" was the scourge of continuing slavery in America. Four score and seven years later that portion of the compromise ripped the nation apart and cost over 600,000 American lives. Although the genius and remarkable accomplishments of the founders cannot be denied, the hypocrisy of their actions in relation to their words will forever compromise their personal legacies.

The problem with compromise in relation to *quid pro quo* is that although one expects it to be some form of win/win, it is not anything like the win/win of *quid pro quo*. In fact, it might even be considered lose/lose if both parties are dissatisfied with the result. In such a case compromise does in fact become conflict management, since there is no genuine conflict resolution.

Based on previous experience, some Owners and Design Professionals think the resolution will always be a compromise of some form of negotiated settlement. In the past they may have received inflated costs on change orders from Constructors. They then had to haggle and ultimately resolve the issue with a compromise and negotiation.

A Contract Administrator must recognize in almost every situation that the party with the most powerful position will win the most in the compromise or negotiation. It is also wise to keep in mind that an Owner holding a payment application and retainage is in a pretty powerful position.

Remember, in some conditions and conflicts, it is the strength of both parties that forces the compromise. This is typically the case in union negotiations with management. Each party believes it has a just position. Often they have inflated their demands just to seek a

Figure 18.3 Peter McGuire. A carpenter, and considered by many the "Father of Labor Day," Peter McGuire was a nineteenth-century labor leader who organized and campaigned vigorously for a standard eight-hour work day

compromise. Neither side is willing to fully concede to the other all of its demands. The matter is resolved, related to how willing management is to take a strike, and how long the union workers are willing to go without pay.

18.1.1.5 All or nothing (win/lose)

This mode is called "competing" by Thomas and Kilmann, but in construction it really is an "all or nothing" approach. The Contract Administrator intends to win at the expense of someone else. it might be a Subcontractor balking over some part of the scope of work. It could be a Supplier not willing to extend a warranty, or it could be the Owner and the Owner's agents not approving extra costs or extending the schedule. In some way, the Contract Administrator has determined that not to get their way on this issue would be disastrous.

Unless extremely well done, with tact and discipline, this conflict resolution mode will almost certainly sever relationships in the future. For that reason the Contract Administrator should probably consult upper management before proceeding too far into this "all or nothing" mode.

The Contract Administrator should make a very thorough examination of the strengths and positions of both parties. Proceeding down an "all or nothing" path may seem fine at first, but by understanding the position of the opposing side, one might realize that the "nothing" portion of the resolution may come to rest on the Constructor.

In this conflict resolution mode, the Contract Administrator must also recognize that the conflict will possibly escalate beyond his or her own control. In such a case the documentation and explanation by the Contract Administrator will be extremely important. It is at this stage that a decision to proceed may again be weighed in relation to the prospect of victory and both the short- and long-term costs of achieving that win.

Although all aspects of the project require good record keeping, it is in a win/lose – "all or nothing" – conflict resolution mode that the Contract Administrator's documentation will be most important, and possibly scrutinized by others. Since it is possible that some conflicts originally presumed to be capable of being resolved under another resolution mode might evolve into a win/lose ("all or nothing") mode, the Contract Administrator needs to always be prepared and up to date with documentation.

It is ultimately from the "all or nothing" that the rest of the discussion of this chapter will lead us. If the Contract Administrator cannot resolve the conflict, and the issue is great enough that it is more beneficial to gamble on winning rather than walking away, then some other party will be required to resolve what the project members (no longer a team) cannot.

* * * * * * * * * *

There is never just one conflict resolution mode available, and often different individuals will pursue issues differently. The Contract Administrator should develop the habit of pausing before each conflict and recognizing the best way to handle that conflict. In deciding what mode to use, the tactics might change, and it is possible the "messenger" might change. Knowing what approach to take might suggest that another person is more capable of handling the situation to a successful resolution.

* * * * * * * * *

18.2 Win/lose dispute resolution

If two friends ask you to judge a dispute, don't accept, because you will lose one friend; on the other hand, if two strangers come with the same request, accept, because you will gain one friend.

Saint Augustine

An ambiguity of some sort first appears, and it turns into a problem, which then can escalate to a conflict, and if not resolved at that stage turns into a dispute. There are several ways win/lose disputes are resolved, almost none of them better than if the conflict had been resolved before it became a dispute.

We will discuss in this section different resolution processes, and then in the next section we will discuss how the Contract Administrator should prepare for the three most common win/lose resolution processes – court trial, mediation, and arbitration.

Rather than court, most contracts call for some other form or a series of dispute resolution processes. In most instances arbitration of some sort is the final method of resolving what cannot be achieved through some other process. These have become known as **Alternative Dispute Resolution** methods or **ADR**. For the Contract Administrator, the formality of using any one method might be burdensome, but the chance of avoiding court or formal arbitration makes the effort, at least initially, worthwhile for all the parties involved.

The following discussion is based on how the different methods generally escalate toward court or formal arbitration. Any or all may be used, and the Contract Administrator might suggest attempting one even if it is not called for in the contract.

18.2.1 Escalation to higher authorities within the companies

This technique is highly recommended, and it can resolve what those in the "trenches" are not able to find agreement on, due to the pressures, personalities, and emotions that have evolved. Good contract administration means keeping upper management advised of difficult issues. Sometimes the Contract Administrator is advised to "get it resolved as best you can," but sometimes that becomes difficult, or even impossible, and it is amazing how different people react when they are doing the negotiation, as opposed to giving directions for someone else to play "hardball."

Often individuals higher in the organizations have a different rapport and relationship than those actually constructing the project. Having them resolve what has been a difficult situation for the field forces can be effective.

18.2.2 Mini-trials

As a further development to getting upper management involved, some projects use mini-trials before upper management. Both sides present their case to the upper managements of all the parties involved in the dispute. From that presentation, "cooler heads" are expected to see which side actually has a grievance and should prevail in the dispute. Of course, such a process might also reveal that both parties are only partially in the right, and both are partially in the wrong. At that point some form of compromise might be in order, even though it generally is not a preferred conflict resolution mode.

In some instances, mini-trials are held before neutral parties acting as the judge and/or jury. They are called in to listen to the different positions. Time is of the essence, so the whole process ideally takes place in less than a day. They are then asked to give a non-binding verdict. The theory is that the verdict from such a judge or jury will be similar to what a drawn-out trial might produce. It gives everyone a chance to reflect on just how worthwhile going further with the dispute will be.

Of course, once outside individuals are involved, the cost of the resolution increases. The cost of project personnel is one cost basically hidden or consumed in the project itself. Nonetheless, the resolution process takes them away from the productive work of overseeing actual units being put in place. Outsiders, moreover, will need to be compensated for their time above and beyond the project costs. Such expenses often are not figured into the compensation of the resolved dispute.

18.2.3 Dispute Review Board

More and more large projects are calling for **Dispute Review Boards**. These can consist of a panel of experts appointed at the beginning of the project, or even a single individual who has expertise in the structure that is being built. The purpose for such a board is to resolve disputes as immediately as possible. This prevents problems escalating into issues and problems between personalities that will continue to fester throughout the remaining project.

Once a problem has reached the level of an unresolved dispute, the outside review board is called in to the project to hear all sides of the issue and make a ruling on a timely basis. Bound by contract, the parties continue with the project, based on the ruling.

18.2.4 Mediation

Generally, the last step before arbitration is **mediation**. If successful, mediation can usually be compared to a compromise or negotiated settlement. In many instances mediation is required by the contract prior to submitting to arbitration. One factor the Contract Administrator might consider, however, is denying to participate in the mediation. If for some reason either party feels that mediation will be a waste of time and money, they can so stipulate, and often the requirement for mediation is waived.

Still, mediation is a worthwhile step in many dispute resolution processes. First, it could work out well and save the more expensive arbitration or court proceedings.

Second, during the mediation process the strengths and weakness of the opposing position become more apparent. If arbitration or court becomes inevitable, then the mediation process will have helped to expose and explore the positions of the other parties involved.

Third, it is possible through the mediation process that a portion of the dispute is resolved. If that is the case, then the arbitration of the remaining issue(s) can often take less time.

18.2.4.1 The mediation process

The mediation process is usually more formal than the previous processes discussed. The first step is to pick a neutral mediator out of a pool that is available. Different institutions and associations such as the American Arbitration Association have a list of trained mediators with expertise in specific areas. The mediator is expected to be impartial, and whatever discussions and offers that might have been put forward during the mediation process cannot be discussed if the process escalates to arbitration.

If travel or distance is an issue for the parties seeking mediation, it is possible to use online mediation or "e-mediation." The purpose and process are similar, but meetings with the mediator are held through internet connections that can include, if desired, video links. This form of mediation is becoming more popular for small mediation cases, since it can save costs in many ways, including travel, time away from productive work, and in some cases reduces or eliminates attorneys' fees.

Whether in person or e-mediation, usually the mediator will have an initial meeting with both parties present. This meeting will reveal the issues of both sides. From that point forward, the mediator will generally meet separately with each party, prompting discussion, points to consider, and usually shuttling ideas and offers back and forth between the parties.

The mediator is not expected to offer compromises or solutions. The better the mediator, the more each party generates its own ideas and offers. There is no requirement that an attorney be present. However, the more complex the mediation issues, the more likely attorneys will be involved.

What the parties have to recognize is that failure to reach a mediated agreement will result in much greater costs during the arbitration process. It is often wise for the Contract Administrator to calculate the cost of continuing with an arbitration versus making some sort of settlement during mediation. In addition to the time of all the parties that will be taken away from productive work during the arbitration, there will be preparation time, lawyers' fees (and lawyers' preparation time), the fees of the arbitration panel (sometimes a single arbitrator, but sometimes a panel of three), and the cost of filing and arbitrating the case through the American Arbitration Association or some other entity.

And, of course, victory is never certain.

Exercise 18.2

Take a moment
How much is enough?

Let us do a quick and dirty calculation on a ten-thousand-dollar claim to be tried before a one-person arbitration panel. (Note: the calculations might be different for court or mediation, and the cost of personnel will vary, depending on salaries and payroll burdens, but the exercise is a start.) We will use round numbers, and they could certainly vary from case to case, in different locations and depending on the issues involved.

Filing fee	$1,000
Payment for Neutral one day	$900
One day preparation – attorney 8 hrs x $300	$2,400
One day of arbitration for attorney	$2,400
Three days Contract Administrator (includes gather information, provide and counsel Attorney, and one day arbitration)	$1,500
Three individuals lose one day to testify	$1,500
Total	**$9,700**

1. On a strictly financial basis, do you decide to press forward with the claim?
2. Now recalculate the above based on higher amounts, such as $25,000 up to $100,000. Certainly the length of the arbitration might increase, and therefore your costs will increase, but figure at what point it becomes financially viable to press forward with the claim.

Discussion of Exercise 18.2

From a financial standpoint, it would be rather foolhardy to spend $9,700 in the hopes of winning back $10,000. It is possible in an arbitration or court case to have the costs of the arbitration, including attorneys' fees, awarded to the winning side, but you have to be the winning side, and usually you have to win significantly. So the hope of winning the $10,000 claim along with an additional $6,700 for the attorney and arbitration fees is a gamble.

What dollar figures you arrived at or how you escalated the costs for larger claims is not as significant as the exercise of looking at what you think it might cost. Against the costs you have to balance not your emotions, but the reality of your prospects for winning. A good Contract Administrator should calculate and express his or her opinions related to cost and the chances of prevailing in any particular claim.

The more the process can be reduced to mathematical analysis, the less likely emotions will be the driving factor in the decision to prosecute a claim.

18.2.5 Arbitration

Unlike mediation, **arbitration** is a binding dispute resolution process, just as a court trial is, but it is usually preferred over court proceedings. Judges prefer arbitration because it removes a lot of claims from their docket, and in most instances judges are not really expert in many of the nuances of the construction industry.

The American Arbitration Association is probably the most widely used and best known of arbitration associations. In addition to arbitration, it does have individuals who will facilitate mediation, but a mediator cannot be called upon to be an arbitrator in the same case.

For years, arbitration was not so universal. Up until 1925, when Congress passed the Federal Arbitration Act, many courts would not enforce an arbitration clause in a contract. Basically, judges looked upon arbitration as less professional than the court system and in fact as infringing on their power and livelihood, even as the dockets of courts were becoming backlogged and congested. Today, almost all states enforce the arbitration clause of a contract, provided it is equitable, meaning both sides have the right to bring a case before an arbitrator.

Today, arbitration is viewed as less expensive and quicker than court proceedings. It is still likely to be quicker than court, but the cost is not necessarily less. There are several reasons for this, but most relate back to the fact that the preparation phase of arbitration is now becoming more and more like the preparation phase for a court case. **Discovery** is the process whereby each party gets to "discover" what the other party has regarding testimony and evidence. It can be a very long and drawn-out process in the court system. Generally, discovery is not expected to be part of the arbitration process, but it is becoming more so in the larger cases. Add to the discovery process the review of the huge volume of e-mails and other electronic communications, and the preparation phase for arbitration balloons from days to weeks, or even months, of billable time for attorneys. Add further to the discovery process the taking of **depositions** from all or at least the key witnesses in some large cases (usually above a million dollars), and gradually the arbitration process becomes as expensive as a trial in court.

Still, arbitration has the reputation of being quicker and less expensive, and for that reason it is found in most contracts to be the method for settling any unresolved claims on a project. The Contract Administrator should check the language of the contract to make sure it includes language stating that all disputes are subject to resolution through arbitration. Some contracts have even included a box to be checked if the signatory parties agree to arbitrate rather than litigate. In such a case both parties to the contract need to agree to arbitrate, or the default resolution process might be litigation through the courts.

A Contract Administrator should also be cautious not to waive the right to arbitrate. This can be done for failure to give notice in relation to the contract, or it may be waived if the parties initially take the case to court. This can often happen if the issue was first presented as a lien against the property.

18.2.5.1 Panel versus single arbitrator

The size of the claim will usually dictate whether one or more arbitrators are appropriate or required for the case. Small claims of under $100,000 are usually heard by a single arbitrator. If the claim is large enough, one or both parties might want a larger panel. Selecting three arbitrators reduces the possibility of a single arbitrator just not understanding the issues. Three arbitrators can probably more effectively see the case from different perspectives. Each of the three can help the others in areas that he or she is more familiar with related to the different aspects of the construction.

While a single arbitrator will cost less, a panel might be worth the additional cost, based on the assumption that getting two out of three to agree with you is better odds than win or lose with one.

18.2.5.2 Overturning an arbitrator's decision

Arbitration is based on the substitution of the judge and jury by neutral individuals who have experience in both the arbitration process and the industry. Attorneys selected as neutrals usually follow closely the procedures of law, while Engineers, Architects, and Constructors rely more upon their expertise related to the industry than on all the technical requirements that would be found in a court proceeding. For example, if an attorney representing one of the parties in the arbitration objects to a particular question or something that is being entered into evidence, the non-attorney arbitrators will often allow the testimony or evidence with a phrase similar to "but we will give it only the weight it deserves." Even if allowing such testimony or evidence would be cause for appealing a verdict in a courtroom trial, it will not be cause for appeal or overturning an arbitration ruling.

Arbitration panels cannot deliberately or blatantly ignore the law, but a decision that is contrary to law cannot be overturned for that reason alone. The primary cause for an arbitration ruling being overturned would be demonstration of prejudice or some form of collusion between the arbitrator and one of the parties.

Because improprieties can happen, **ex parte** (away from the other party) communications or meetings are not to take place between the arbitrator and either of the parties to the arbitration without the prior knowledge and consent of the other party. This is also true regarding witnesses for either side in the case.

Examples of improper and ex parte actions:

1. An arbitrator seeks inclusion on an upcoming bid list from an Architect who is testifying on behalf of the Owner in a Contractor vs. Owner arbitration.
2. An arbitrator sees one party at a ball game and they decide that chatting over a couple beers is more entertaining than watching the game.

The exception to meeting without the other party present is when the other party does not communicate or acknowledge notices regarding the arbitration. In such a case the entire case may be heard without the presence of the other party. Such cases are sometimes "heard" or decided on a documents-only basis, without testimony from witnesses.

18.2.5.3 Selecting a panel of arbitrators

Typically, a group of names of neutral arbitrators is made available to both sides. They are expected to determine, based on the resumes of the arbitrators, which one or ones would be most beneficial to their case, while being able to understand the issues. For that reason a Constructor might want a Contractor for a single arbitrator, while an Architect might want an Architect, and an Engineer might want an Engineer. Choosing just one arbitrator can be a roll of the dice. In larger cases with three arbitrators sitting on a panel, there might be three different disciplines, such as an attorney, a Contractor, and an Architect or Engineer. It is also possible that every one of the panelists would come from a single discipline.

If, upon selection, an arbitrator recognizes there is a potential conflict, due to previous or even expected future relationships with one or more of the parties involved in the arbitration, then the arbitrator is required to disclose such relationships. It is then up to the two parties to determine if such disclosures would require another neutral arbitrator to replace the current one.

Table 18.1 Comparative dispute resolution processes

Type	Time	Appeal	Formal	Cost
Mini-trial, Dispute Review Board or other project site resolution	Usually very quick, normally within 30 days.	Almost always appealable, but could be binding by contract or project site agreement.	No.	Fairly inexpensive.
Mediation	Shorter than arbitration, and usually a time limit of a month is set from the time both parties agree to mediation.	Only if the parties do not settle the dispute. Binding on what is agreed to and settled.	Yes, to the extent a process will be followed and the final agreement set forth in writing, signed by both parties.	Modest, but a cost that has not been expected, and the settlement is usually not all that either party would have wanted.
Arbitration	Under $100,000, usually resolved within 30 to 90 days. Larger cases can take months, even years.	Only under very narrow circumstances, such as impropriety of an arbitrator.	Yes, but actual law and legal proceedings may not be followed as closely as in a court trial.	Moderately expensive in money, expensive in time for preparation and presentation.
Court trial	Usually months before case is called and sometimes years before final resolution.	Yes, from a lower court to a higher court, but good cause must be shown.	Very.	Expensive in both time and money.

In fact, this need for disclosure is so important, in order to maintain the validity of the eventual decision and not give grounds for overturning the decision as discussed above, arbitrators must continue to make any disclosures throughout the proceedings if witnesses or other issues arise that might be construed as affecting the impartiality of the arbitrator.

Before actually proceeding to prepare for a specific dispute resolution process, from project site resolution to the ultimate "day in court," a review of the time, cost, and ultimate finality of the potential processes might be helpful.

18.3 How to prepare for dispute resolution

This is a court of law, young man, not a court of justice.

Oliver Wendell Holmes

When some formal process of strangers deciding the outcome of a dispute seems inevitable, the Contract Administrator will be required to prepare for such a possibility, and in some cases the preparation involves different issues.

A Contract Administrator must also recognize that at the end of a dispute process a court trial rather than a less formal arbitration process may be the "end-game." In such a case, the rules of law will definitely apply, and the need to obtain competent counsel to guide the company and the individuals through the process will be important. Unfortunately, every Contract Administrator must also recognize that at the end of a court trial the law will have spoken, but justice may not have been served.

Those who trade a court proceeding in lieu of arbitration have the right of appeal, but the process is lengthy and not always successful. It becomes more and more financially draining, and the whole process is also a major drain on time for the participants. The Contract Administrator will have to be prepared for lengthy discussions and the collection of voluminous evidence if the proceedings are to be successfully concluded.

18.3.1 Preparing for a court trial

If the case is brought before a jury, the likelihood of the majority of the jurors understanding the construction industry is remote. Scheduling procedures, payment requisitions, construction sequencing and erection practices, and even basic vocabulary and expressions will not be familiar to the average juror.

It will be important to create **demonstrative evidence**, such as charts, photographs, and even videos showing the issues involved. Remember that the picture that is worth a thousand words is far better than the thousand words that will put a juror to sleep. Even the best-intentioned, including a judge, will not be able to follow the details the same way that a Constructor, the Owner, or a Design Professional who has lived with the project for months is able to understand the issues readily.

In the end, what seems obvious to a Contract Administrator may not be so obvious to someone listening from the judge's bench or the jury box. The Contract Administrator must assemble evidence that is simple to understand. Similarly, the witnesses need to address the issues in laymen's terms, simply and honestly.

One of the most important parts of the presentation will be the pricing of the claim. Jurors will need to be carefully shown the intricacies of the billing process, including the schedule of values, the percentages of completion, and the concept of retainage. They might also need to understand the issues of Liquidated Damages, as well as late or withheld payments and the reason for the same. While such items may be obvious to those who submit the documents, jurors will need to be carefully guided through any disputed applications and the accounting behind them.

For the Contract Administrator, imagine explaining a difficult subject, such as physics, to a middle school student, and you are well on your way to finding an approach for a typical jury.

A CASE IN POINT

Once upon a time I did find myself crossing the bar and sitting at the defendant's table. It was no fairy tale. Big dollars and bigger principles were at stake.

The judge kept us waiting for over forty-five minutes. That only increased all of our tensions, and looking back, I wonder if that is exactly what he intended.

Once he appeared, he surveyed the room, and that was the last he looked at any of us. He recited a blistering prologue he may have used many times before, advising us that if we were still foolish enough to roll the dice in his court in two weeks, we could return and he would hear the case.

Otherwise he advised us strongly to settle, because he was certain neither party would like his judgment when he got done. With that, he left us.

We settled.

And although one might think in a win/lose court trial situation somebody is happy, the truth is that the judge was right in most construction cases. For Constructors, court is not the Super Bowl or the World Series. There are very few outright winners.

18.3.2 Preparing for mediation

As mentioned above, prior to arbitration, mediation might be required, or at least suggested. Preparing for mediation actually requires some further thought than would be necessary for either court or arbitration. In both court and arbitration a third party or group (such as a jury or panel of neutrals) will be deciding the ultimate outcome. In mediation, the decision still rests with the parties involved in the dispute.

Basically, mediation is a chance to settle before incurring more costs in court or arbitration, so the Contract Administrator needs to calculate the cost of proceeding past mediation, factor the chance of prevailing (or even winning a partial settlement), and then recognize what would be the best outcome possible and compare that to what would be the least acceptable outcome that would still make mediation worthwhile.

Calculating the least acceptable settlement not only establishes a floor for the mediation process, but it will also help the Contract Administrator, as well as those backing the mediation process, to recognize when efforts toward a settlement have become futile.

While preparing for the mediation process, it is useful to consider what will come from an arbitration process not only from a cost point of view, but also from a time and potential of success perspective. Some refer to such an appraisal as considering one's **BATNA** (Best Alternative to a Negotiated Agreement). If you have not been able to reach an agreement, what comes next is worth considering. A Contract Administrator with a good attorney might win the battle and lose the war.

Generally, the weaker party has the least control in such a situation. The stronger party can dictate the terms with more authority, with the conviction that their side will prevail if the mediation offer is not accepted.

In the process of preparing one's case, therefore, the Contract Administrator should be as objective as possible in analyzing just how good are the prospects for success. Recognizing the strengths and trying to mitigate the weaknesses of one's own position are very important. For instance, contract language and the documents might clearly demonstrate an obvious ambiguity. There might be little doubt the **Spearin Doctrine** should work in favor of the Contractor. The Owner, however, might be satisfied that some exculpatory clause will be sufficient to negate the Spearin Doctrine. The Constructor recognizes that the weakness of their position is lack of timely notice and documentation of the ambiguity as soon as it was known. The Owner, however, realizes that on an **estoppel** basis payments and changes had been carried out on similar issues without the proper notice or documentation.

At this point the Contract Administrator should gather everyone on the project team to do a **SWOT** (Strengths, Weakness, Opportunities, Threats) **analysis**. If it is done well, the project team should have a clear understanding of what will arise in the mediation. Spending as much time as possible on analyzing the other party's position is vital. The weaknesses and threats that the project team faces will help to guide or point toward what type of settlement is reasonable or advisable.

Exercise 18.3

Take a moment
Take a SWOT at it

Take any team you are familiar with, such as a school team, sports team, lab team, project team, or perhaps even your family. Do a SWOT analysis of the entire team, based on what you know about the individuals.

Strengths:

Weaknesses:

Opportunities:

Threats:

Discussion of Exercise 18.3

As you go forward in the construction industry you will work more and more in teams. Usually they will be project teams, or estimating teams, or other business units. No team is perfect, and yet you will be expected to strive constantly for perfection. In order to do that you will have to honestly appraise the SWOT elements of your team, using the assets as best as you can while doing whatever is necessary to eliminate any weaknesses and ward off any threats.

Eventually you will become so good at SWOT analysis that you will do it instinctively and constantly, depending on the situations you are involved in, but until that time, keep the acronym handy and refer to it often.

A CASE IN POINT

No one is perfect. Nick is a great Engineer, and he has saved many buildings from structural defects. But he has told me he has on at least three occasions gone out into the field and seen problems with the steel erection that he could not understand.

Consulting the shop drawings and approvals, he realized that his own office had made some significant errors. His rapport and respect, however, among all who have worked with him is such that the solutions were never in doubt.

Nick spoke with the Contractors, acknowledging the error and his company's responsibility. He asked for the following terms to fix the problems. All actual costs for material and labor would be paid for by his company. He asked only that no markup be put on the costs.

Nick's career has spanned over fifty years, and just three such mistakes is really an excellent record, but the greater record is fifty years of respect and admiration that the entire construction community has for him. When people give so much of themselves, others do not hesitate to give back. That is a form of accommodation that ultimately leads to future win/wins.

Just one of Nick's accomplishments has been to save the tower of Independence Hall with minimally invasive structural support so that generations to come will be able to appreciate the great symbol and birthplace of the nation's founding without ever knowing it could have fallen over years ago. I know it could have, because my grandfather was asked a couple times to check on it well before Nick performed his structural steel surgery.

Figure 18.4 Independence Hall, Philadelphia
Photo credit: Charles W. Cook

If other parties, such as bonding or insurance companies, are involved, the project team and Contract Administrator might find they are under additional pressure to find a settled solution. In most instances, non-constructors involved in a mediation process are more inclined to settle rather than roll the dice in the court or arbitration on a construction dispute they might not fully understand. They usually approach each dispute on a dollars and cents basis, and to them settling the dollars makes *sense*.

18.3.3 Preparing for arbitration

More and more contracts are requiring mediation before arbitration. As mentioned above, however, mediation can be useful only if both parties want it to be. If one party refuses to mediate in good faith, then the whole process is often dropped and both parties go immediately to arbitration.

Although a SWOT analysis in preparation for arbitration is a good idea, it is not as imperative as it is in mediation. Nonetheless, preparing for arbitration should not be a "back burner" item. The Contract Administrator must consider the arbitration process as though it were a court case. In arbitration, there may actually be more leniencies regarding what will be admitted as evidence and what testimony will be allowed.

In some instances, contract language, such as exculpatory clauses, might be considered differently or even ignored as unfair by the neutral panel. An attorney will certainly caution against over-optimism or inflated expectations, but there is often far more latitude in the presentation of a case before arbitrators than in a court.

The Contract Administrator, therefore, should prepare thoroughly, including on issues and language which the attorney might state cannot be used. Careful consideration should be given to the attorney's observations, but the Contract Administrator should also know when an issue deserves some weight in the arbitration process. There may be issues of estoppel (the bar preventing a party from actions contrary to what they have previously done), industry practice, common sense, safety, or even impossibility that could override the language of the contract.

Attorneys understand the law. It is the Contract Administrator's challenge to make sure they also understand the vagaries of construction that have made the dispute inevitable. If the arbitrator or panel of neutrals can be shown that a reasonable interpretation by a competent Constructor was made, then success is not guaranteed, but it is a possibility. At the same time attorneys for the Owner and Design Professionals will have Contract Administrators helping them do the same – to present a case that shows their reasonable and competent approach to the project and the contract language.

Such a sobering reality brings us once again full circle to the final and ultimate conclusion that avoiding the dispute resolution process in the first place was probably the better solution.

INSTANT RECALL

- For the Contract Administrator, although disagreeing might be inevitable, continuing to disagree over an issue is not at all beneficial.
- The Contract Administrator should always look for ways to resolve conflict so that the team is stronger after the resolution than it was before.
- *Ignore or avoid the conflict* – this is the territory of procrastinators, but it is also the realm of self-preservation.
- *Concede* – it is sometimes beneficial to give in on some matters.
- *Quid pro quo* – this is Latin for "this for that." Everyone gets something out of the resolution, but it is usually not the same thing.
- *Compromise* – unfortunately, this mode is usually considered by many Constructors, Owners, and Design Professionals to be the most useful and easily reached in a conflict. Although that may be true, it is also true that compromise is usually the least satisfying and effective resolution mode.

- Win/lose – "all or nothing" – the Contract Administrator intends to win at the expense of someone else.
- Escalating the conflict resolution to a higher authority within the company is highly recommended, and it can resolve what those in the "trenches" are not able to find agreement on due to the pressures, personalities, and emotions that have evolved.
- Often individuals higher in the organizations have a different rapport and relationship than those actually constructing the project.
- Mediation is a worthwhile step in many dispute resolution processes.
  - First, it could work out well and save the more expensive arbitration or court proceedings.
  - Second, during the mediation process the strengths and weaknesses of the opposing position become more apparent. If arbitration or court becomes inevitable, then the mediation process will have helped to expose and explore the positions of the other parties involved.
  - Third, it is possible through the mediation process that a portion of the dispute is resolved. If that is the case, then the arbitration of the remaining issue(s) can often take less time.
- What the parties have to recognize is that failure to reach a mediated agreement will result in much greater costs during the arbitration process.
- Unlike mediation, arbitration is a binding dispute resolution process, just as a court trial is, but it is usually preferred over court proceedings.
- For years, arbitration has been viewed as less expensive and quicker than court proceedings. It is still likely to be quicker than court, but the cost is not necessarily less.
- The size of the claim will usually dictate whether one or more arbitrators are appropriate or required for the case.
- Arbitration is based on the substitution of the judge and jury by neutral individuals who have experience in both the arbitration process and the industry.
- In a court trial, the rules of law will definitely apply, and the need to obtain competent counsel to guide the company and the individuals through the process will be important.
- Preparing for mediation actually requires some additional thought than would be necessary for either court or arbitration.
- While preparing for the mediation process, it is useful to consider what will come from an arbitration process, not only from a cost point of view, but also from a time and potential of success perspective.
- Attorneys understand the law. It is the Contract Administrator's challenge to make sure they also understand the vagaries of construction that have made the dispute inevitable.

YOU BE THE JUDGE

Commerce Bank (Commerce) had contracted with DiMaria Construction, Inc. (DiMaria) to renovate offices. During the course of the construction, Commerce removed DiMaria from the project, claiming delay. DiMaria sued for the amount of construction work done, along with attorneys' fees. DiMaria was late in accordance with the contract filing for arbitration, but the arbitration went forward. The arbitration panel of three awarded DiMaria $250,000 plus attorneys' fees of $64,554.

Commerce appealed the award on the basis that the arbitrators had no authority to rule on a case that had not followed the initial legal requirements. This was clearly an error of law made prior to the panel hearing the case. In addition, Commerce argued that the award was "outrageous." It should not be held responsible for paying such a sum.

How do you rule?

IT'S A MATTER OF ETHICS

In the above exercise you calculated the financial issues related to the cost of pursuing conflict resolution in court or through arbitration. Under what circumstances, if any, would you be prepared to seek court or arbitration on the basis of principle rather than cost?

Going the extra mile

1. Complete the "You be the judge" exercise and then go to the Companion Website for a discussion of the actual case.
2. **Discussion exercise I** (with friends or as part of a classroom discussion): In Plato's *Republic*, Socrates is depicted as arguing with a sophist named Thrasymachus, who contends that justice is nothing more than the interest of the stronger. Is Thrasymachus correct? Are there circumstances under which Thrasymachus is correct even though it might not be right? Do the interests of the strong by necessity become evil or bad? If not, could the interests of the strong be good for all?
3. **Discussion exercise II** (with friends or as part of a classroom discussion): Discuss situations where only one conflict resolution is appropriate. For example, anyone not doing things safely on the jobsite has to lose to your win. Safety is an "all or nothing" resolution mode where the nothing means that the individual or company not complying leaves the project. Sometimes it is imperative to avoid the situation. As Douglas MacArthur pointed out, "We are not retreating. We are just advancing in the opposite direction." Come up with your own list of single conflict resolution mode situations.
4. **Discussion exercise III:** Discuss situations in which different conflict resolution modes could be used effectively, and the reasons why. For instance, in the negotiation process with a client, some might wish to accommodate all the requests of the Owner to show how receptive and responsive the company will be. Others might want to be polite but firm, to show confidence and set the tone for how the project will go forward with deliberate and effective management.

5. Go to the Companion Website and watch Part V of the Ethics video.
6. Go to the Companion Website for additional study aids, including flash cards and sample questions.
7. Effective use of the correct vocabulary improves communication and promotes respect among colleagues. The reader should become familiar with the key words listed below that have been used in the chapter. Definitions can be found in the Glossary. Additionally, the student is encouraged to use the flash card exercise in the chapter's Companion Website as practice for possible quiz or exam questions.

Key words

Alternative Dispute Resolution (ADR)

Arbitration

BATNA

Demonstrative evidence

Depositions

Discovery

Dispute Review Boards

Estoppel

Ex parte

Mediation

Spearin Doctrine

SWOT analysis

CHAPTER 19

Ethics

For some, "business ethics" is an oxymoron. Unfortunately, that might be more accurate than humorous.

For the first eighteen chapters of this workbook, we have been focused on legal right and wrong. In addition to civil law, based on our common law system, we have also looked at tort law and statutes and regulatory laws. This chapter will deal with another set of right and wrongs. Ethics involves what is right, regardless of whether there is a law defining it as right.

Ethics is a system of directed conduct that helps individuals and groups attain success through specific behavior, but such behavior cannot deliver benefits to one party at the expense of another. Many associate ethics with morals, laws, and codes of conduct, but the most effective application of ethical behavior is determined by how well it improves the interaction of those who practice it.

Student learning outcomes

Upon completion the student should be able to. . .

- **understand the importance of ethics in a professional environment**
- **recognize how personal ethics and group ethics are distinguished but dependent upon each other**
- **interpret ethical behavior in regard to group interaction**
- **analyze personal ethics in order to improve personal success**
- **create approaches to help with organizational ethical transformation.**

19.1 Ethics

When skating over thin ice, our safety is in our speed.

Ralph Waldo Emerson

For many in construction, as well as in business and in life itself, the standard operating procedure is "whatever one can get away with is fine." Ethics is for the other person. But sooner or later we skate too slowly or find ourselves stopped on thin ice. The result can be minor or catastrophic. Unfortunately, in an industry as dangerous as the construction industry, with such narrow margins of profit, the likelihood of "catastrophic" is almost a certainty at some time. Cheating others is a good way to assemble a large following of those who will cheat you back. Expecting someone to rescue you after you have stepped all over them on your selfish rise to what you thought would be the top is a good way to discover that everything is now up to you. Trying to stay ahead of disaster while dealing unethically throughout your career is a sure way to be constantly looking over your shoulder until you finally fall over the cliff you did not see in front of you.

Each of us has limited time and multiple prospects. We all must, therefore, make choices, and it is the result of the choices we make that affects our own success and our relationships with others. The Contract Administrator has a responsibility to see that the contract is fulfilled. That places additional responsibility upon the Contract Administrator in relation to Owners, Design Professionals, Subcontractors, Suppliers and the workers involved in the project. It also, however, places a responsibility upon the Contract Administrator for personal success.

Exercise 19.1

Take a moment
To take a look in the mirror

Rank on a scale of 1 to 3 the following traits and skills in relation to how important and how well they are part of your actions. Use 1 to indicate important, 3 to indicate less important (but not necessarily unimportant). In the same way, rank on a scale of 1 to 3 always part of your actions, and or only sometimes a part. A 2 will represent middle of the scale.

Trait	*Important*	*Part of my actions*
Fair		
Dependable		
Reasonable		
Trustworthy		
Honest		
Caring		
Hardworking		
Helpful		
Team Player		
Organized		
Punctual		
Opportunistic		

Discussion of Exercise 19.1

It is possible that you gave each trait a 1 in relation to how important it is, but hopefully no one gave a 1 to each of the traits in relation to how much a part of your actions that trait is. For example, I am sure that if we are honest with ourselves, procrastination does affect how *punctual* we sometimes are, and if we are late or ill-prepared on some important matters that we have put off too long, then that will also affect how *dependable* we are.

None of us is perfect, and it is recognizing what we can and need to improve that allows us to move forward to better days and more success. Although good fortune can play a role ("opportunistic" is last on the list for a reason), we all must realize that we cannot wait for fate to deal us a good hand. It is up to each of us in contract administration to apply our best characteristics to fulfill our obligations.

An honest analysis of traits in Exercise 19.1, therefore, will be a start for the reader to recognize what traits might be improved. Using the method suggested by Aristotle in his *Nicomachean Ethics* (discussed in the Ethics video on the Companion Website) can be an excellent way to implement improvement in one or two traits. The reader might also apply one of the above traits in Exercise 19.2 later in this chapter.

The word *ethics* comes from the Greek word ("ethos") for "character." It was Heraclitus who pointed out that "character (*ethos*) is destiny." And thus, we might assume, "ethics is destiny." How we conduct ourselves through life and our careers will have a major impact on our destiny. Certainly there will be networking opportunities, mere chance acquaintances, perhaps a single "glance across a crowded room" that might change our fortunes and our lives, but more than anything, the relationships we make and keep will be determined by our character – our *ethos*.

In this Greek derivation, ethics becomes personal, but we must recognize as well that ethics is also a system of behavior that a group might impose upon itself. The guiding ethic of medicine comes from the Hippocratic oath of "First, do no harm." From that principle many others are derived for those who practice medicine, but just because there is a principle, there is no guarantee that it will always be adhered to by every individual within the group.

The same is true within the construction industry. Professional organizations related to the construction industry have codes of ethics or business conduct codes. Owners and Designers may want to see a written code of ethics as part of the qualifying process for Constructors on a bid list. The **Federal Acquisitions Regulations Act (FAR)** requires Contractors working on Federal projects to have a written code of ethics that is both distributed to and complied with by the employees of the firm.

Contractors are expected to:

1. promote compliance with the code of business ethics and conduct that is suitable for the size and business the company performs in relation to government contracts;
2. train employees about the code;
3. have an internal control system to monitor training and compliance and to disclose in a timely manner improper conduct;
4. ensure corrective actions are taken to eliminate recurring issues of non-compliance.

For individuals, compliance with group ethics can sometimes be a personal challenge. Each member of the industry also needs to recognize early on in their career how personal responses are compatible with the group ethics they work with on a daily basis.

Examples:

1. It is a challenge to be honest with other members of the project team when others around you are not treating everyone the way you believe they should.
2. If you are faced with an incompetent or dishonest member of the project team, the ability to deal with such an individual is usually quite difficult because schedule or even budget may be dependent on successfully working with that person. Such a relationship will certainly create challenges for your own behavior in relation to that person.
3. Pressures placed on you by others not performing as needed or expected can add to your own burden to accomplish more than expected or within your concept of what is your duty or responsibility.

It is important, therefore to understand both the personal aspects of ethics as well as the group dynamics of ethics.

A CASE IN POINT

Things don't change overnight.

When I first started working for my father, I began in the field. When I was moved into the office, I soon experienced the "mayhem" of bid days and the "buying-out" of projects. Bid shopping in all its forms was rampant among the estimators. I am sure it was part of the corporate culture before my father bought the company, and I also believe my father condoned it, because, coming from larger construction firms, he had witnessed it as the rule rather than the exception.

I was uncomfortable with the whole process. I never believed a company with a reputation of bid shopping got the best efforts or prices initially from those bidding to us. As the young and naive just-out-of-college kid still finding his way in the company, let alone the industry, I had no chance to change the "corporate ethic." Estimators with twenty to forty years of experience were not about to listen to me.

I believed my approach had to be guided by what was possible. As I became more involved in the estimating, I established among those bidding to us that I did not shop prices. At first a few Subcontractors got irate with me when I awarded projects to other companies whose prices I had used to submit the bid. The irate Subcontractors complained I had never called them back to "sharpen their pencil" or give them last look. Slowly and surely, the word got out that when you bid to me, give your final best price.

Eventually, after years of experience and considerable success, the corporate policy became "no bid shopping." I believe it ultimately improved our success ratio, but I do know for certain that it reduced stress and improved relations.

19.1.1 Personal ethics

The more a Contract Administrator's time is consumed in the details and challenges of the project, the less time one has to take a moment and contemplate how good things can come from ethical conduct. Personal ethics is meant to define specific behaviors that will guide us through successful interactions with the people and situations that we confront throughout our lives, but it cannot be a "spur of the moment" thought sequence. For this reason we can understand how it must be closely associated with our own character. Something as simple as whether we are aggressive or shy will have an effect on how we relate to others.

On the surface, being aggressive or shy may not seem to have anything to do with ethics, but it might ultimately have an important influence. Someone who is overly aggressive might step across the boundaries of cheating in order to get their way in a situation. Winning might become more important than doing the right thing. On the other hand, someone who is shy might not point out an important issue and the project schedule, cost, quality, or, even worse, safety might be affected.

Examples:

1. Bid shopping is an example of how overly aggressive individuals get a thrill out of manipulating others to give them lower prices.
2. Seeing a row of uncapped vertical rebar and hoping someone else will take care of the safety issue can seem minor, since no one probably will fall and impale themselves on it, but safety is everyone's duty in the construction industry.

It was with personal behavior that Aristotle first began to study ethics. He disagreed with his teacher, Plato, who had followed his own teacher, Socrates, in believing that by knowing right from wrong one would do what is right.

Aristotle thought that following right from wrong, for the average human being, was far more complex than simply that, knowing right from wrong, one would or could always do right. He believed that we can know right from wrong, but we still might do wrong. He recognized that people are driven to succeed in their own perspective. Such a perspective can, of course, be positive or negative.

Aristotle then wanted to develop an argument in favor of the positive. He wrote in his *Nicomachean Ethics* that people can improve their personal virtues until they become habits. If we practice positive habits they will ultimately lead to greater success and happiness. For some, this might seem a little too good to be true, but, following Aristotle's concept just a little bit further, we can see that there is a greater chance of positive interaction with others by being virtuous than there is by being a scoundrel.

About seventeen centuries after Aristotle, Machiavelli pointed out that scoundrels have great success, but in a crisis they are on their own. Everyone deserts those who mistreat them as soon as the opportunity presents itself.

In business, particularly one as difficult and dangerous as construction, building friends that will stick by you in the difficult times can be very important. Virtuous behavior, therefore, can have a very important and positive effect on the career and success of the Contract Administrator.

19.1.1.1 Aristotle's method for improving ethics

An important aspect of Aristotle's concept of ethics is that each human being seeks happiness. We are guided by what we find desirable. We seek being happy, and we try to avoid pain or what makes us unhappy.

The essence of Aristotle's method is to grow virtuous personal behavior by first finding the ideal mean between two excesses. Make that mean into a virtue. Practice that virtue until it becomes a habit, and then enjoy the success that the habit will bring.

Mean

↓

Virtue

↓

Habit

↓

Excellence (happiness and success)

Figure 19.1 Aristotle's Path to Excellence

In order to find the mean to turn into a virtue, we must first discover the two extremes of what prevents individuals from achieving the mean. He considered one extreme to be an excess of the intended virtue, and the other extreme to be a deficiency or total lack of the virtue.

In the "Going the extra mile" video, the mean leading to the virtue of courage is used as an example. In this discussion, let us use another of Aristotle's examples: the mean of being generous. Such a virtue could lead to improved relations, but of course it could also be abused by others who would take advantage of someone who is too generous.

Aristotle defined the two extremes of such a virtue as being wasteful with one's resources at the one extreme and being stingy with what one has at the other extreme.

Wasteful Generous Stingy

Going back to the conflict resolution modes, the Contract Administrator might recognize how this type of mode fits with the accommodation mode. At one extreme, not helping others when you can do so will lead to poor morale and bad attitudes toward you among the team members. At the other extreme, if you keep giving things away, you will waste both time and profits on those who will continue to take advantage of you.

For each of us, however, we must also recognize that even though each virtue that we turn into a habit is personal, it is best to understand its potential in relation to how it will improve our ability to successfully work with others, to complete our projects, and to grow our companies.

Exercise 19.2

Take a moment
Making a habit

Decide on a new virtue or, more likely, an existing one you would like to improve.
Determine the two extremes of that virtue, which should be the "Golden Mean" between the two excesses. Take time to think where you actually fall in relation to the extremes, and determine what you will need to do to move toward the "Golden Mean."

Insufficient extreme Golden Mean Excess extreme

What do you need to do to move toward the Golden Mean and make it a habit?

Discussion of Exercise 19.2

It is sometimes difficult to really identify the two extremes, so be sure that you have taken the time to know what they are so you can better tell where you are.

For instance, it would be easy to say the insufficient extreme of honest is dishonest, but that does not help as much as actually identifying the extreme trait as "lying." Now that we see that as the extreme, we might recognize that we do not lie, but we sometimes do not tell the truth, or the whole truth, or we bend the facts to meet our purposes, and so we do have some distance to go to reach the Golden Mean of honesty.

However, at the opposite extreme of excessive honesty there is the trait of being too outspoken. Molière wrote a comedy entitled *The Misanthrope*, in which the lead character never knows when to "keep his mouth shut." He tells everyone exactly what he thinks, and in the process not only hurts everyone's feelings but destroys friendships.

> **There is a virtue in knowing when to engage the brain before motoring the mouth.**
>
> **As you do review your strategy for developing a new virtue, realize that change does not come easily and new habits take time to develop. Some believe we must consciously apply ourselves for about a month before we develop a new, positive habit.**
>
> **With a sound understanding of what we want to accomplish, along with diligent practice whenever the occasion arises, and sometimes with the help of others, improving on a virtue until it becomes a new habit for our own success is possible.**

The importance of identifying the virtue is to put it into practice as a habit. Without the virtue being a habit, it is not likely to be effective. Knowing what to do is fine, until the mountain begins to fall upon you. Knowing your mean for courage, generosity, responsibility, or other virtues is not enough when you are confronted with a great challenge to that virtue. Making it a habit, an automatic response, is far more likely to lead to success when the situation becomes difficult. It is easy to be virtuous when the stakes are low, but much harder when the price goes up.

19.1.1.2 John Locke and the pursuit of happiness

The debate about knowing good from bad, and doing what is right with that knowledge, continued for centuries. Aristotle's observation that people do what makes them happy was what stood in the way of those who wanted to impose universal ethics (or morals) upon individuals who could easily understand right from wrong.

Eventually the English philosopher John Locke would pose an important observation that would become a foundation for American thought and ethics. He believed that ethics were related to experience, and therefore they might shift, depending on experience, but the foundation of all human experience was three fundamental rights – life, liberty, and the fruit of one's own labor in the constant pursuit of one's happiness. In close agreement with Aristotle's concept of motivation, Locke's premise would be set into beautiful prose by Thomas Jefferson, and it would become the foundation not only of the United States Declaration of Independence, but the ultimate argument by Abraham Lincoln for ending the horrors of slavery in the United States. Finding no legal justification to end slavery in the Constitution, Lincoln would rely upon Jefferson's ". . . all men are created equal, that they are endowed by their Creator with certain unalienable Rights, that among which are life, liberty, and the pursuit of happiness."

Unfortunately, one person's pleasure might be another person's pain. While the words of Jefferson may inspire the citizens of the United States, and even the world, it is also sobering to realize that the very struggle that consumed Lincoln's presidency was part of Jefferson's own world and the practice of slavery was how Jefferson's beloved Monticello estate was maintained.

Locke may have crystalized Aristotle's concepts of how humans are motivated to seek their own happiness, but as long as the perspective of each person's happiness differs, personal ethics connected strictly to an individual happiness will always be a moving target. Although Aristotle can point us in a direction to improve our own success through improving our virtues, there comes a time when we must go back to Socrates' concept that we can know right from wrong, and in so knowing we should choose doing what is right, even though we may not see how it makes us happy in the short run.

It was Lincoln, himself, who might provide us with a greater insight into the dilemma of what constitutes personal ethics. Criticized for not belonging to a specific church, Lincoln responded: "I will join any church that truly practices the Golden Rule."

19.1.2 The Golden Rule

Exercises to improve our virtues that in turn are expected to improve our success and happiness are all valid, but perhaps time consuming. The Contract Administrator faced with a difficult situation during the progress of a project may not have time to develop a new virtue. We cannot download virtues as an app every time we need one.

In such a case, the one universal that everyone can turn to is the Golden Rule. Underlying the judgment of all human interaction is the concept of fairness. We probably learned this from our parents or our kindergarten teacher, but along the way we all came to understand the simple phrase:

> Do unto others as you would have others do unto you.

It still works. On a good team every member of the construction team, from Owner through Design Professionals to the Subcontractors and Suppliers, will appreciate and respond to how you treat them in the same way that you will respond to them based on how they treat you.

Every Contract Administrator may have difficult issues to handle, but how they are handled will be the key to how well the difficulty is resolved. Everyone on the project team should want to find solutions. If that is not the case, why are they on the project team? When a problem arises, think of how you would want to be treated, and then proceed as quickly as possible to find a solution.

Combining the best of Aristotle's understanding of human nature with the correctness of Socrates in knowing what is right, we merge the pursuit of personal happiness with the practice of the Golden Rule. In so doing, perhaps we come to the final conclusion that personal ethics is the striving for the good life by striving to do what is good.

This then, is not only the strength of personal ethics, but the foundation of all group ethics and codes of conduct.

19.2 Group ethics and codes of conduct

What is moral is what you feel good after, and what is immoral is what you feel bad after.

Ernest Hemingway

In business there may be many ways to feel good without doing good. In fact, we have been indoctrinated that some things that seem natural or justified are the way to proceed. We all understand that "competition is good – it makes us all better." But there is a time and place for us all to be better together. More often than not, particularly on a project team, my winning does not depend on your losing.

Too often, companies publish a mission statement, a statement of values, a code of conduct, or some other form of inspirational thoughts and ideals that few people read and almost no one ever remembers. The primary purpose for most companies is to include their code of conduct and/or mission statements in their advertising and sales pitches. This is unfortunate.

A CASE IN POINT

Figure 19.2 The Great Wall of China
Photo credit: Kristian Reid and Cory Schmitt

Certainly the greatest "horizontal" structure ever conceived and built is the Great Wall of China. It was meant to keep marauding hordes coming out of the north from entering the land of China. As an obstacle, it was indeed formidable, but as a solution, it was a failure.

On three occasions those who intended to invade China did so without engaging in battle to cross the wall. They merely bribed a guard each time and walked through the wall, unopposed.

No matter how many safeguards we build into the codes of conducts, mission statements, and ethics of our companies, we are only as strong as the commitment of everyone to doing the right thing.

19.2.1 Types of group ethics

Complicating the issue or subject of ethics is that there are different views on what truly constitutes ethical behavior. As simple as the Golden Rule might sound, it can become more complicated once money or other assets of extreme value are involved. Groups and companies establish practices within their own "inner circles" that might never be put into writing, and probably would not be advertised to Owners and Designers.

19.2.1.1 The utilitarian approach

John Stuart Mill and Jeremy Bentham were English philosophers of the nineteenth century. Their approach to ethics and decision making in relation to business and life was based on doing the most good for the most people. This became known as **utilitarian ethics**. However, it also has an immediate built-in flaw. That is, who determines what is best or "most good," and who decides what are enough people to constitute the "most people?" Under a utilitarian approach, different individuals might have very different ideas.

19.2.1.2 Shared moral values

In most cultures, workers have a common set of work ethics that go back through their parents and even grandparents. These are often founded on moral principles.

As construction becomes more and more a global industry, Contract Administrators must also recognize that different countries do not share the same moral values, at least not in business dealings. There are stories of how, in some countries, bribery is the only way to get something accomplished. Or Design Professionals may wish to retain future rights to designs, but copyrights in some countries are seldom enforced.

Expecting business values to be universal is not going to be a successful approach in a global industry.

19.2.1.3 Categorical imperative

Immanuel Kant was a German philosopher who was adamant about everyone following a common set of rules. This has become known as the categorical imperative. He firmly believed, as Socrates had pointed out centuries before, that everyone could know right from wrong. Once right is known, it was essential that everyone should follow and comply with what is right. He insisted that what was right for one would be right for all.

This sounds very reasonable until one actually looks at the details. When is it right to exceed the speed limit? If you have a person about to give birth in your car at 3:00 AM, do you ignore the red light when you determine no one is coming? Taking it to a more complex level, we recognize that many prisoners have been set free from death row after DNA analysis of the evidence. However, DNA analysis of evidence has not been available until recently. Is it possible that one or more innocent persons were killed for a crime they did not commit? If that is true, do we reconsider capital punishment?

One of the mistakes of the twenty-first-century mind over that of the twentieth-century thinker is that we tend to think we now know it all. Unfortunately, we are each limited by our own perspective. We all have our unique experiences, and they form the basis of our reaction to the world around us, including what is right and wrong. Someone in the tropics would not understand through experience how easy it is to walk on water, but someone in Wisconsin in the winter would not give it a second thought.

Who determines what is right and wrong can also have devastating effects. A government might impose rules and codes that are not wrong but, without them, the government itself might suffer. This is basically true of any government. Benjamin Franklin conceded that taxes were as certain in life as is death; but how much taxation is fair, and who really determines the level of taxation, particularly if it is without representation?

Even something as basic as the ethics of environmental concerns can vary considerably from one nation to another. How can one balance the need to protect our planet's resources in a country of plenty with the same restrictions in a country where the immediate needs of the majority of the population are such that they are malnourished, underemployed, and lack adequate healthcare?

19.2.1.4 The Platinum Rule

Some have rewritten the Golden Rule to become the **Platinum Rule**, presumably more precious than gold. The Platinum Rule states "Do unto others as they would have you do unto them." This, of course, becomes disastrous and complete nonsense. If you would like to be continually taken advantage of, follow the Platinum Rule, but that will quickly lead to bankruptcy as others take advantage of you, and there are always those who will.

Being overly generous and giving value away without a return may be one way to get to heaven, but in business it will mean the poorhouse. It is well known that greed is considered "one of the seven deadly sins," and no one would propose that that should be a motivating force behind the actions of anyone, but self-interest is a key motivator. We do what we do because it is good for us. As long as what is good for us is not bad for others, then self-interest is reasonable. In fact, if what is good for us also leads to what is good for others, then self-interest is excellent.

19.2.2 Effective group codes of ethics

The purpose of group ethics or codes of conduct is to ensure that the group virtues will lead to ever greater success with those customers who recognize the value of such virtues. Businesses can use codes of ethics or conduct to communicate to employees the standards the company expects from their behavior. There are several major areas or categories that can be addressed in codes of ethics, including:

- honesty
- safety
- quality
- relationships with clients, vendors and the public
- fair practices in relation to products and actions
- financial goals and expectations, especially in relation to reporting
- contracting practices
- sharing information within and outside the company
- environmental concerns.

If the purpose of business ethics could be contained in a single word, it would be "trust." The greater the trust a Contract Administrator can build with and for each member of the team, the more effective and successful the actions and interactions of the team will be. Of course, the opposite is also true. Once trust is lost, the teamwork disintegrates and the project is in deep trouble.

In "Going the extra mile," the reader will be encouraged to check codes of conducts and ethics of several associations within the construction industry.

Examples:

ASCE – In viewing its immense responsibility not just to the Construction team, but to the entire general public, the American Society of Civil Engineers has put out a code of ethics that includes fundamental principles, along with several canons of conduct, and then each canon includes several guidelines for conduct to fulfill the principles of that canon.

CMAA – The Construction Management Association of America provides first-person pledges that each member is expected to adhere to in issues regarding client service, qualification and availability of services, fair competition and fair compensation, conflicts of interest, and professional development.

AIA – The American Institute of Architects has established different levels of behavioral goals and conduct to include canons (broad principles related to behavior and conduct), ethical standards (specific goals for member to guide their behavior and conduct), and rules of conduct (required conduct, for which a violation will lead to disciplinary procedures).

In the best companies, ethics is not a sometimes thing, nor is it merely a notice upon the wall. A good ethics program combines many aspects of corporate involvement and all levels of the workforce.

The Business Ethics Leadership Alliance (BELA) is an organization created to share best practices. The Ethisphere award for Most Ethical Conduct is presented to companies throughout the world based on weighted criteria regarding ethical codes and conduct:

1. compliance with a defined ethics program (25%);
2. industry reputation, leadership, and innovation, including legal and ethical track record in regard to litigation (20%);
3. corporate governance and ability to monitor compliance with standards (10%);
4. corporate good citizenship in regard to such issues as community service and environmental concerns (25%);
5. culture of ethics fully recognized throughout the organization (20%).

A CASE IN POINT

For a code of conduct or ethics program to be truly successful it must be a top-down initiative that everyone recognizes and adheres to, but it does not have to be imposed as a burden. When I first met Bob Bowen, CEO of Bowen Engineering, I said to myself "nobody can be this happy and positive all the time and still survive in the construction industry." Over the years, I have found that I was wrong.

A written goal on every project of Bowen engineering is "To have fun." That might seem a stretch, but I was working with Bob's people in one of their offices to help AGC of America put together a video entitled *Continuous Planning Improvement*, based on the techniques that Bowen Engineering uses. Close to the end of the project, one of the video crew came up to me and asked if I knew whether he could get a job with Bowen Engineering. I asked him why someone in video production would want to work in construction. He answered, "Because everybody is having fun around here."

Having received recognition and awards from several sources as "best place to work," the people at Bowen Engineering do have fun in a challenging industry.

And it starts with Bob Bowen insisting that everyone who works for him joins his club, and it's not his country club. It is "The Compliment a Day Club." Working at Bowen Engineering, you are expected to give a compliment to someone every day.

I can tell you it works, because, beginning with Bob Bowen, there is someone that happy and positive every day in this incredibly challenging industry.

Exercise 19.3

Take a moment
Where do your priorities lie?

On a scale of 1 to 3 (1 = very important, 2 = somewhat important, and 3 = "not me at all"), rank the following:

1. I seek leadership status over others: _____
2. I rank enjoying life above all else: _____
3. Higher education is also a time for vacation: _____
4. I study harder than I have to: _____
5. I am planning well for my future: _____
6. I would rather help than get ahead: _____
7. I always practice the Golden Rule: _____
8. For all my hard work I expect great returns: _____
9. If it is to be, it is up to me: _____
10. I am always prepared to criticize when necessary: _____

Take a look at the list and write below any one or two that you think might help you develop a better approach toward a more successful future.

Discussion of Exercise 19.3

In doing the above exercise, you may have noticed there are no right or wrong extremes to the listed priorities. In one sense, the 1 to 3 ranking may be compared to Aristotle's extremes at both ends of a Golden Mean. What we must recognize is that how we approach different priorities is just as important as whether we have them as a priority or not.

For instance, priority #1 on Leadership is important for a Contract Administrator, but it should not be the sole motivation. Great leaders are focused on the goal, but great leaders are also humble and team centric rather than self-centered. They are not leaders for their own egos. At the other extreme, doing an excellent job for all the other team members is critical. If one does not wish to be a leader, contract administration will become a heavy burden, and managing tasks and personnel will seem to be the tiresome burden of always pushing others to do even the minimum.

The customer, of course, can be the Owner, but in the broad sense of service, customers can also be Design Professionals, Subcontractors, Suppliers, craft workers, bonding and insurance agents, or the public in general. By serving others in their needs we will be more likely to have others serve us in our needs. Continually being positive and working well with one another grows the circle of success.

When one looks at the history of the world, one is usually drawn by the horrors of the wars that have moved borders and given rise to super powers. Even the rise to success of the United States as a super power has many dark sides to it (e.g. treatment of the Native Americans, slavery, internment of Japanese Americans in World War II).

However, when one looks at the progress of civilization, one is struck by a different driving force. Tribes have worked together to become villages, towns, cities, nations. Science, technology, and let us not leave out the construction industry, have made our lives more and more comfortable. We have ultimately achieved more in civilization than we have lost in wars.

By working together rather than fighting each other, we have put humans into space, found incredible cures for the scourges of several diseases, and have made worldwide and universal communication possible on an ever smaller planet.

The foundation of these codes of ethics is a simple principle – reciprocity. What one wants is that the service given will be returned in some way, and in the return, the recipient receives more than the cost of the original service. If this becomes the case, then with each exchange the two parties will continue to grow. The receiving party will benefit from what is originally given, and the giving party will benefit from what is received in return for the original service. Everything is based on what each party receives being of greater value to themselves than what they had to give.

Of course, this is the essence of capitalism. But it is also the story of how early hominids decided "you scratch my back and I'll scratch yours." From that simple code of conduct we can see the purpose of the more advanced and refined codes of conduct that companies put forward. The success is not in the wording but in the practice of scratching each other's back – or, in the more dangerous world of the military (and why not construction), "I've got your back."

Perhaps we do not accept the fact that we can do good simply by knowing the difference between right and wrong. Based on the second law of thermodynamics, systems move from order to chaos. For a cynic, this can often summarize the construction process on a daily basis, where each day starts with everyone understanding what to do and eventually evolves to multiple forms of crisis control. Yet it is possible, in the examination of teams and, indeed, civilization, that humans uniquely stand in opposition to the chaos into which the universe is bound to evolve, based on physical laws. In such a case, creating and implementing a code that enables everyone to work together toward the same success can be a great advantage.

Ethics, therefore, may improve the success of the group as much as it can improve the success of the individual. The challenge is to get the group to actually "buy in to" the ethics, and this can often be difficult if the code is long, never referred to, and not practiced on a universal basis throughout the group.

Examples:

1. The boss insists that everyone come to work on time, but he is always late.
2. The Contractor hangs a code of ethics in the lobby, but seldom practices what it preaches when it comes to paying Subcontractors or Suppliers.
3. Only the clients who read the marketing brochure, but none of the workers, know what the mission and code of ethics of the company are.
4. The motivational poster reads "The Customer Comes First!" but the bonuses are given out to those who can make the most profit, without regard to client retention.

A CASE IN POINT

Sometimes short is better than sweet. Back in the early 1970s everyone was writing mission statements for their companies. Ours was quite verbose and, frankly, no one ever read it more than once. In fact, it was rather long, and if someone started to read it, I am not sure they ever finished it.

The mission statement collected dust over the years, framed on an office wall that everyone walked by without pausing.

Finally, it occurred to me that the mission statement was ineffective. I sat down and wrote what I believed could be our vision and mission for all our team members to not only know by heart but also believe we could do it.

Every first-day hire would be told it, and all of us were to ask ourselves, in whatever challenge we had, if we were actually practicing it.

The new mission statement simply read:

To Build for the Betterment of All.

19.2.3 Establishing a code of ethics

The Contract Administrator may not be in a position to establish a corporate code of ethics, but there is the possibility of presenting the project team with such a code. Whether for an entire company or a single project team, some steps should be taken to make sure everyone understands expectations when it comes to ethical conduct:

1. Recognize why you want to establish a code of ethics in the first place and determine if what you want is really a code or a mission statement.
2. Determine what the team seeks to achieve together.
3. Discuss what interactions will achieve the desired results for all the team members.
4. Determine who will formalize the code or mission statement, and how it will be kept current and active in the dealings of all the members of the team.
5. Continue to refer to the code once it is established (e.g. at project meetings) and seek the input of others related to conduct throughout the project.

Perhaps the final and ultimate resolution about ethics for the Contract Administrator must always come back to what is in it for themselves and the team. We all need to balance the short-term pull of duties and responsibilities with the long-term benefits of doing what is right for everyone. Ethical behavior should benefit everyone. The tragedy of unethical behavior is not merely that someone gets hurt. It is that, in the long run, everyone gets hurt. Those who think there is nothing for them in doing good are not sufficiently focused on the fact that there is a huge, although sometimes delayed, penalty in not being ethical.

A CASE IN POINT

Figure 19.3 Citigroup Center is one of the taller skyscrapers in New York City. As an engineering challenge with ethical implications, however, it perhaps ranks at the top.
Photo credit: Charles W. Cook

Structural Engineer William LeMessurier found, in recalculating for wind loads in relation to bolted joints that had been changed from his previously designed welded joints, that Citicorp Center would be in danger of serious structural damage, even collapse, with winds hitting at a 45-degree angle. LeMessurier claimed that he had not initially known of the change from welded to bolted, but the seriousness was no less imminent.

LeMessurier was not certain at first how to proceed. He did not want to alarm the public, and, of course his reputation was at stake, but with hurricane season approaching, there was major cause for concern. The building needed to be repaired. He approached Citicorp and asked for it to have two-inch-thick steel plates welded over all the bolted connections. He also asked that the entire process be kept secret while it was being done.

For almost twenty years, the secret was never reported. Some have criticized LeMessurier for not revealing the problem as well as the solution. They also feel he had not demonstrated effective oversight, based on having to recalculate the wind loads for 45 degrees and also not knowing that the welded joints had been changed to bolted. Half way through the repairs a hurricane was headed straight for Manhattan. Fortunately it veered out to sea, but what might have happened had it hit Manhattan dead on and brought into the city the 70 mph winds at a 45-degree angle that could have caused tremendous damage and even loss of life from the toppling of the Citigroup Tower? Others have felt that LeMessurier's quick response in approaching Citicorp, knowing his own errors and omissions were on the line for the costs, was the ethical thing to do. They recognize that correcting the error was paramount. If one's reputation could be maintained without public exposure, then there was no harm done.

(In "Going the extra mile" see Discussion exercise I to review how you and your classmates feel about how this situation was handled.)

19.2.4 Staying true to the ethical path

There is no certainty that ethical behavior will be perfect. It would be rare to find that it ever is. Commitment is not a guarantee. However, confronting failure, analyzing issues, and recommitting to our code is a good beginning to improving ethical behavior whenever our actions fall short of our words or intentions.

Just as improving project delivery is a forever ongoing process, continued ethical behavior will always be a challenge. The preceding chapters have been only a beginning for the Contract Administrator. The chapters end. The course concludes, but the journey for each of us goes on. For each reader who has come this far, may your journey be long upon this earth, filled with good health, great project teams, and demanding challenges to make the journey worthwhile.

INSTANT RECALL

- Ethics is a system of directed conduct that helps individuals and groups to attain success through specific behavior, but such behavior cannot deliver benefits to one party at the expense of another.
- The choices we make affect our own success and our relationships with others.
- The Contract Administrator has a responsibility to see that the contract is fulfilled.
- How we conduct ourselves through life and our careers will have a major impact on our destiny.

- Personal ethics is meant to define specific behaviors that will guide us through successful interactions with the people and situations that we confront throughout our lives, but it cannot be a "spur of the moment" thought sequence.
- Aristotle wrote in his *Nicomachean Ethics* that people can improve their personal virtues until they become habits.
- Making a virtue a habit or an automatic response is far more likely to lead to success when the situation becomes difficult. It is easy to be virtuous when the stakes are low, but much harder when the price goes up.
- John Locke believed that ethics was related to experience, and therefore ethics might shift, depending on experience, but the foundation of all human experience began with three fundamental rights – life, liberty, and the fruit of one's own labor in the constant pursuit of one's happiness.
- Although Aristotle can point us in a direction to improve our own success through improving our virtues, there comes a time when we must go back to Socrates' concept that we can know right from wrong, and in so knowing we should choose doing what is right, even though we may not see how it makes us happy in the short run.
- Underlying the judgment of all human interaction is the concept of fairness.
  - "Do unto others as you would have others do unto you."
- On a good team every member of the construction team, from Owner through Design Professionals to the Subcontractors and Suppliers will appreciate and respond to how you treat them, in the same way that you will respond to them based on how they treat you.
- When a problem arises, think of how you would want to be treated, and then proceed as quickly as possible to find a solution.
- Negative group ethics is not as effective as a positive set of standards.
- If the purpose of business ethics could be contained in a single word, it would be "trust."
- By serving others in their needs we will be more likely to have others serve us in our needs.
- Reciprocity in serving others is based on what each party receives being of greater value to themselves than what they had to give.
- John Stuart Mill and Jeremy Bentham were English philosophers of the nineteenth century. Their approach to ethics and decision making in relation to business and life was based on doing the most good for the most people.
- If what is good for us also leads to what is good for others, then self-interest is excellent.
- Whether for an entire company or a single project team, some steps should be taken to make sure everyone understands expectations when it comes to ethical conduct:
  - Recognize why you want to establish a code of ethics in the first place and determine if what you want is really a code or a mission statement.
  - Determine what the team seeks to achieve together.
  - Discuss what interactions will achieve the desired results for all the team members.
  - Determine who will formalize the code or mission statement, and how it will be kept current and active in the dealings of all the members of the team.
  - Continue to refer to the code once it is established and seek the input of others related to conduct throughout the project.
- Ethical behavior should benefit everyone.

WHAT WOULD YOU DO DIFFERENTLY?

For our final "You be the judge" exercise we are going to change the format. This chapter has been a discussion about ethics, and ethics are not part of an adjudicated case before a judge, jury, or panel of neutrals.

The tragedy of the building collapse mentioned in a previous chapter has had many looking for answers, along with many looking for blame, but we cannot now turn back the hands of time. In a communication dated Friday, July 19, 2013, over a month after the incident had taken place, Michael A. Nutter, Mayor of Philadelphia, stated, "The collapse of the building at 22nd and Market was the kind of tragedy that should never recur."

He chose the word "recur" correctly. Now we need to understand how we can avoid such situations in the future. After a tragic incident, in addition to finding who is to blame, many governments, from municipal to Federal, pass new regulations in an attempt to make sure such an incident never takes place again (yet they seem to happen over and over).

Think to yourself or discuss with friends whether more regulations are required. Is there some other issue that needs to be addressed? If new regulations are passed, who should pay for the costs? Will they just be handed down to the Owner/Developer? What about projects already under construction when the new regulations are passed? Who pays for the extra costs on those projects? Do exculpatory clauses protect the Owner in such a case and does the Constructor have to carry the burden of the new regulation costs?

IT'S A MATTER OF ETHICS

Have you thought about your price? It is better to know your limits before you get into a difficult situation. Take some time to think about how far you would actually go for a job. To keep a job? How much money would it take for you to do something you would not want your friends or relatives to read about if you got caught? How much would you want to make in order to take a risk yourself? What if the risk was to someone else? Is there something more than money that would cause you to do any of the above? Being honest with yourself may be more important than discussing with others, but feel free to discuss your thoughts on the matter with friends if you believe it might help.

Going the extra mile

1. **Discussion exercise I** (with friends or as part of a classroom discussion): Go back to the LeMessurier story and discuss how you believe it was handled. Would you have done differently?
2. **Discussion exercise II** (with friends or as part of a classroom discussion): Bid shopping is considered a way of life by some, while others do not consider it appropriate. Some consider bid shopping unethical, but it is difficult to prove that it is illegal. Sometimes an overly enthusiastic "shopper" might cross the boundaries in tort law and commit some form of fraud, but it would still be difficult to bring such a case to court.
3. Let us turn our attention for a moment outside the construction industry. Well before the days of advertising circulars and the internet, in the 1920s my mother worked as a "shopper" for a large department store in Providence, Rhode Island. As a "shopper" she

would go out to other stores and see what items in the store were selling for, and then she would bring that information back to her own store so it could match or beat the price. Today, most national retail stores have what is called "price matching." They will "meet or beat any competitor's price." Discuss whether this is another form of bid shopping. If so, is bid shopping that wrong? Why would someone not bid shop? Is it unethical? Is price matching unethical? Are there some aspects of doing business that are unethical?

4. **Discussion exercise III** (with friends or as part of a classroom discussion): It may not seem obvious at first, but take any or all of the priorities in Exercise 19.3 above and discuss how any one taken to an extreme may not be worthwhile. How does one find the Golden Mean within the priority and turn it into a virtue?
5. **Discussion exercise IV** (with friends or as part of a classroom discussion): In the text and in "Instant recall," it was stated: "How we conduct ourselves through life and our careers will have a major impact on our destiny." How firmly do you believe this? Are there other elements of life that will affect our destiny? How do you handle them? What do you think you should do now that will most affect your destiny in a positive manner?
6. If you would like to research further the topic of ethics you might start with the Ethics Research Center at www.ethics.org. Also consider going to the websites of any of the professional organizations associated with the construction industry, including ASCE, NSPE, AIA, CMAA, PMI, AACEI.
7. Go to the Companion Website and watch Part VI of the Ethics video.
8. Effective use of the correct vocabulary improves communication and promotes respect among colleagues. The reader should become familiar with the key words listed below that have been used in the chapter. Definitions can be found in the Glossary. Additionally, the student is encouraged to use the flash card exercise in the chapter's Companion Website as practice for possible quiz or exam questions.

Key words

Federal Acquisitions Regulations Act (FAR)

Platinum Rule

Utilitarian ethics

APPENDIX I

Sample contract

R. S. Cook and Associates, Inc.

1234 Spring Street
Philadelphia, PA 19100

Subcontract

Date: Julember 18, 2013

Contract's Job # 9999
Subcontract No.: 2

Subcontractor: XYZ Electric
9876 Main Street
Philadelphia, PA 19100

Principal Contract:
Owner: Happy Industries
Date: Julember 15, 2013
Location of Work: Riverfront Plant
123 River Avenue
Philadelphia, PA 19100

R. S. Cook and Associates, Inc., herein called "Contractor" and the above named Subcontractor, herein called "Subcontractor," hereby agree that the portion of the Principal Contract Work below specified shall be performed by Subcontractor in strict compliance with all the requirements of the Principal Contract applicable thereto and the provisions hereinafter stated, including the General Terms and Conditions included with this subcontract and any special conditions attached hereto, which by this reference are incorporated herein. Subcontractor acknowledges that it is familiar with the requirements of the Principal Contract applicable to the work to be performed hereunder.

1. (a) Work to be Performed: All Electrical Work

(b) Plans, drawings, and specifications: Proposed New Office Building, Riverfront Plant, Architect LMNOP Associates, Drawings A-1 through A-12, S-1 through S-5, M-1 through M-6, P-1 through P-5, and E-1 through E-8, Specifications dated Julember 1, 2012, addendums 1, 2, and 3.

2. Items to be furnished by Contractor: None

3. (a) Time of Commencement: Julember 31, 2013

(b) Time of Completion: Julember 31, 2014

4. (a) Compensation: $1,000,000 (one million dollars)

(b) Invoices to be submitted in duplicate on Subcontractor's request for payment form (sample attached) by the 25th of current month, signed and notarized.

Terms of Payment: Within seven days of payment by Owner.

5. Insurance Required:

Coverage	***Amounts and Limits***
Workers' Compensation	Statutory
Public Liability	3,000,000
Property Damage	2,000,000
Automobile	1,000,000
Automobile Property Damage	1,000,000

6. Bonds Required: None

Special Provisions: Liquidated Damages $350/day

GENERAL TERMS AND CONDITONS

1. **CONTRACTUAL RELATIONSHIP:** In the performance of the Subcontract, Subcontractor shall perform as an independent contractor and not as agent of the Contractor or Owner.
2. **ITEMS TO BE FURNISHED BY SUBCONTRACTOR:** Subcontractor shall supply and furnish at the location where the work is to be performed all items, including labor and materials, necessary for the complete and satisfactory performance of the subcontract work, except such items as the Contractor in the Subcontract specifically agrees to supply or furnish to or for the use of the Subcontractor. As used in these General Terms and Conditions, "materials" shall refer to and include all substances, supplies, equipment, machinery, apparatus and other things used or useful in the performance of the Subcontract work.
3. **PERFORMANCE REQUIREMENTS AND SPECIFICATIONS:** Work shall be performed in accordance with all the requirements of the Principal Contract and its specifications and drawings applicable thereto, the provisions of this Subcontract, and such specifications and drawings as may be referred to in the Subcontract. Anything mentioned in the specifications and not shown on the drawings, or shown on the drawings and not mentioned in the specifications, shall be of like effect as if shown and mentioned in both. In case of conflict with the drawings, the specifications shall govern. In case of ambiguity or obscurity the matter shall be submitted immediately to Contractor for determination.
4. **CHANGE IN SPECIFICATIONS AND DRAWINGS:** Contractor reserves the right by written notice to correct any errors or to make any changes in the specifications and/or drawings. If such changes cause a material increase or decrease to the cost of performing the work or the time of performance and written notice thereof is given to either party, within the terms of the Principal Contract with the Owner, an equitable

adjustment in the contract price and/or the time of performance shall be made. If the parties cannot agree upon such adjustment within ten days after receipt of such notice, the matter will be submitted to arbitration as hereinafter provided, but the Subcontractor shall proceed immediately with the work as changed.

5. **EXTRA WORK:** Subcontractor shall not be entitled to any compensation in addition to that specified in this Subcontract for the performance of any work not required under this Subcontract, unless prior to the performance of such work, it shall have received from Contractor written authorization to perform such work and additional compensation shall have been agreed upon in writing, provided, however, that in the event work is being performed as provided in Article 4, this Article does not apply thereto.
6. **INSPECTION AND TEST OF EQUIPMENT:** All materials and work furnished or performed by Subcontractor shall be subject to final acceptance, inspection and tests by Contractor and Owner upon completion of all Subcontract work, whether or not previously paid for by Contractor. At any and all proper times during manufacture or performance of the work, all materials and work furnished or performed by Subcontractor shall be subject to inspections, tests, and approvals by inspectors of the Contractor, or Owner, at any and all places where such manufacture or performance shall be carried on. Failure of such inspectors to make inspections or tests or to discover defective work or material shall not prejudice the rights of Contractor or Owner on final inspection and test. If facilities of Contractor are not available, Subcontractor shall furnish such facilities as may be necessary for the making of such inspections and test. If upon any such inspection or test any material or work shall be found to be defective or not to conform to the requirements of this agreement it shall be promptly rejected and the Subcontractor shall be notified thereof. Subcontractor, at its own expense, shall promptly correct work which does not comply with such requirements by making the same comply therewith and shall promptly replace any material (except such as may have been furnished by Contractor) which does not conform to such requirements. If Subcontractor shall fail to replace or correct rejected material or work promptly, Contractor, at its option may replace or correct the same and all costs and expenses of Contractor in connection therewith shall be borne by Subcontractor, including costs of tests.
7. **TIME OF PERFORMANCE:** Time is of the essence of this Subcontract. Subcontractor shall indemnify and hold harmless Contractor from and against any penalty or liability incurred by Contractor to Owner because of Subcontractor's failure to perform the work within the time agreed upon.
8. **EXTENSION OF TIME AND SUBCONTRACTOR'S WAIVER OF DAMAGES FOR DELAY:** In case of any delay caused by Contractor or the Owner which was not reasonably ascertainable by Subcontractor at the time this Subcontract was entered into, written notice thereof and of the anticipated results shall be given promptly to Contractor by Subcontractor. Failure to give such written notice promptly shall be deemed sufficient reason for a denial of an extension of time by Contractor. Contractor shall notify Subcontractor if, in its opinion, the cause of delay specified is such as not to entitle Subcontractor to an extension of time. After such cause of delay has ceased to exist, Subcontractor shall file with Contractor a statement in writing, of the actual delay resulting from such a cause. If, in the opinion of Contractor, the cause of delay was beyond the reasonable control of Subcontractor and was not reasonably ascertainable by Subcontractor at the time this Subcontract was entered into, the duration of delay shall be determined by Contractor and the time of performance of the work, the performance of which has been delayed thereby, shall be extended, in writing, by Contractor for such additional period as shall be required to allow for such delay. Subcontractor shall not

be entitled to and hereby waives, any and all damage which it may suffer by reason of Contractor or Owner hindering or delaying Subcontractor in the performance of the work, or any portion thereof, from any cause whatsoever.

9. **ORDER OF PERFORMANCE OF WORK:** Contractor reserves the right to direct the Subcontractor to schedule the order of performance of the subcontract work in such manner as not unreasonably to interfere with the performance of work by Contractor, the Owner and other contractors or subcontractors.
10. **REMOVAL OF DEBRIS AND WASTE MATERIAL:** During performance of work under this Subcontract and upon termination or completion thereof, Subcontractor shall promptly remove all debris and waste material and keep and leave the site of the work clean and in good condition.
11. **DESIGNATION OF SUPERINTENDENT:** Subcontractor shall designate a competent superintendent who, on behalf of Subcontractor, shall have complete charge of all work under the Subcontract. Subcontractor shall promptly advise Contractor in writing of the name, address and telephone number (day and night) of such designated superintendent and of any change in such designation.
12. **LIENS AND CLAIMS:** Subcontractor shall indemnify and save harmless Contractor and Owner from all claims, liabilities, suits, demands, causes of action, losses, damages, costs and expenses (including court and attorney's fees) of whatsoever nature arising out of or in connection with any and all of the acts or representations of Subcontractor, its subcontractors, agents or employees in the performance of the subcontract work or in furnishing of any labor or materials by Subcontractor, or its subcontractors, agents or employees. Subcontractor for itself, its subcontractors, agents, employees and all parties acting through and under it, covenants and agrees that no mechanics' claims or liens shall be filed or maintained by it, them or any of them against the buildings, improvements or the work to which this subcontract pertains or against the ground appurtenant thereto or any part thereof, for or on account of any work done or materials furnished by it, them or any of them, under this subcontract or otherwise, and subcontractor for itself, and its subcontractors, agents, employees and all parties acting through or under it, hereby expressly waives and relinquishes the right to have, file or maintain any mechanic's lien or claim. Subcontractor shall indemnify and save harmless Contractor and Owner from any and all mechanics' liens or claims filed, asserted or maintained by Subcontractor, its subcontractors, agents, employees or anyone acting through or under it, including all damages and expenses incurred by Subcontractor and/or Owner with relation thereto (including attorney's fees). Contractor may, as a condition precedent to any payment hereunder, require Subcontractor to submit complete waivers of the lien and releases of any and all claims of itself and any person, firm or corporation supplying labor, services or material to Subcontractor.
13. **PATENTS AND ROYALTIES:** All royalties, license fees, or other charges, payable with respect to any patented materials, process or method of construction furnished or used by Subcontractor, or its subcontractors, or agents, or with respect to the title, possession, use and/or sales thereof by Contractor and/or Owner, shall be the sole responsibility of Subcontractor; and Subcontractor shall defend all suits relating to and hold Contractor and Owner harmless from any and all royalties, claims, liabilities, suits, demands, causes of action, losses, damages, costs and expenses (including court costs and attorney's fees) of whatsoever nature arising out of or in connection with any infringement or alleged infringement of any patents, as well as for the misuses of any patented article by Subcontractor, or its subcontractors, agents or employees, in the performance of the subcontract work.

14. **INJURY OR DAMAGE TO PERSONS AND PROPERTY:** Subcontractor shall be solely responsible for and shall indemnify and hold harmless Contractor and Owner from any and all claims, liabilities, suits, demands, causes of action, losses, damages, and expenses (including court costs and attorney's fees), of whatsoever nature arising out of or in connection with injuries (including death) or damages to any and all persons, employees, and/or property in any way sustained or alleged to have been sustained in connection with or by reason of the performance of the subcontract work by Subcontractor, its subcontractors, agents or employees or the use by any such persons, whether due to negligence or otherwise, of tools, dies, materials, or other items furnished by Contractor, its subcontractors, agents or employees. Subcontractor shall assume full responsibility for compliance with all applicable regulations issued by the Secretary of Labor of the Contract Hours and Safety Standards Act, commonly known as the Construction Safety Act, and/or the Williams-Steiger Occupational Safety and Health Act of 1970, and shall indemnify and hold harmless Contractor from any and all claims, liabilities, suits, demands, causes of action, losses, damages and expenses (including court costs and attorney's fees) of whatsoever nature arising out of or in connection with Subcontractor's failure to comply with such regulations.
15. **LABOR CONDITIONS:** Contractor may require Subcontractor to discharge any incompetent or undesirable employee serving on the site of the contract work.
16. **RESPONSIBILITY FOR WORK, TITLE AND LIMITATIONS ON USE OF PROPERTY:** (a) Subcontractor shall be responsible for all materials furnished by Subcontractor or its subcontractors or agents and for all work performed under the Subcontract until completion of all subcontract work and acceptance after final inspection and tests, as referred to in Article 18(c), below, and upon such completion and acceptance, the same shall be delivered and turned over complete and undamaged. Title to materials so furnished shall not pass to Contractor or Owner until the same have been so accepted, except as to materials the title to which passes by reason of becoming affixed to realty. Notwithstanding the foregoing, Contractor shall have and retain title to materials furnished by Contractor to Subcontractor, whether or not incorporated in or annexed to materials furnished by Subcontractor, or its subcontractors or agents, and Contractor shall also have title to materials furnished by Subcontractor, or its subcontractors or agents, which are incorporated in or annexed to materials furnished by Contractor. Regardless of the party in which ownership rests or who furnishes the materials, risk of loss and damages with respect to all of the foregoing materials and work shall rest upon Subcontractor until acceptance.
(b) All plans, drawings, designs, blueprints, specifications, and other delineations and writings, furnished by Contractor to Subcontractor, or prepared and furnished by Subcontractor, or its subcontractor, or its subcontractors or agents, to Contractor, shall be and remain the property of Contractor, and shall be carefully protected and preserved by Subcontractor and returned to Contractor upon demand; and any information derived therefrom or otherwise communicated by Contractor to Subcontractor in connection with the subcontract shall constitute strictly confidential information and be treated as such by Subcontractor, its subcontractors, agents and employees, and shall not be disclosed to or used for the benefit of any third person, without the prior written consent of Contractor.
(c) All tools, dies, jigs, welds, furniture, gauges, patterns and other aids furnished by Contractor to Subcontractor, or prepared by Subcontractor, its subcontractors or agents, at Contractor's expense, shall be and remain the property of Contractor, and shall be carefully protected and preserved by Subcontractor and returned to Contractor

upon demand; and if not so returned Subcontractor shall be liable for the value thereof to Contractor. No such items shall be used by or for Subcontractor for the benefit of any third person without the prior written consent of the Contractor.

17. **INSURANCE AND BONDS:** Subcontractor at its own expense, shall procure, carry and maintain on all its operations hereunder the bonds and policies of insurance in the amounts specified in the Subcontract. The bonds and policies of insurance shall be in such form and shall be issued by such company or companies as may be satisfactory to Contractor. Subcontractor shall cause to be furnished to Contractor certificates of insurance from the insuring companies which shall include the following clause: "ten (10) days advance notice shall be given in writing to R. S. COOK AND ASSOCIATES, INC. on cancellation, termination, or any alteration of the policy or policies evidenced by this Certificate."

18. **COMPENSATION, INSPECTIONS AND PAYMENTS:** (a) Subcontractor agrees to accept the compensation specified in this Subcontract as full compensation for doing all the work and furnishing all the materials contemplated by and embraced in the Subcontract, for all loss, damage or expense arising out of the nature of the work or from the action of the elements or from any unforeseen or unknown difficulties or obstructions which may arise or be encountered in the prosecution of such work until final delivery and acceptance and for all risks of every description connected with such work.
(b) Unless otherwise provided in the Subcontract, Contractor shall make partial payments as the work progresses, as follows: At the end of each calendar month or as soon thereafter as practicable, Contractor shall estimate or cause to be estimated the Subcontract value of all work performed hereunder. Such estimate shall be conclusive upon Subcontractor for purposes of this paragraph. Contractor shall pay Subcontractor in accordance with the terms of payment specified in this Subcontract less the aggregate of all payments made or charged to Subcontractor.
(c) As soon as practicable after completion of all work hereunder, final inspection and tests shall be made by Contractor and Owner. When such inspection and tests prove satisfactory, the work shall be accepted and the amount then remaining due to Subcontractor shall be paid; provided that Subcontractor shall have furnished Contractor and Owner with a waiver and release of all claims against Contractor and Owner arising under or by virtue of the Subcontract, together with the waivers and releases referred to in article 12, above. Such claims, if any, as may be consented to by the Contractor and Owner may be specifically exempted from the operation of such waivers and releases in stated amounts to be set forth therein.
(d) Under no circumstances will a Subcontractor sue for payment the Contractor or Surety Company until all conditions of the Prime Contract have been fulfilled.
(e) Payments otherwise due may be withheld by Contractor on account of defective materials or work not remedied, claims filed, or reasonable evidence indicating probability of filing claims, failure of Subcontractor to make payments properly to subcontractors or for materials or labor, or a reasonable doubt that the Subcontract can be completed for the balance then unpaid. If the foregoing causes are removed, the withheld payments shall promptly be made. If the said causes are not removed on written notice, Contractor may rectify the same at Subcontractor's expense. Should any valid indebtedness arise after final payment is made, the Subcontractor shall reimburse the Contractor for any amount that it may pay in discharging any lien therefor on any claim affecting title to the work or Owner's property.

19. **UNEMPLOYMENT INSURANCE AND TAXES:** Subcontractor shall accept full and exclusive liability for the payment of any and all taxes and contributions for unemployment

insurance, workmen's compensation and disability benefits which may now or hereafter be imposed by the federal government or any state or municipality, whether measured by the wages, salaries, or remuneration paid to persons employed by Subcontractor or otherwise, for the work required to be performed hereunder. Subcontractor shall comply with all federal, state, and municipal laws on such subjects, and all rules and regulations promulgated hereunder, and shall maintain suitable forms, books, and records and indemnify and save Contractor harmless from the payment of any and all such taxes and contributions, or penalties. Subcontractor shall likewise pay any and all taxes, excises, assessments or other charges not specifically imposed by law on Contractor, levied by any governmental authority on or because of the work to be done hereunder, or any equipment used in the performance thereof.

20. **TAKING OVER PERFORMANCE – TERMINATION OF CONTRACT:** (a) Should Subcontractor at any time refuse or neglect to supply a sufficiency of properly skilled workmen or materials of the proper quality or quantity Contractor may, after forty-eight (48) hours written notice to Subcontractor, provide any such labor or materials and deduct the cost thereof from any money due or thereafter to become due Subcontractor under this Subcontract.

(b) Should Subcontractor at any time refuse or neglect to supply a sufficiency of properly skilled workmen or materials of the proper quality or quantity, or fail in any respect to prosecute the work or any separable portion thereof with promptness and diligence, or fail in the performance of any of the agreements on its part contained herein, or should Subcontractor make any assignment for the benefit of creditors or make admission in writing of its inability to pay its debts generally as they are due, or should a receiver or trustee of the Subcontractor be appointed for any of its property in any insolvency or bankruptcy proceeding, or should any bankruptcy, arrangement, insolvency or liquidation proceedings under any federal, state or municipal law be instituted or commenced by or against the Subcontractor, Contractor may, after forty-eight (48) hours written notice to Subcontractor, terminate Subcontractor's right to proceed with the work or such part of the work as to which such defaults have occurred. In the event of such termination, Contractor may enter upon the premises and for the purpose of completing the work, take title to and possession of and remove all materials thereon belonging to or under the control of Subcontractor, whether or not such materials are in completed or deliverable state, and may finish the work by whatever method it may deem expedient, including the hiring of another contractor or contractors under such form of contract as Contractor may deem advisable. In such case Subcontractor shall not be entitled to receive any further payment until the work is finished. If the unpaid balance of the amount to be paid on this Subcontract shall exceed the expense of finishing the work, compensation for additional managerial and administrative services and such other expenses, costs and damages as Contractor may incur or suffer, such excess shall be paid to Subcontractor. If such expenses, compensation, costs and damages shall exceed such unpaid balance, Subcontract and its sureties, if any, shall be liable for and shall pay the difference to the Contractor.

(c) Should the Principal Contract terminate or be cancelled, pursuant to the terms thereof, or should conditions arise which, in the opinion of Contractor, make it advisable to cease work under this Subcontract, Contractor may terminate this Subcontract by written notice to Subcontractor. Such termination shall be effective in the manner specified in said notice and shall be without prejudice to any claims which Contractor or Owner may have against Subcontractor. On receipt of such notice, Subcontractor shall, unless the notice directs otherwise, immediately discontinue the work and placing

of orders for materials in connection with the performance of this Subcontract, and shall, if requested, make every reasonable effort to procure cancellation of all existing orders or contracts upon terms satisfactory to Contractor, and shall thereafter do only such work as may be necessary to preserve and protect work already in progress and to protect material, plant, and equipment on the work or in transit thereto. In such event Subcontractor shall be entitled only to prorate compensation for the portion of the Subcontract already performed, including material for which it has made firm contracts, it being understood that Contractor shall be entitled to that material.
(d) Contractor's remedies shall be cumulative and the Contractor may exercise any remedies at any time and from time to time either concurrently or successively. Failure of Contractor to exercise any remedy or right under this contract should not be deemed a waiver thereof nor excuse Subcontractor from compliance with the provisions of this Subcontract nor prejudice the right of the Contractor to recover damages for any such default.

21. **WARRANTIES:** Subcontractor represents and warrants that all materials furnished by Subcontractor, or its subcontractors or agents shall be new and of first class grade and quality and conform strictly to all specifications, drawings, samples or other designations furnished or adopted by Contractor and/or Owner, and shall be free from all defects in workmanship and material; and from all defects of design if furnished by Subcontractor, or its subcontractors or agents; and in addition to other legal remedies available, and notwithstanding acceptance after final inspections and tests, Contractor shall have the right to require Subcontractor to repair or replace, at Subcontractor's cost and expense, and in a manner satisfactory to Contractor, any material which is found, within one (1) year after such acceptance, to be defective or not as warranted.
22. **LAWS AND REGULATIONS:** Subcontractor, its subcontractors, agents, servants and employees shall comply with all laws, ordinances, rules and regulations, federal, state and municipal, applicable to materials supplied by Subcontractor and its subcontractors and agents, and shall certify to Contractor that such materials were produced in compliance with all applicable requirements of the Federal Fair Labor Standards Act of 1938, as amended, including the requirements as to records.
23. **NOTICES:** Any notice required hereunder may be served by delivering it personally to the superintendent of either party at the job site or may be served by prepaid registered mail directed to the address shown in the Subcontract. Any notice served by mail shall be deemed to be given 48 hours after mailing.
24. **ARBITRATION:** In case of any dispute between the parties as to the interpretation of this agreement or the performance of the same, either party may demand in writing that the dispute be submitted to arbitration, in which event said dispute shall be submitted to arbitration by an arbitrator appointed by the American Arbitration Association whose decision thereon shall be final and binding on the parties hereto. The cost of arbitrations shall be divided evenly between the parties hereto. In no case shall submission of a matter to arbitration be cause for delay or discontinuance of any part of the work under this Subcontract.
25. **NON-ASSIGNMENT:** Subcontractor shall not subcontract or assign the work, or any part thereof, nor any moneys to become due hereunder, without first obtaining written consent of Contractor; and any such subcontracting or assignment without such prior written consent shall be void and of no effect.
26. **SUBCONTRACT INCLUDES ENTIRE AGREEMENT:** This Subcontract embodies the entire agreement between Contractor and Subcontractor. Subcontractor represents

that in entering into this Subcontract it does not rely on any previous oral or implied representation, inducement or understanding of any kind or nature.

27. This from of Subcontract and these General Terms and Conditions are designed for general use in any of the states of the United States of America where Contractor is engaged in the performance of work; and any provisions which contravenes the laws of any such state or of the United States of America shall not be considered to be part of this Subcontract.

IN WITNESS WHEREOF, the parties hereto have executed this Subcontract the day and year first above written.

Contractor:	Subcontractor:
R. S. Cook and ASSOCIATES, INC.	
By: ______________________	By:__________________
Date:__________________	Date:__________________

(Sign, date, and return both copies to R. S. Cook and Associates, Inc.)

Daily work report

Daily Work Report – XYZ Company

Report No. _____ Date:__________

Labor on Project

Classification	Number	Hours

Weather

AM

Temperature:

Conditions:

PM

Temperature:

Conditions:

Work Performed:

Deliveries and Material Received:

Tools and Equipment Used on Site:

List Visitors to the Project:

Meetings and Issues Discussed/resolved:

Requested Information:

Is anything currently delaying any part of the project?

Report submitted by:

Glossary

Acceleration – a requirement to increase production on a project, usually due to a delay, to meet the original contract completion date, or, in some instances, because the Owner's need to complete has been moved earlier than originally planned.

Acceleration clause – a clause requiring the Constructor to increase personnel, material, and/or equipment on the project in order to meet a specific schedule.

Acceptance – in contract law the act of agreeing to the offer (promise) made. Without acceptance there is no contract.

Acts of God – in some instances referred to as ***force majeure***, these are conditions brought about by natural occurrences that the Owner and Contractor have no control over.

Actual Authority – in construction, the contractually defined authority of a particular agent of the Owner (*see* **Expressed Authority** and **Implied Authority**).

Additional insured – other parties named and protected under the term of an insurance policy other than the one that has taken out the insurance coverage.

Adversarial system – dispute resolution whereby two or more parties present opposing positions before a third party that will pass judgment.

Affirmative Action – regulations that require taking positive actions in regard to portions of the population that have previously suffered from discrimination practices.

Age Discrimination in Employment Act (ADEA) – a 1967 law prohibiting discrimination of anyone on the basis of age.

Alternative Dispute Resolution (ADR) – processes or techniques of resolving conflict without resorting to litigation

Ambiguities – discrepancies between different parts of the contract documents or between the documents and the actual conditions.

Americans with Disabilities Act (ADA) – law passed in 1990 prohibiting under certain circumstances discrimination on the basis of disabilities.

Apparent Authority – the authority that arises out of the accepted actions of an individual(s) rather than the Expressed Authority contained within the language of the contract.

Arbitration – an **Alternative Dispute Resolution** process in which the parties outside of a court submit the conflict to one or more neutral parties to determine a resolution to the conflict.

"As-built" drawings – documents detailing changes in the actual in-place construction from what was originally shown on the contract documents. Today more and more "as-builts" are being done electronically.

Assumption of Risk – a defense put forward by defendants to show that the plaintiff agreed to take on the risk that resulted in the damages.

Attractive Nuisance – a law placing responsibility for damages upon the Owner of property if children (or sometimes adults) trespass and are injured due to unprotected and dangerous material or equipment on the property.

BATNA – "Best Alternative To a Negotiated Agreement."

Bid Bonds – the guarantee by a surety that the bidder will undertake the contract under the conditions of the proposal.

Bid shopping – the act of using one Subcontractor or Supplier's price to negotiate a better price from another Subcontractor or Supplier.

Boilerplate – language in a contract that is considered standard and used generally without change from one contract to another.

Bonds – guarantees by a surety that the entity being bonded is capable of fulfilling the conditions for which the entity is being bonded.

Breach – an action not in conformance with the contract requirement or conditions.

Breach of contract – a failure to comply with one or more parts of a contract's requirements.

Broad Form Indemnification – a requirement within the contract for Party B to hold Party A harmless and protect against damages even if those damages are caused by Party A.

Builder's Risk Insurance – a type of property insurance that indemnifies against loss while the structure is under construction.

Building Information Modeling Technology (BIM) – computer rendering technology of 3-Dimensional imaging of a project, allowing each space to be viewed from different angles, and identifying where installation of different equipment and materials interfere with other objects in the structure. The 4th Dimension of Building Information Modeling is considered schedule, and the 5th Dimension is cost.

Cardinal Changes – a change or changes so large or numerous that the total scope of the project is significantly altered.

Cash on Delivery (COD) – a requirement of the seller to receive payment for the material when it is passed on to the buyer.

Causation – the act or process of making something happen.

Certificate of Occupancy (C/O) – issued after inspection by governing authority to recognize a structure can be occupied for its intended use.

Certificate of Substantial Completion – verification, usually by the Designer, that the Constructor has completed the structure to the extent that it can be used by the Owner for its Intended purpose.

Change order – a modification to the contract documents, usually made by the Owner or by a Design Professional and then subsequently authorized by the Owner. It can result in an increase, decrease or no change in cost.

Civil Rights Act – major piece of legislation passed in 1964 outlawing discrimination on the basis of race, religion, national origin, and gender.

"Claims made" – covers incidents that occurred during or before the policy so long as they are reported during the term of the policy.

Clean Air Act – passed in 1970 and subsequently updated, the legislation gives the Environmental Protection Agency the authority to write regulations to eliminate airborne contaminants.

Clean Water Act – enacted in 1972, the act governs the release of pollutants into the waterways.

CM at Risk – a Construction Management firm that assumes responsibility for final cost of the project, usually based on pricing developed through early involvement during the design process.

Commercial General Liability (CGL) – protects a company against loss due to accidents that injure another's person or property.

Commissioning – the process of maintaining and verifying that all systems of a new structure continue to function in the manner intended over a specific period of time.

Comparative Negligence – the principle whereby a claim in which both parties contributed to the damages will be shared, based on the amount of damages caused by each party.

Compensable delay – an extension to the completion date for which the Contractor will receive compensation.

Compliance Officer – OSHA inspector charged with making sure projects are complying with current standards of safety.

Comprehensive Environmental Response, Compensation, and Liability Act (CERCLA) – also known as "superfund," this act is intended to clean up contaminated sites as well as assess costs for doing so to appropriate parties.

Concurrent delay – a delay that takes place simultaneous to or overlapping another delay.

Condition precedent – a clause within a contract requiring certain conditions to be met or occur before other conditions are required.

Conscionability – in law, a legal expression stating the person was acting in good faith within prudent, conscientious, scrupulous, and ethical boundaries.

Consent of Surety for Final Payment – formal document from surety to Owner agreeing that final payment can be made to Constructor, but acknowledging all legal obligations of the surety remain in place.

Consideration – what the party accepting the original promise of the contract must give in return for the delivery of what is promised by the first party of the contract.

Construction Manager – an individual or entity acting as an agent of the Owner to oversee the construction processes of a project, particularly as they relate to schedule, budget, quality, and safety.

Constructive Acceleration – in the absence of an approved **change order** for accelerating the work or extending the schedule, the Contractor incurs additional costs to meet the existing schedule of the Owner and, in all likelihood, to avoid Liquidated Damages. This is usually done with the intention of seeking reimbursement for the additional costs through some post-project means.

Constructive changes – informal directives by the Owner that might not initially be recognized as extra work, but constitute additional costs based on differences in interpretation of the plans and specifications.

"Construed against the drafter" – an important concept that, barring a patent defect that a prudent Constructor should have recognized, the contract documents will be interpreted in favor of the person who did not create the documents (and thus any ambiguity).

Contract – an agreement between two or more parties for which if there is a breach of the agreement a court can determine a remedy.

Contract modification – an addition or clarification to the contract documents. Although they can be informal, courts prefer contract modifications to be in writing and that they stand the test of ***quid pro quo.***

Contributory Negligence – a concept of defense for negligence when both parties contributed to their own damages.

Cost Plus – a form of compensation in which the Constructor receives payment for the cost of doing the work plus an agreed-upon markup to cover overhead and profit.

Crash – a point or moment in the installation by one or more trades that interferes with another part of the installation. The 3-D modeling of **Building Information Modeling Technology (BIM)** makes the revelation of such points possible that generally were not found on 2-D drawings.

Critical delay – a delay that will affect the scheduled completion date.

Critical path – a schedule of activities that must be completed in sequence to determine the earliest completion of the project.

Critical Path Method (CPM) – a scheduling technique to show the relationship of independent activities to the duration, execution, and completion of the project. The entire project is divided into the events necessary to be executed in order to complete the project, and then durations to complete those events are assigned to each task. These events are then sequenced in the order in which they must be completed, and the dependency of the events is determined in order to understand what events must be completed prior to other events beginning. Based on the sequencing, a "critical path" can be determined, which will be the sequence of events that will require the longest time to complete without delaying or affecting other events. Events not on the critical path are also assigned beginning and ending dates, which will determine along with the assigned duration of such events the earliest and latest dates such events can begin without affecting the critical path.

Davis Bacon Act – 1931 law passed by Congress requiring workers to be paid a prevailing wage for each craft employed on a project receiving Federal funding.

Deep pockets – the party with the best financial position and thus considered capable of carrying the costs for all the parties that are responsible for certain damages.

Demonstrative evidence – exhibits, such as photographs, videos, or diagrams, that "show" the judge, jury, or arbitrator(s) the evidence.

Depositions – testimony of a witness taken outside of the court during the discovery process to be used later as the basis of testimony.

Differing Site Condition – a situation or circumstance different than what has been shown on the drawings or described in the contract documents.

Directed acceleration – the Owner has agreed to compensate the Constructor for additional costs that will be required to get the project back on the original (or another approved) schedule.

Disadvantaged Business Enterprises – a for-profit business that is at least 51% owned by an individual or individuals who are socially and/or economically disadvantaged.

Discovery – pre-trial exchange of evidence and witness information, including the potential to collect depositions.

Dispute Review Boards – an **Alternative Dispute Resolution (ADR)** process that puts the issues before a panel of experts to decide the most appropriate outcome of a conflict.

Eichleay formula – a special formula on government contracts to allow a portion of home office overhead to be included in delay or sometimes Termination for Convenience cases.

Emotional Intelligence – the ability to control one's own emotional response to situations or stimuli.

Employee Retirement Income Security Act (ERISA) – law passed in 1974 to establish rules and regulations regarding retirement plans.

Equal Employment Opportunity Commission (EEOC) – the Federal Agency created to enforce workplace anti-discrimination laws.

Equal Pay Act – passed in 1963 and signed by John F. Kennedy to guarantee equal pay for equal work and abolish disparity of wages between male and female workers.

Errors and Omissions Insurance – coverage, usually for professionals, such as architects or engineers, that insures against acts of negligence.

Estoppel – a legal bar preventing a person from asserting or acting something contrary to previous statements or actions.

Ethics – a discipline that deals with what is good and bad, right or wrong.

E-Verify – Federal free internet resource that compares information from an applicant for employment's I-9 report with facts collected by the government.

Ex parte – Latin for "from one party," which is a term meaning the other party in a dispute was not present during the discussion.

Excess insurance – protects the purchaser from the "excess" costs not covered by an underlying policy.

Exclusive remedy – a method of recovery that eliminates further action for additional damages.

Exculpatory clause – a clause relieving one party of responsibility for damages due to actions or lack of actions that are not the fault of the other party to the contract.

Excusable delay – a delay for which the Owner recognized the Contractor is entitled to at least an extension to the completion date.

Experience modifier – a factor related to previous claims that is applied to the standard costs of some insurances, such as workers' compensation insurance.

Expressed Authority – the authority granted a party by contract.

Fair Labor Standards Act (FLSA) – established in 1938 as a Federal statute to govern work-week hours and minimum pay.

Family Medical Leave Act (FMLA) – law passed in 1993 requiring covered employers to provide job-protected and unpaid leave for medical and family-related reasons.

Fast tracking – a process of starting some parts of the construction process (e.g. foundations, steel) before the entire plans for the project are completed.

Federal Acquisitions Regulations Act (FAR) – rules and regulations established by the Federal government to provide uniform procedures for acquiring products so that the customer receives a quality product, with the highest standards of integrity and honesty involved in the process.

First or second tier subcontractors or suppliers – relative to the closeness to the prime contract, a first tier is one contract below the prime, and a second tier is another contract removed from the prime contract.

Float – the time between early start and late finish times of a specific event not involved in the critical path of a project that can either be delayed or completed early without affecting the overall completion of the project.

Flow down clause – wording within a contract that makes a lower party responsible for the conditions required in a higher contract. In construction a General Contractor will usually obligate a Subcontractor (lower party) to the conditions the General is bound to by Owner's contract (superior).

Force majeure – actions that can affect the time and/or cost of the project in damages that are out of the control of the Owner and Constructor.

Free on Board (FOB) – the place or vehicle where the buyer takes possession and responsibility for the item purchased.

Front-end loading – the act of putting a greater value on events that will happen early in the project by taking away actual costs from work that will happen later in the project.

General Contractor – the entity responsible for construction of the whole project, or those parts of the project that are not contracted to other Prime Contractors (e.g. Mechanical, Electrical, Plumbing).

Guaranteed Maximum Price (GMP) – a set cost based on plans and specifications that will not be exceeded unless significant changes take place in the scope of the project.

Hold Harmless Clause – a statement within a contract in which the party signing the contract will not hold the entity that created the contract responsible for injuries sustained as a result of the contract operations. It is usually included in contracts where there is some element of risk.

Immigration Reform and Control Act – 1986 law passed to reform immigration law, granting citizenship to some immigrants while requiring employers to verify legal status of workers.

Implied Authority – authority that is the reasonable extension of Expressed Authority.

Impossibility – in the legal context of a construction issue, the conclusion that a contract or a portion of a contract could not be performed by any Contractor.

Indemnity Clauses – a part of a contract that requires one party to assume responsibility for the loss of another party.

Insurance – the transfer and protection from risk by one entity to another in return for a financial payment.

Integrated Project Delivery – a collaborative group of individuals and/or companies working toward a common goal, sharing skills and assets to deliver the complete project.

Intentional torts – damages that occur to other parties or property that the party causing the damages clearly understood would or could happen.

Joint venture – two or more entities forming a separate corporation to share personnel and assets in order to more effectively execute the obligations of a contract.

Jurisdictional disputes – conflicts involving which union will perform certain work on a project or in a region.

Latent Ambiguity – a discrepancy in the contract documents that could not be known at the time the bid proposal was put together.

Lean Construction – a process of improving flow and final quality based on principles of manufacturing that add to successful scheduling and management of the construction process.

LEED – Leadership in Energy and Environmental Design. Developed by the Green Building Council to provide a set of standards for improving the environmentally beneficial structures.

Libel – remarks made in writing that are false and defamatory.

Lien – an encumbrance on a property to secure payment for a debt owed by the property owner

Liquidated Damages – a contractual requirement placed upon the Constructor to pay for the estimated costs to the Owner for not being able to use the structure for its intended use by the completion date of the project. Note: Liquidated Damages need to be a reasonable and accurate assessment of the real costs rather than a penalty.

Little Miller Acts – based on the Federal Miller Act, a statute by each state that requires state contracts to include bonding for payment and performance.

Look-ahead schedules – a short time-frame schedule (usually two or three weeks) often developed by field personnel to analyze the specific needs and requirements to complete the scheduled events during the upcoming time period.

Lump sum – a firm and total price for the identified scope of work.

Machiavellian – character or actions tending toward the principles described by Machiavelli, including cunning, treachery, and duplicity.

Means and methods – a recommended way to perform certain work on a project. Often a Design Professional may have an understanding of a best practices method of executing work, but sometimes for liability reasons the means and methods of doing so are not given.

Measured Mile – a method of costing the effect of lost productivity by measuring units put in place during a period of time unaffected against a period that was adversely affected.

Mechanics Lien or construction lien – an encumbrance placed upon a property to secure interest in the property for work or material supplied to the property but not paid for.

Mediation and/or Arbitration – forms of dispute resolution that have become popular alternatives to litigation.

Miller Act and **Little Miller Acts** – Federal (and state – Little Miller) legislation requiring Prime Contractors to furnish bonds for payment and performance on contracts above $100,000.

Modified Total Cost Method – a method in which the Constructor subtracts from the total increased cost of the project some factor for Constructor error or acknowledged fault.

Multi-prime Contracting – an Owner contracts with several different entities (e.g. separate General, Electrical, Mechanical, Plumbing Contractors) to do the work on a single project.

National Labor Relations Act – guarantees basic rights for private sector employees to form a union and protections based on actions that they may take if represented by a union.

National Labor Relations Board – an independent agency of the Federal government to oversee labor practices between management and unions.

Negligence tort – claims against another party due to actions or inactions that caused damages to persons and/or property of others through failure to take prudent or appropriate steps to prevent the damages.

No Damages for Delay Clause – a part of a contract that obviates all responsibility of a party for any costs associated with a delay, even if that delay is caused by that party.

Non-compensable delay – a delay for which the Constructor will not receive any financial compensation for damages.

Non-critical delays – delays associated with events that are not on the critical path of the project and are short enough that they will not affect the completion date.

Non-excusable delays – delays caused by the Contractor or its agents (e.g. Subcontractors) for which the Owner will neither grant a time extension nor financial compensation.

Occupational Safety & Health Administration (OSHA) – the 1970 agency established under the Labor Department to oversee safety in the workplace.

Occurrence policy – insurance that covers parties against claims made during the term of the policy regardless of when the claims were made.

Open shop – also known as merit shop, an employer that does not place the condition of being part of a union for employment.

Pareto Principle – Originally meant to be a management tool to show 80% of results come from 20% of the effort, it has been changed variously to express any 80:20 ratio, often as a demonstration of frustration. In construction, the concept that 80% of the effort is consumed on the last 20% of the project. It was named after Vilfredo Pareto, who observed 80% of Italy's land was owned by 20% of the population.

Parol evidence – usually oral testimony that contradicts the written contract, and is thus not admissible so long as the written contract appears to be whole.

Patent Ambiguity – a discrepancy in the contract documents that a reasonable and experienced Constructor should recognize and therefore should not be cause for an increase change order.

Patent flaw – similar to a **Patent Ambiguity**, a discrepancy in the contract documents a reasonable person could have foreseen or recognized.

"Pay under the table" – the practice of paying cash for services rather than including the local, state, and federally mandated payroll insurance and tax obligations.

Pay when Paid/Pay if Paid Clauses – a condition within a contract that attempts to shift the payment to a Subcontractor or other entity to a time after the Owner has paid the Contractor.

Peculiar Risk – Owners can assume the damages caused by Contractors working for them if they understand the dangers and do not protect against them.

"Pennies on the dollar" – an expression that refers to the general outcome of receiving less from a bankrupt party than is owed.

Performance specifications – a contract or part of a contract that sets the functional requirements of installed equipment or material.

Platinum Rule – "Do unto others as they would have you do unto them."

Portal to Portal Act – defines what is and is not compensable under the Fair Labor Standards Act (FLSA) related to travel to and from employment.

Potentially Responsible Party (PRP) – an entity that can be held responsible for clean-up of hazardous waste.

Precedence Clause – a part of the contract that states in what order different sections or documents in the contract will be relied upon. This becomes particularly important when dealing with ambiguities.

Prime Contractor – an entity that has a direct contractual relationship with the Owner or Owner's agent for all or a portion of the work involved in completing the project. A Prime Contractor may subcontract portions of work to other contractors.

Private Public Partnership (PPP) – a business venture between a public body and one or more private entities to accomplish a project.

Progress payments – payments made, usually on a monthly basis, for the percentage of work accomplished up until the application for payment was submitted.

Promissory estoppel – a situation in which the full requirements of a contract may not be in place, but one party is prevented from withdrawing from obligations another party acted upon in good faith.

Punitive damages – an amount of additional money added to the award made by a jury or judge above and beyond the actual costs of the damages suffered by the injured party.

Quantum meruit – a term used to describe what compensation an individual should receive for work done. Literal meaning from the Latin is "as much as he deserves."

Quid pro quo – Latin, meaning "this for that." The essence of a contract, "I promise this if you promise that."

Remedy – an imposed settlement, usually by a court, to return an injured party to the approximate condition before the injury occurred.

Request for Information (RFI) – an application, usually to a Design Professional, requiring clarification or direction on a condition encountered during construction.

Resource Conservation and Recovery Act – 1976 Act passed to govern the disposal of solid and hazardous waste.

***Respondeat superior* or "reputed negligence"** – Latin for "let the master answer." In this principle the employer is responsible for the actions or inactions of the employees.

Responsible – a bid submitted by a Contractor that can in fact do the work the project will require.

Responsive – a bid that has included all the responses required in the bid proposal.

Retainage – an amount (usually 10%) subtracted from a pay application until project completion or other milestones have been achieved in the schedule of the work.

Risk shifting clauses – a clause or section of a contract that places responsibility onto a party that may not normally be responsible for the risk or the damages resulting from action or inaction related to contract requirements.

Scope clause – a section of the contract that details all the documents that are to be included as part of the whole contract.

Sheriff's Sale – a public auction conducted by the sheriff on a property being sold to satisfy a court-imposed judgment of debt.

Shop drawing – a rendering, usually provided by a Supplier or Subcontractor, of specified material, showing details of manufacturing and installation.

Shop steward – an individual on a jobsite appointed by the union to monitor, and resolve if possible, disputes regarding working rules on the project.

Slander – remarks made orally that are false and defamatory.

Spearin Doctrine – the concept that if a Constructor relied upon the documents supplied by the Owner, the Constructor cannot be held responsible for design errors.

Special Purpose Vehicle (SPV) – in the case of **Public Private Partnerships**, an entity created solely for the execution of the work required to construct and maintain the project.

Specialty Contractor – *see* **Subcontractor/Specialty Contractor**.

Stop Notice – a directive to a public authority to withhold payments to a Contractor until certain debts owed by the Contractor are paid.

Subcontractor/Specialty Contractor – a party that contracts to perform part or all of another's contract. In recent years the term "Subcontractor" has been changed to "Specialty Contractor" on the basis they do specialized parts of a project (e.g. electrical, ceiling, excavation) as opposed to the Construction Manager or General Contractor that may not do any or very little of the work with its own forces.

Subrogation – the right of a party (such as an insurance company) to seek restitution for payments made from the party responsible for the loss.

Substantial Completion – the date on which the Owner can move into the building or use the structure for its intended purpose.

Sub-subcontractors – Contractors twice removed from the prime contract (e.g. the plumber is the sub-subcontractor to the Mechanical Subcontractor to the Prime General Contractor).

Superfund – the more popular name for the Comprehensive Environmental Response, Compensation, and Liability Act.

SWOT analysis – a technique to assess or analyze a team or group's "Strengths, Weaknesses, Opportunities, and Threats."

Termination for Cause – a provision within a contract to terminate the contract under certain specified circumstance.

Termination for Convenience – a provision within a contract to end the contract without cause.

Three Cs of bonding – character, competence (or capacity), and capital.

Time and Material – a form of compensation for work done based on the costs expended plus an agreed-upon markup for "overhead and profit."

"Time is of the essence" clause – a phrase (often overused) that stresses the contract completion date is of critical importance.

Tort law – a civil wrong for which a governing body will impose a remedy.

Total Cost Method – a method to compute the difference in cost of a project in which the original contract amount is subtracted from the final total cost of the project. In a slightly more difficult way this can sometimes be done for specific parts of a project.

Toxic Mold – potentially dangerous spores usually developed due to moisture penetration or lack of adequate ventilation.

Trade stacking – usually done during the final days of a project to get whole sections of the work done with different crafts working in the same area at the same time.

Treble damages – three times the amount of actual costs of damages.

Type I Differing Site Conditions – the actual conditions encountered differ substantially from what was indicated in the contract documents.

Type II Differing Site Conditions – the actual conditions differ from what one would expect under normal circumstances to exist.

"Umbrella" Insurance – additional insurance that provides additional coverage in any areas where other insurance limits are exceeded by a claim.

Uncertainty – the legal recognition that a contract was void, because one or more parties could not fully know the scope or cost required to complete the project.

Unemployment insurance (UI) – a form of insurance that compensates workers when they become unemployed through no fault of their own.

Uniform Commercial Code – a set of rules established in 1952 to establish a set of standards regulating commerce within and between the fifty states, as well as the District of Columbia.

Unit Cost – a form of compensation for work done on the basis of specified units at the cost of so much per unit (e.g. cubic yard of concrete, lineal foot of six-inch pipe, square foot of paint).

Utilitarian ethics – a form of ethics based on doing the most good for the most people, thereby reducing suffering while increasing happiness.

Waiver – the cancelling of an existing right from one party to another.

Warranty – the assurance by one party to a contract that the installation will remain in operation and without flaw for an agreed period of time or corrective measures will return the installation to its proper and original condition or working order.

Wildcat strike – a work stoppage taken by union employees against management without the official authorization of the union officials.

Workers' Compensation Insurance – insurance that covers workers against injury and any resulting loss of income without the need to bring suit against the employer for such compensation.

Further Reading and Research

Construction claims

Kimberly, Gilbert (Managing Editor), *Construction Claims Monthly*, Business Publishers, Inc., published monthly since 1949.

Written more from and for the perspectives of attorneys, this is still an excellent resource for updates on how different cases have been decided in various jurisdictions around the United States.

Dougherty, Jay, *Claims, Disputes and Litigation Involving BIM*, Routledge, 2015.

This first book to investigate case studies and case law relating to BIM projects gives us a crucial insight into how liability and other factors can be different under BIM.

Ethics

Mirsky, Rebecca and Schaufelberger, John, *Professional Ethics for the Construction Industry*, Routledge, 2014.

Talks about ethics in a construction industry context using case studies, broken down by activities such as codes and compliance, and discrimination. Written to meet the ACCE's requirements for ethics teaching on construction courses.

Dalla Costa, John, *The Ethical Imperative*, Harper Business, 1999.

Covers the topic of ethics from a business perspective, developing the concept that the practice of good ethics is good business. The author relates how to establish a working model for good business ethics within a company.

Johnson, Craig E., *Ethics in the Workplace*, Sage Publications, 2007.

Describes an approach to assert ethical practices from within the organization. There is a strong emphasis on how to take personal approaches to ethics and apply them to the broader organizational business model.

Design issues

Beard, Jeffrey L., Wundram, Edward C., Loulakis, Michael C., *Design Build*, McGraw Hill, 2001.

This was one of the first and still best publications covering the resurgence of the Design-Build movement in the construction industry. It carefully and clearly covers the needs and relationships of architectural, engineering, and construction services involved in the challenges of the construction process.

Levy, Matthys and Salvadori, Mario, *Why Buildings Fall Down*, W. W. Norton, 1992.

An excellent description of many causes of failure of structures. This book provides the reader with numerous warnings of what can happen based on design flaws.

Petroski, Henry, *To Engineer is Human*, Vintage Books, 1992.

A more general analysis of engineering issues than in Levy and Salvadori above, the work gives critical analysis of some important failures, including the horrific Kansas City Hyatt Regency Hotel disaster.

General

Chinowsky, Paul S. and Songer, Anthony D., *Organization Management in Construction*, Routledge, 2011.

Looks at ten key strategic management issues and their impact on the construction organization, drawing on the experiences of leading international authorities.

Strange, Peter S., *Steerageway*, Keen Custom Media, 2013.

A personal perspective by a very successful CEO of how employee Owners set direction of a construction company amid the currents of change.

Project management

Fewings, Peter, *Construction Project Management: An Integrated Approach, 2e*, Routledge, 2012.

Takes a holistic view of the project manager's role, in how it can influence communication, ethics, and efficiency across the team, as well as cost, time, and risk.

Negotiations and teambuilding

Fisher, Roger, and Ury, William, *Getting to Yes*, Houghton Mifflin, 1981.

A recognized standard on how to negotiate and resolve situations for the best interests of all parties.

Napier, Rod, and McDaniel, Rich, *Measuring What Matters*, Davies-Black Publishing, 2006.

Offers excellent tools and techniques to build and reinforce teams based on assessments and analysis of measurement models.

Important websites

www.agc.org

A leading construction organization in America, representing Contractors on important issues critical to the industry, as well as providing resources on education, training, and opportunities for employment. AGC was the driving force behind the creation of ConsensusDOCS.

www.aia.org

The leading website for architects in America. This is an important resource for perspectives on the industry from the architectural viewpoint. AIA contracts are available for purchase and use through agreements with AIA.

www.cpp.com

A source for the Thomas–Kilmann conflict mode measurement instrument. Using thirty questions, the instrument analyzes a person's tendencies related to conflict resolution.

www.dbia.org

The Design Build Institute of America is the main organization promoting Design Build practices and techniques through education, training and networking.

www.ejcdc.org

An important source of information from the engineering perspective. Engineering contracts are described and available through the Engineers Joint Contract Documents Committee.

www.osha.gov

Important source for research and updates on safety regulations affecting the construction industry, as well as all industries.

Index